Smart Ship Technology 2017

London, United Kingdom
24-25 January 2017

ISBN: 978-1-5108-8299-7

Printed from e-media with permission by:

Curran Associates, Inc.
57 Morehouse Lane
Red Hook, NY 12571

Some format issues inherent in the e-media version may also appear in this print version.

The Royal Institution of Naval Architects
8-9 Northumberland Street
London, WC2N 5DA
United Kingdom

Phone: 020 7235 4622
Fax: 020 7259 5912

publications@rina.org.uk

Additional copies of this publication are available from:

Curran Associates, Inc.
57 Morehouse Lane
Red Hook, NY 12571 USA
Phone: 845-758-0400
Fax: 845-758-2633
Email: curran@proceedings.com
Web: www.proceedings.com

CONTENTS

TOWARDS SHIPPING 4.0

Ø.J. Rødseth, SINTEF Ocean AS, Norway

SUMMARY

Shipping 4.0 can be seen as the fourth technology revolution in modern shipping, closely aligned with the corresponding "Industrie 4.0" or "Industry 4.0" evolution on land. It is characterised by a huge increase in the use and availability of digital information, but also by new service and operation concepts made available with better connectivity between ship and shore. Examples are e-navigation and the increasing use of shore based services for the ship, including more advanced on-board decision support combined with remote assistance. The eventual deployment of autonomous ships is conceptually the culmination of Shipping 4.0. This paper will give an overview of the four technology revolutions in shipping; examine the different components of Shipping 4.0; and how these are already being picked up by the shipping industry. The paper will also point out some key elements in the new technology that will need attention by the community, particularly in the area of international standards.

1. INTRODUCTION

In parallel with land industry, shipping has experienced three technical revolutions: steam engines (~1810), diesel engines (~1910) and digital control systems (~1970). Still in parallel with industry, it is facing its fourth revolution based on the so-called "Industry 4.0" revolution, integrating Internet of Things, Cyber Physical Systems, Big Data and other technologies. As with previous revolutions, we expect that "Shipping 4.0" will have a profound effect on shipping operations and on how the business will be put together. The society's focus on energy efficiency and reduced emissions is expected to amplify the uptake of the new technology.

Shipping 4.0 is mainly characterised by a huge increase in available digital information, but also by new service and operation concepts made available with better connectivity between ship and shore. Examples are already emerging in e-navigation, third party analytics and administrative shore services. Supporting evidence is, e.g. in the amendment to the IMO FAL convention which will require electronic port clearance of ships from 2019 and the strong interest in cyber security in IMO and elsewhere. The interest in, and eventual deployment of autonomous ships, can be seen as the culmination of Shipping 4.0.

In this paper we look at Shipping 4.0 primarily as a revolution for the ship systems and for the operation of the ship, but it will obviously also have implications for shore services in port, in owner's office and elsewhere.

2. THE TECHNICAL REVOLOUTIONS

One can discuss what technological breakthroughs in ships and shipping were the most important over the ages. In this paper, we have selected the ones that were also aligned with and driven by similar major technology changes on land. This means that we have omitted the initial developments of oars and sails, the introduction of electronic navigation instruments and global positioning systems as well as the introduction of shipping containers around 1970. Our approach is in line with the definitions used by the developers of the Industrie 4.0 (or Industry 4.0) concept [1], when it was introduced for the first time at the Hanover Fair in 2011. The following sub-sections will give a brief overview of these main technology changes.

2.1 FROM SAIL TO STEAM (1810)

The first industrial revolution was represented by the introduction of mechanized power and particularly in the form of steam engines that could provide rotational power. This happened around 1800 and the technology was rapidly picked up by the shipping industry. Sea-going ships started to appear from around 1810 and the "S/S Great Western", the first purpose-built transatlantic steam ship, set out on her maiden voyage in 1838. This heralded a revolution in international shipping where one of the significant developments were scheduled liner shipping due to the much more dependable departure and arrival times for these ships.

2.2 FROM STEAM TO DIESEL (1910)

The second industrial revolution was the introduction of serial and mass production. This principle was in use already from the 1800s when mechanized power first became available, but did not really change the general production industry before around 1900. The introduction of electric power was one important factor enabling this, as power sources could be much more easily distributed. This replaced the single steam engine with a line shaft and belts to drive the individual machines. In the shipping industry, one can argue that the development of the internal combustion engine, i.e. the diesel engine, was the corresponding major change. One of the first motor ships was "M/S Selandia", built in Copenhagen and launched in 1911. The diesel engines allowed the ship to operate much more efficiently, both in terms of overall power efficiency per volume and weight of fuel as well as in much fewer people needed to service the engine. This increased the ship's range and reduced the necessary crew substantially.

2.3 INTRODUCTION OF COMPUTERS (1970)

The third industrial and maritime technology shift was the introduction of computerized control systems around 1970. This was preceded by different types of electronic and other types of normally analog control systems. The first ship to sail with a fully automated engine control system was the "M/S Kinkasan Maru", launched in 1961. This allowed all important engine control functions to be operated from the bridge. The introduction of computers in the control systems enabled a much higher degree of automation with significant lower complexity in wiring and control system construction. The shipping sector was very fast to adopt this and the first computerized control system on sea was on board the "M/S Taimyr", launched in 1969. Taimyr had a fully computerized engine room as well as a computer controlled anti-collision system (ARPA) on the bridge. The two computers used on this ship were also the two first commercial deliveries from the now defunct Norwegian computer manufacturer Norsk Data [2].

2.4 CYBER-PHYSICAL SYSTEMS (2010)

The fourth major change in industrial automation, or "Industry 4.0" as it is commonly called [3], started around 2012. Industry 4.0 includes extensive use of new technology like Big Data, Cloud Computing, Internet of Things (IoT) and Internet of Services (IoS) as well as the central concept of "Cyber-Physical Systems" (CPS). Cyber-Physical Systems are integrations of computation and physical processes where embedded computers and networks monitor and control the physical processes, usually with feedback loops where physical processes affect computations and vice versa [4]. Industry 4.0 implies much more intelligent physical equipment that generate a plethora of new data and information as well as enabling new services that make use of the data.

The corresponding term "Shipping 4.0" was, as far as we know, first proposed in [5]. It was observed that the shipping industry was adopting many of the Industry 4.0 technologies and for the same reasons as the shore based industry. Shipping 4.0 promised, among other things, improved monitoring of shipboard systems, increased use of shore-based services, increased instrumentation and more integrated ship-control while using data analytics to optimize operations. This is arguably the start of a fourth major technologies change in the shipping industry, hence the proposed name "Shipping 4.0".

3. COMPONENTS OF SHIPPING 4.0

There have been a number of publications discussing what technology components Industry 4.0 encompasses, see, e.g. [6], [7], [8] and [9]. In [10], the concept is also put into a maritime context. However, there is no general agreement of what Industry 4.0 or Shipping 4.0 means or contains, but an attempt to highlight the most important technologies have been made in Figure 1. This has been slightly adapted from the concept presented in [5]. In the following sub-sections, we will give an overview of these technologies and their application in Shipping 4.0 as we see it. We have left out some commonly mentioned technologies from Industry 4.0. The reasons for this are briefly discussed in Sec. 4.

Figure 1: Shipping 4.0 Main Components

3.1 CYBER SECURITY

The digitalization of shipping will require a high emphasis on cyber security for all facets of Shipping 4.0. Ships and shore operations will increasingly become dependent on secure connectivity and the correctness of transferred data. Many data exchanges will also be business or privacy related and sensitive to eavesdropping. We also see that important business transactions are becoming digitalized. This requires trust between the parties, which can be provided with electronic signatures and similar mechanisms [11].

This has been recognized by most organisations active in the maritime domain and there are currently hectic activity in IMO and all shipping interest organisations to analyse the problem, provide guidelines for improved security and define new technical standards to protect the systems and the data exchanges.

Apart from the communication infrastructure itself, cyber security is perhaps the most important component of the successful digitalization of shipping.

3.2 OPEN SYSTEM INTEGRATION

The open and integrated access to data on board the ship is critical for Shipping 4.0. This applies to Internet of Services, cyber-physical systems as well as real-time optimizations and other forms of decision support. The ability to add new support tools on board the ship and on shore, as easily as possible, is crucial to the effective implementation of Shipping 4.0.

This means that one needs an open integration framework on the ship that covers all important ship functions and all layers of the on-board data networks as well as data format specifications [12]. This is illustrated in Figure 2 where the four main layers have been drawn in the vertical direction and where most layers have been subdivided horizontally into several sub-networks.

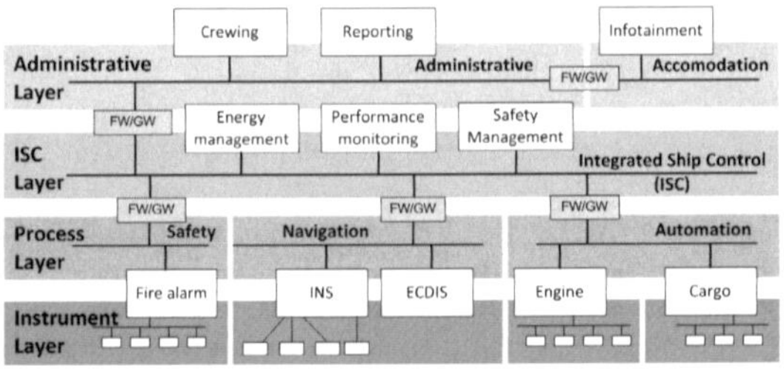

Figure 2: Typical Ship Network Subdivisions [13]

The integration problem is most pronounced on the process layer, where the different main ship control functions are located, e.g. navigation (INS – Integrated Navigation System and Electronic Chart Display and Information System – ECDIS), engine automation, cargo control, fire and safety.

An important obstacle is that the individual networks normally are isolated and have limited connectivity to each other. If there is connectivity at all, this will typically be via special purpose firewalls or gateways (FW/GW in the figure). The firewalls and gateways will also reduce connectivity to whatever is pre-programmed into the respective gateways.

Another related obstacle is the use of many specialised and normally proprietary communication protocols in the different ship systems. This makes it difficult to connect third party equipment to any of the networks and get full access to all the data it can provide. As an example, connecting an advanced instrumented compressor to the automation network will normally only result in a few new tags on the automation network, controlling start and stop, output pressure and retrieving any error signals form the unit. To retrieve more extensive operational or diagnostic data, one will normally have to go on-board and connect a computer directly to the device.

The open integration should also apply to connectivity between ship and shore. In this case, one can probably rely on reuse of standards from the Internet domain, although adjusted for the lower quality of service that ship communication currently suffers from.

3.3 INTERNET OF THINGS AT SEA

The basic idea of Internet of Things (IoT) is the pervasive presence of a variety of things or objects that through unique addressing and communication schemes are able to interact with each other and to cooperate with their neighbours to reach common goals [14].

IoT is an integral part of Industry 4.0 and necessary for the realization of Cyber-Physical Systems and Internet of Services. IoT on ships is generally more complicated than on land and for this reason, we prefer the term “Internet of Things at Sea”. Ships can be large and the use of steel as construction material makes general ship-wide wireless connections more difficult than on land. The weather-exposed external sensors and actuators also require special measures for being securely connected to the ship data networks.

International shipping is making use of a variety of free and internationally regulated data communication systems, typically based on medium or very high frequency digital data links with limited bandwidth. AIS (Automatic Identification System) is the most common example of this. Finally, the dependence on generally low capacity and long latency satellite communication when at high seas also makes connectivity to the general Internet more challenging than on land.

It is clear that a conventional centralized approach, where high-level application platforms control the whole information flow, is not feasible on ships. However, we still want to make use of mainstream IoT architectures like the ones developed by IEEE [15] or ETSI [16]. To fit the use at sea, these architectures have to allow more efficient use of both intelligence and provisioning of services at the edge networks as well as include adaptive and dynamic collaborations between application programs. In addition, IoT enabled units need the ability to retrieve, process, combine, and provide information and services to other units or applications without necessarily being in continuous contact with them. The future IoT at Sea solution should support applications where both networks and traffic are sparse and a continuous internet connection is not always available or often constrained.

3.4 CYBER-PHYSICAL SYSTEMS

Cyber-Physical Systems (CPS) are systems of collaborating computational entities which are in intensive connection with the surrounding physical world and its on-going processes providing and using, at the same time, data-accessing and data-processing services available on the Internet [9]. For ships, this is very important, as there are a high number of physical processes on-board. One can claim that a ship is a self-sufficient small village on the sea, including systems to support navigation, energy production, life-support and so on. CPS can been seen as a further development of the automatic control systems already common on board ships in the form of track-pilots, autopilots, engine control systems and many others control systems.

The first CPS are already in place on ships: Many new engines and other complex systems have embedded microcontrollers with sensors and actuators that can monitor, control and adapt physical systems to changing operational conditions or demands. The computers will collect information that can be used in diagnostics of the physical systems to better plan maintenance or repairs as well as being invaluable for the manufacturer or user to assess the performance of the equipment.

A major problem in increasing the deployment and usefulness of CPS is the lack of an open system integration mechanisms (see sec. 3.2). Connecting the CPS as one or more IoT devices requires full connectivity to the main ship networks and to shore. This is difficult or costly to realize today.

3.5 DATA ANALYTICS

Data analytics is already playing a large role in digitalization of shipping. Common applications are traffic analysis with AIS data (Automatic Identification System) or ship performance analysis based on engine, navigation and environment data.

There are, however, important challenges in using data from on-board systems. One is the connectivity problem that was discussed in the previous sections that may make some data difficult to obtain. Another problem is the various data quality issues that apply to the data that is available [17]. This means that special methods may have to be employed to extract reliable information from the data. Many ship systems are also highly non-linear and clustered data analysis may be necessary [18].

Another problem for data analytics on ships is the limited bandwidth and relatively costly transfers of data via satellite. This normally means that real-time analyses have to be performed on-board. Alternatively, it can be done after the ship has called in port and transferred the data via lower-cost and faster shore data links.

3.6 AUGMENTED REALITY

A significant percentage of ship accidents are at least partly caused by human error. Various sources site different numbers and EMSA (European Maritime Safety Agency) reports that in 908 accidents investigated in the period 2011 to 2014, 67% were attributed to human errors [19]. Human error is, however, a vague term that covers anything from direct operator errors, via fatigue and boredom to badly designed bridge systems [20].

However, it is commonly acknowledged that there is a need to look into how the bridge crew can use the different ship sensors and navigation information more efficiently and safer. This includes full or part automation of the ship control (autonomy) which will be returned to later.

An important problem is to find better ways of presenting both chart information and sensor data in a more useful format than in the conventional exocentric information format used by the traditional chart displays [21]. With even more information being made available to the mariner through the new e-navigation services, the need to investigate alternative display methods becomes more acute. Various forms of augmented reality, e.g. in the form of head up displays or even virtual reality goggles have been proposed and are being tested. Exactly how the future ship's bridge will look is still not decided, but different visions of the future have been published by most large ship bridge manufacturers and it is quite certain that new display technologies will be an important part of the future digital shipping scenario.

Augmented reality may also play a role in shore control centres for fully or partly unmanned ships. The challenge is to monitor several ships at a time, while retaining situation awareness for each ship when intervention is necessary. One will also need to consider how the shore control and the crew cooperate in cases where the ship is partly autonomous. Augmented reality technology can play a role also here.

3.7 INTERNET OF SERVICES AT SEA (IoSS)

The Internet of Services (IoS) describes an infrastructure that uses the Internet as a medium for offering and selling services [22]. IoS at Sea (IoSS) represents a development where supporting ship operations are being moved to shore, either implemented by the ship operator or bought from third party providers. Typical services are fuel quality control, performance monitoring and weather routing. Many shore authorities are considering providing nautical services to the ship as part of the implementation of e-navigation. If one looks at the 16 currently defined "Maritime Service Portfolios" [23], one will find that all can be characterized as IoSS.

In our definition, we prefer to classify shore services directly related to the physical operation of the ship, such as traffic management and nautical monitoring or control services, under "robotics and autonomy" rather than as IoSS. This is because these services are closely related to and necessary for the implementation of autonomous ships.

As is the case for IoT, the realization of IoSS will need to cater for variable quality of communication links. This means that the concept of IoS for ships may have to be implemented somewhat differently from that on shore. This may include more use of store and forward type data exchanges and avoiding reliance on high bandwidth and low latency. This does in particular apply to services that are provided by coastal or port states, that may use special purpose, low bandwidth and cost free data connectivity. The developing VHF Data Exchange System (VDES) [24] is one example of the latter.

3.8 SIMULATION AND OPTIMIZATION

Simulation and optimization is conventionally employed in the ship design phase, but data used in these processes as well as the tools will also be important in the operational phase. This may, e.g. include use in voyage optimization and weather routing.

There has always been a strong focus on making ships as efficient as possible and design simulation and optimization is not a new concept in shipping. There is also an increasing focus on life-cycle operational optimization that can be based on simulations of ship operations over several years [25], [26]. Simulations and optimization during ship design is also part of a life cycle use of digital descriptions of ships and ship systems, a concept that has been called the "digital twin" by DNV-GL [27].

3.9 ROBOTICS AND AUTONOMY

Increased automation and use of robotics are important components of Shipping 4.0. This is also necessary for implementation of autonomous and unmanned ships.

It is likely that many ship operations will be automated to save time and to reduce unwanted incidents. This includes automated berthing, lock passages and other operations that can be improved by better ship control as well as improved positioning and sensor systems. Robotics, in the form of automated cargo handling, auto-mooring and automatic hook-up to shore power supplies, will support this and are already being deployed.

One can also imagine robots being used on board ships to do maintenance that today are performed by the crew, possibly in combination with additive manufacturing such as 3D printing to produce spare parts on-board. However, our opinion is that it may be more cost effective to improve reliability, maintenance planning and redundancy of ship systems (see sec. 4.2).

Partly autonomous ships can provide various types of "driver assistance" or decision support systems that improve lookout, voyage efficiency, heavy weather operation, engine performance and so on. Together with shore based monitoring and assistance functions, this can be used to, e.g. implement partly unmanned bridge operations. During deep-sea passages and in reasonable weather it should be fully possible to let the bridge be controlled by partly autonomous systems, e.g. at night. Shore personnel will stand by to handle the more complex situations, until it is necessary to muster one of the crew to take over.

Shore supervision and intervention in nautical operations on sailing ships can be seen as one possible realization of the Internet of Services at Sea, but in our opinion, it can more usefully be seen as part of the autonomy and robotics technology. At least today, these types of services are probably necessary for the implementation of partly or fully autonomous ships [28]. We also include emerging real-time traffic management services from coastal authorities, such as those demonstrated in the STM Validation project [29] or even more centralized approaches as suggested in [30], in this category.

Full autonomy is the culmination of ship automation and increased robotics. It means that the ship can sail independently and without a crew controlling the ship. The MUNIN project [28] was the first major research project in this area. It started in 2012 and was followed by a number of similar projects from 2013 and onwards. MUNIN investigated a partly unmanned bulk carrier, but later research shows that the ship has to be fully unmanned to reap the most benefits from the technology [31]. This mainly has to do with the possibility to build the ship differently and to create new transport systems.

This particular aspect, that one can design completely new transport systems, is perhaps the most exciting with full autonomy. It can release traditional constraints such as economy of scale and can make smaller ships more competitive with trucks. It also enables new designs, such as new power systems, e.g. based on batteries.

4. NON-PRIORITIZED COMPONENTS

The concept of Industry 4.0 have been analysed in many papers and contexts, e.g. [6], [8] and [10]. Each paper highlights different set of components that are claimed to define Industry 4.0 and in the previous section, we have described those that we believe are most relevant for Shipping 4.0. However, some commonly mentioned Industry 4.0 components have not been included and this section will give a brief rationale why this is so.

4.1 CLOUD COMPUTING

Cloud computing is a model for enabling ubiquitous, convenient, on-demand network access to a shared pool of configurable computing resources (e.g., networks, servers, storage, applications, and services) that can be rapidly provisioned and released with minimal management effort or service provider interaction [32].

Cloud computing is often listed as one of the components of Industry 4.0. However, satellite data connectivity for ships is not normally of high enough quality to utilize cloud computing as defined above. For ships, one will probably see this technology implemented more as a form of IoSS where specific services rather than the general computing resources are made available. Thus, cloud computing has not been included in the key technologies for Shipping 4.0.

Another type of "cloud computing" is the so-called maritime cloud [33]. This is a proposed mechanism by which information can be transferred seamlessly between system and over different communication link in the nautical domain. This typically includes ships, shore installations and port services. By its nature, this is more akin to IoT, and the developers of the maritime cloud do stress that it is not related to cloud computing.

4.2 ADDITIVE MANUFACTURING

Additive manufacturing, most commonly exemplified by 3D printing, is commonly mentioned as an important component of Industry 4.0. It is expected that this technology can complement traditional manufacturing techniques for complex or short series products and have a significant impact on the future production industry.

In shipping, one can imagine that this technology could get a position in spare part or consumables production on-board ships. This has already been considered and is being investigated by the US Navy and the Maersk shipping company [34]. Currently this is limited to printing of relatively small plastic parts, but metal sintering techniques and larger printers could increase the use of the technology.

3D-prining is currently an expensive technique and extension into metal printing will not make it cheaper. Operation of precision printers on-board a ship moving in the waves and with vibrations from large engines and propellers is also challenging. The trend in shipping, as we see it, may be towards even fewer crew and less possibilities for doing maintenance during the voyage. For maintenance, it will probably be more important to develop new condition monitoring and maintenance planning systems to provide better plans for service of the ships during short stops in port and planned dry-dockings [35]. Thus, we have also left additive manufacturing out of Shipping 4.0 key technologies.

4.3 BIG DATA

Big data is also a common component of the different visions of Industry 4.0. Big data is commonly understood as data sets that cannot easily be handled by conventional data processing hardware and applications. It is often associated by the four Vs: Volume (amount of data), Variety (high number of sources and types of data), Velocity (Speed of new data generation) and Veracity (variable quality and reliability of the data) [36]. Well-known examples of big data are medical and diagnostic data from large patient groups, data from search engines and social networks and traffic data from road traffic (Intelligent Transport Systems).

Data analytics (sec. 3.7), will be important in Shipping 4.0. However, the relevant data is rarely of the type described above. One may argue that large area and long time-frame Automatic Identification System data may qualify, but this data, although large in volume, is not particularly variable in format or quality. Other data related to the ship operation may have high variety and veracity, but rarely high volumes.

Thus, we prefer the more general term "data analytics". We do, however, expect that technologies from big data can and will be applied in Shipping 4.0 data analytics.

5. HIERACHY IN SHIPPING 4.0

The technical components included in Shipping 4.0 can be seen as forming a hierarchy of capabilities, where higher layers are building on the lower. This is illustrated in Figure 3. The layering does not mean that enabling technology in some cases cannot appear as stand-alone applications – or vice versa, but rather indicates the general relationship between technologies that we expect will become the most common in Shipping 4.0.

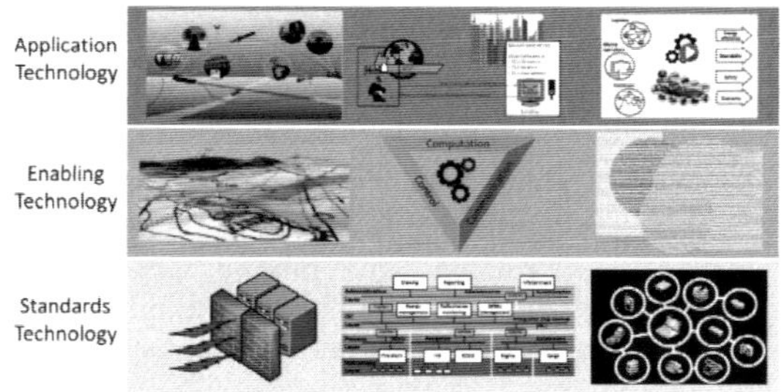

Figure 3: Hierarchical capabilities in Shipping 4.0

5.1 STANDARDS TECHNOLOGY

The foundation level represents the technology components that need to be developed as common and open standards for the rest of Shipping 4.0 to be realizable. This includes, most importantly, standard technology for cyber security, open system integration and Internet of Things at Sea.

These technologies are needed to enable safe, secure and cost-effective connectivity between all components of the ship, physical and computers, as well as between the ship and shore or other ships. The innovations needed to realize Shipping 4.0 will depend on a cost-effective and safe infrastructure and unless these technologies are open and standardised, it will be much more difficult to get enough innovation momentum to create the new enabling and application technologies (see sec. 8).

5.2 ENABLING TECHNOLOGY

The next layer represents technologies that build on the standard technologies and realizes important central functions in the Shipping 4.0 framework, without necessarily being applications themselves. This includes data analytics technology, augmented reality and the cyber physical systems.

5.3 APPLICATION TECHNOLOGY

In our opinion, the main types of on-board applications we will see coming out of Shipping 4.0 can be illustrated as shown in the top layer.

Increased use of automation, robotics and eventually full autonomy, represent a central part of the future of the physical and nautical handling of ships. This includes new real-time decision support systems for manned ships, including new augmented reality based navigation systems. These systems will be supported by new traffic management services from coastal authorities and other forms of real-time and shore based assistance to the ship.

Internet of Services at Sea (IoSS) represents the trend to move more administrative and supporting services off the ship and to land. This will increase the quality of the services, provide better decision support for crew and owners and will relieve sailing crew of many time consuming non-nautical tasks. IoSS will be a crucial component in energy efficient voyage planning, enabling better coordination between ports, ships and cargo owners. It will also be an important part of improved maintenance and condition monitoring schemes, improving system efficiency and reducing unexpected repairs. Cyber-Physical Systems will be major component in realizing improved system supervision and management on-board, based on shore services.

Simulation and Optimization is already playing a significant role in ship design and operation and we expect this role to increase in importance. Weather routing and voyage optimization is important for saving ship fuel and greenhouse gas emissions and simulations during the design can lead to more optimized ship designs for the intended trade and operational profile. Advanced data analytics will feed ship responses back to the simulation and optimization tools, both for design and operation. Simulation and optimization will also be an enabling component of IoSS services.

6. SHIPPING 4.0 IN THE SHIP LIFE-CYCLE

6.1 SOLUTIONS FOR ALL PHASES

Shipping 4.0 as it has been described here is mainly targeting the operational phase of the ship, but it is obvious that some applications like simulation and optimization also can be used in design of ships and ship systems as illustrated in Figure 4.

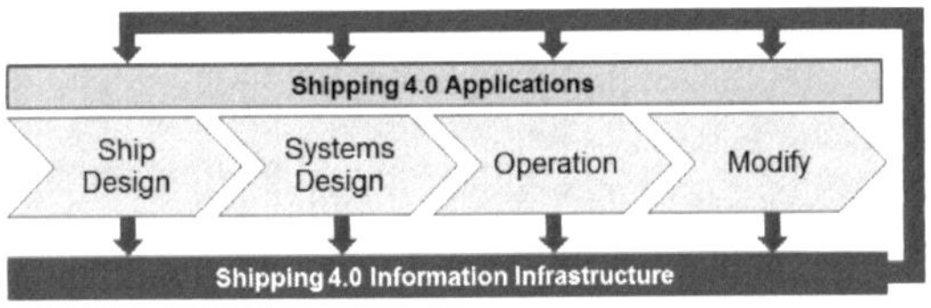

Figure 4: Life-cycle information use

Communication, monitoring and control technologies like IoT, CPS and open data networks can also be very useful during commissioning of IT-systems and physical equipment on the ship. This can be used effectively during the building and modifications of the ship.

6.2 DATA ACCESS THRUGH THE LIFE-TIME

Figure 4 also shows an infrastructure for data collection and access that can be used during the lifetime of the ship. Data collected and analysed during ship operations and after modifications can be used to improve Shipping 4.0 applications used in all phases of the ship's life cycle and to improve designs for future ships.

The simulation and optimization tools can reap the most benefit from improvements based on real-life operational measurements and data analysis. This applies both to on-board optimization tools, e.g. for routing, speed control and ballasting of ship as well as simulations or optimization tools used during the design of new ships.

6.3 DATA AVAILABILITY AND OWNERSHIP

The access to information on-board will be critical for realizing many Shipping 4.0 applications as well as the data infrastructure. One prerequisite is open and standard access to the data sources on-board. This is related to the discussion on open standards in sec. 5.1 and sec. 8. Today, proprietary solutions and closed interfaces make this more difficult than desired.

It would also be very useful if it was possible to compare and benchmark each single ship against other ships of similar type or in similar trade, independent of ownership of ships. This would require some form of third party data base management, e.g. of the type found in Shipping KPI [37], or direct cooperation between different owners so that data can be exchanged between them.

The possibility to reuse data through the ship's life or between ships raises the issue of data ownership. While most people tend to be of the opinion that the ship owner also owns data generated on-board, the reality is not that simple. One example is advanced data acquisition systems built into new engines and other complex systems, i.e. through the realization of cyber-physical systems, where the system manufacturer may be the only one able to access or understand all data elements. Another problem is the proprietary data networks that are commonly used on board which effectively hinders the owner in accessing the data flowing on that network (see sec. 8). These problems need to be addressed to realize the full potential of Shipping 4.0.

7. SOME EXISTING ACTIVITIES

7.1 IMO AND DIGITALIZATION

E-navigation is defined as "the harmonized collection, integration, exchange, presentation and analysis of marine information on-board and ashore by electronic means to enhance berth to berth navigation and related services for safety and security at sea and protection of the marine environment" [38]. E-navigation has defined a number of Maritime Service Portfolios (MSP) [23] that all are examples of IoSS or IoT at Sea applications, typically provided by port and coastal authorities. The strategy development for e-navigation is mostly done in the Maritime Safety Committee (MSC) of IMO while much of the technical work is coordinated by IALA (The International Association of Marine Aids to Navigation and Lighthouse Authorities) and IHO (International Hydrographic Organization).

The Facilitation Committee (FAL) in IMO as well as MEPC (Maritime Environment Protection Committee) are also working on electronic reporting formats for ships that can be said to be part of Shipping 4.0. These formats will enable the use of automated services on shore for receiving arrival reports and mandatory ship documents in conjunction with port calls.

For all, maritime cyber security is a critical element.

7.2 EU AND E-MARITIME – MARITIME ITS

Another example of an international initiative to increase digitalization in the maritime area is e-Maritime in Europe, which aims at extending e-navigation with elements that also focuses more on the commercial issues in shipping, such as cost and efficiency of ship and port operation. This has sometimes been called the maritime variant of the Intelligent Transport System (ITS) [12].

7.3 IN THE COMMERCIAL FIELD

Many commercial parties are highlighting the need for further digitalization of shipping. This ranges from headlines like the "Connected ship" and "Digital Shipping" to several highly profiled and large projects on autonomous ships. There is also a very high interest in data analytics or "big data" in the maritime businesses.

Maritime technology forecasts from both Lloyd's Register [39] and DNV GL [40] highlight the same developments towards smarter ships, connected ships, autonomous ships and so on. Thus, there is a clear consensus that digitalization in shipping will continue and that the future will look very similar to what we have outlined as Shipping 4.0.

8. NEED FOR STANDARDS

Standards are already being developed to support Shipping 4.0. This includes standards for ship data networks and for information transfer between ship and shore [41]. However, developments are relatively slow and much more effort is needed to provide the necessary foundation for the innovations that Shipping 4.0 can provide.

As has been pointed out before, open standards are perhaps the most important factor to speed up the development of Shipping 4.0. The shipping business is relatively small with around 50 000 ships in regular international trade. This is a correspondingly small market with very limited room for innovations if the current regime with proprietary interfaces and networks continues. There will of course be innovations from the different manufacturers of systems for use on their networks, but much of the promise of Shipping 4.0 lies in the integration of information from across systems and between ships where, more often than not, control systems are delivered by different manufacturers.

It is probably necessary with a stronger involvement from the ship owners, where they require more openness in system integration and data interfacing. This will come at a cost, but this will be much lower than the gains one can get from better integration of systems and simple access to data for analysis.

On a similar subject, one should also investigate the implications of data ownership. There is a tendency today that some advanced monitoring and control systems supply performance and maintenance related data that is not accessible for the ship owner. This may hinder deployment of Shipping 4.0.

Finally, one needs to encourage data sharing. There are relatively few ships with similar construction, on board equipment and comparable trades. Being able to pool data from several ships, independent of owner and operator, gives much more information to build new analysis functions on. However, data sharing has implications for privacy and data ownership that need to be solved for this to work properly.

9. CONCLUSIONS

We have given an overview of how we believe the next revolution in ship operations – "Shipping 4.0" – will manifest itself. We have given an overview of the main components, how these interact and how the components are linked to the corresponding "Industry 4.0" developments in the shore based industry. We have given some brief references to already ongoing work and some rationale for selecting the technology components that we have included in Shipping 4.0.

Shipping 4.0 extends the historical connection between shore technology developments and uptake in the maritime sector.

We believe that Shipping 4.0 contains crucial elements that are needed to realize further efficiency gains and reductions in fuel consumptions on sailing ships. The introduction of Shipping 4.0 can reduce greenhouse gas emissions and increase the competitiveness of the sector, which is particularly important in short sea trades.

Finally, we have emphasised the very important role that open standards need to play in these developments. This includes issues related to data ownership and sharing.

10. ACKNOWLEDGEMENTS

This work has in part been supported by the Norwegian Research Council through the projects SESAME Straits (contract 235414/O70) and CySiMS (contract 256508/O80).

11. REFERENCES

1. KAGERMANN H., LUKAS W.D., WAHLSTER W. 'Industrie 4.0: Mit dem Internet der Dinge auf dem Weg zur 4. industriellen Revolution'. *VDI Nachrichten*, (13) 2011. Retrieved December 2016 from: http://www.vdi-nachrichten.com/Technik-Gesellschaft/Industrie-40-Mit-Internet-Dinge-Weg-4-industriellen-Revolution

2. STEINE, T. O.. The founding, fantastic growth, and fast decline of Norsk Data AS. *IFIP Conference on History of Nordic Computing*. Springer Berlin Heidelberg, 2010. p. 249-257.

3. WAHLSTER, W. 'Industry 4.0: From smart factories to smart products'. *Forum Business Meets Research BMR*. 2012. 22nd May 2012, Luxembourg – Kirchberg.

4. LEE, E. A. 'Cyber physical systems: Design challenges'. *11th IEEE International Symposium on Object and Component-Oriented Real-Time Distributed Computing (ISORC)*. IEEE, 2008. p. 363-369.

5. RØDSETH Ø. .J. 'Sustainable and Competitive Cyber-Shipping through Industry 4.0'. *Singapore Maritime Sustainability Forum 2016: Smart Maritime Solutions and Overcoming Challenges*, April 19th 2016, Suntec City, Singapore.

6. HERMANN M., PENTEK T., OTTO B., 'Design Principles for Industrie 4.0 Scenarios' *49th Hawaii International Conference on System Sciences (HICSS). IEEE, 2016. p. 3928-3937.*

7. LEE, J.; KAO, H.-A.; YANG, S. 'Service innovation and smart analytics for industry 4.0 and big data environment'. *Procedia CIRP, 2014, 16: 3-8.*

8. BRETTEL, M., et al. 'How virtualization, decentralization and network building change the manufacturing landscape: An industry 4.0 perspective.' *International Journal of Mechanical, Industrial Science and Engineering, 2014, 8.1: 37-44.*

9. MONOSTORI, L. 'Cyber-physical production systems: roots, expectations and R&D challenges.' *Procedia CIRP, 2014, 17: 9-13.*

10. ANG, J. H.; GOH, C.; LI, Y. 'Smart design for ships in a smart product through-life and industry 4.0 environment.' *Evolutionary Computation (CEC), 2016 IEEE Congress on. IEEE, 2016. p. 5301-5308.*

11. RØDSETH, Ø.J. LEE, K. 'Secure Communication for e-Navigation and Remote Control of Unmanned Ships.' *In: Proc. of the 14th Conference on Computer and IT Applications in the Maritime Industries-COMPIT.* Technische Universitat Hamburg-Harburg 2015, ISBN 978-3-89220-680-4

12. RØDSETH, Ø. J. 'A maritime ITS architecture for e-navigation and e-maritime: Supporting environment friendly ship transport.' *IEEE ITS Conference. 2011. p. 1156-1161.*

13. RØDSETH, Ø. J.; CHRISTENSEN, M. J.; LEE, K. 'Design challenges and decisions for a new ship data network.' *International Symposium Information on Ships 2011 – ISIS 2011, Hamburg, September 2011.*

14. ATZORI L, IERA A, MORABITO G. 'The internet of things: A survey.' *Computer networks. 2010 Oct 28;54(15):2787-805.*

15. IEEE, 'IEEE IoT Architecture' (Retrieved Nov. 2016): h*ttp://standards.ieee.org/develop/wg/IoT_Architecture.html*

16. ETSI, 'ETSI IoT Architecture' (Retrieved Nov. 2016): *http://www.etsi.org/technologies-clusters/technologies/internet-of-things*

17. RØDSETH, Ø. J., PERERA, L. P., MO, B. 'Big data in shipping-Challenges and opportunities.' *15th International Conference on Computer and IT Applications in the Maritime Industries - COMPIT '16.*

18. PERERA, L. P., MO, B. 'Marine Engine Operating Regions under Principal Component Analysis to evaluate Ship Performance and Navigation Behaviour.' *IFAC-Papers On Line, Volume 49, Issue 23, 2016, Pages 512-517*

19. EMSA. 'Annual Overview of Marine Casualties and Incidents 2015', *Report from European Maritime Safety Agency, Lisbon 2015.*

20. LÜTZHÖFT, M; GRECH, M. R.; PORATHE, T. 'Information environment, fatigue, and

culture in the maritime domain.' *Reviews of human factors and ergonomics, 2011, 7.1: 280-322.*

21. PORATHE, T; PRISON, J. 'Design of human-map system interaction.' *In: CHI'08 Extended Abstracts on Human Factors in Computing Systems. ACM, 2008. p. 2859-2864.*

22. CARDOSO, J., VOIGT, K., WINKLER, M., 2008, June. 'Service engineering for the internet of services.' *In International Conference on Enterprise Information Systems (pp. 15-27). Springer Berlin Heidelberg.*

23. IMO. 'Draft E-Navigation Strategy Implementation Plan', *Report to the Maritime Safety Committee, Annex 7, NCSR 1/28,* 10th September 2014.

24. ITU. 'Technical characteristics for a VHF data exchange system (VDES) in the VHF maritime mobile band.' Recommendation ITU-R M.2092-0. October 2015.

25. FATHI, D. E., et al. 'Integrated decision support approach for ship design.' *OCEANS-Bergen, 2013 MTS/IEEE.* IEEE, 2013. p. 1-8.

26. HASSANI, V., et al. 'Virtual Prototyping of Maritime Systems and Operations.' *ASME 2016 35th International Conference on Ocean, Offshore and Arctic Engineering. American Society of Mechanical Engineers, 2016.*

27. DNV GL, 'Making your asset smarter with the digital twin', Retrieved December 2016 from: *https://www.dnvgl.com/article/making-your-asset-smarter-with-the-digital-twin-63328*

28. RØDSETH, Ø. J., BURMEISTER, H. C. 'Developments toward the unmanned ship.' *In Proceedings of International Symposium Information on Ships–ISIS 2012 (Vol. 201, pp. 30-31).*

29. http://stmvalidation.eu/ [Accessed December 2016].

30. VAN WESTRENEN, F.; PRAETORIUS, G. 'Maritime traffic management: a need for central coordination?' *Cognition, technology & work, 2014, 16.1: 59-70.*

31. RØDSETH Ø.J. 'Design Challenges and Opportunities for Future Unmanned Ships', *Autonomous Ship Technology Symposium 2016, Amsterdam 21-23 June.*

32. MELL, P; GRANCE, T. 'The NIST definition of cloud computing.' *NIST Special Publication 800-145.* September 2011.

33. IMO. Development of an E-Navigation Strategy Implementation Plan, Overview of the Maritime Cloud concept, NCSR 1/INF.21. 25 April 2014.

34. https://3dprint.com/9021/maersk-ships-3d-printers/ [Accessed December 2016].

35. RØDSETH, H., MO, B. 'Integrated Planning in Autonomous Shipping—Application of Maintenance Management and KPIs.' *In Proceedings of the 10th World Congress on Engineering Asset Management (WCEAM 2015) (pp. 497-504).*

36. HU, H., et al. 'Toward scalable systems for big data analytics: A technology tutorial.' *IEEE Access, 2014, 2: 652-687.*

37. https://www.shipping-kpi.org/. [Accessed December 2016].

38. IMO, 'Report of the Maritime Safety Committee on its Eighty-Fifth Session, Strategy for the Development and Implementation of E-Navigation.'. *MSC 85/26/Add.1, Annex 20.* 6 January 2009.

39. LLOYD'S REGISTER ET AL., 'Global Marine Technology Trends 2030', *Report from Lloyd's Register, QinetiQ and University of Southampton.* Published August 2015.

40. DNV GL, 'A Broader View, The Future of Shipping.' *Report published by DNV GL,* Høvik, 2014.

41. RØDSETH, Ø. J. 'Integrating IEC and ISO information models into the S-100 Common Maritime Data Structure.' *E-navigation Underway International 2016 – Printed as MARINTEK report 302000778, 2016-02-14.*

12. AUTHOR BIOGRAPHY

Ørnulf Jan Rødseth holds a master's degree in electronics engineering from the Norwegian Institute of Technology (now NTNU) in 1983. He has worked with maritime computer and data communication systems since 1990. He is currently a senior scientist at SINTEF Ocean AS in Trondheim, Norway and is, in addition to research on Shipping 4.0 technologies, mainly working with autonomous merchant ships. He is the author of numerous technical papers on these subjects. He is also active in international standardisation and has been editor of and contributor to several international technical standards.

WILL THE SMART SHIP ALSO BE THE LIABLE SHIP?: AN ANALYSIS OF THE APPLICATION OF LIABILITY TO THE SHIP ITSELF

H Stones, Institute of Maritime Law, University of Southampton, UK

SUMMARY

As engineers develop the future of shipping: unmanned ships, and autonomous operating systems, which do not need a master or crew on-board, it is asked: 'who shall be held responsible?' Research into autonomous systems has considered holding the owner, the manufacturer or programmer responsible by considering the autonomous system as their instrument. Another suggestion is to make the ship liable.

The problem is that the idea of the system itself being held liable has not been considered in relation to shipping. It is concluded that it would not be possible to hold the ship liable in international maritime law. Previous research found that it is problematic to impose remedial measures on a system; additionally, in maritime law the system is based on the owner being liable and minimal change being required is needed to ease the introduction of unmanned ships. Therefore, liability will not be imposed on the ships themselves.

1. INTRODUCTION

Unmanned ships are the next big revolution in the shipping industry, but how they operate will vary greatly. The aim is for them to operate without a master or crew on-board. This could be achieved through smart systems, such as remote-control, or autonomy. Therefore, it is without a doubt that an unmanned ship will be a smart ship (as will manned ships that combine these systems with crew on-board).

However, what is in doubt is whether changing how a ship is controlled, and operated, will change the liability system. Therefore, this paper will explore the entities involved in the operation of an unmanned ship that could be held liable. The current liability laws will also be explained, and it will be discussed whether the current system will be appropriate. Then alternative systems of imposing liability will also be considered, importantly including whether the ship itself can be held liable. This builds on the current research on models of liability by applying them to smart shipping. This will include a discussion of the challenges of imposing remedial measures on a ship.

2. ROLES OF PEOPLE AND SYSTEMS

Although smart ships are considered to be unmanned they may still be manned, and the systems complement the role of the crew. Smart ships will vary in manning, and in how their control systems operate (see figure 1). Therefore, smart shipping does not mean the instant and utter end of seafaring: some ships may need to be manned, but will make use of the technology, and other ships may represent a phased introduction of smart technology (especially during the early stages, until reliability and trustworthiness are established).

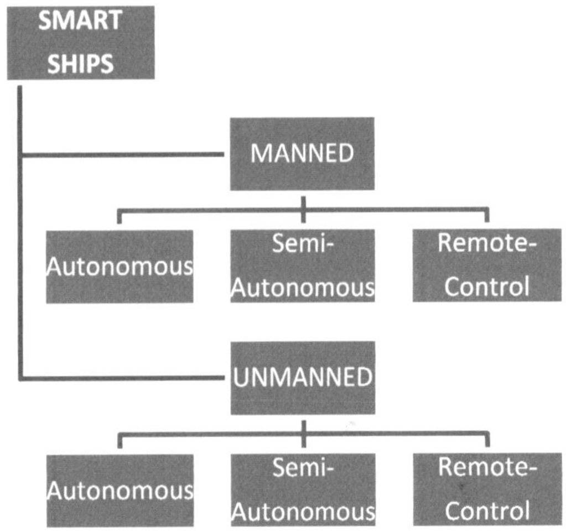

Figure 1: showing that both manned and unmanned ships could use similar smart technology to operate.

2.1 PEOPLE AND SYSTEMS ON UNMANNED SHIPS

On unmanned ships the ship will operate either autonomously or through remote-control, or a combination of the two. The main human involvement with unmanned ships is during the creation of the ship (even more so when autonomously controlled) to ensure that the smart technology is capable and reliable. However, humans will be more directly involved when remote-control is used. There will be greater reliance on the smart technology on an unmanned smart ship than on a manned smart ship.

2.2 PEOPLE AND SYSTEMS ON MANNED SHIPS

Manned ships could be autonomous for large parts of a voyage. For example, autonomous mode could be used when performing deep sea shipping then the crew could take control when near the coast or in busy shipping lanes. In the Maritime Unmanned Navigation through Intelligence in Networks project it was suggested that the ship could be unmanned when in the deep sea, but a pilotage crew come on-board when approaching the coast and return manned control [1]. The same approach could be taken, but instead of getting them on-board they are waiting on-board. This would be especially useful and more cost-effective for ships that are in port frequently.

Manned ships could utilise the technology from remote-control systems, so instead of a simulated bridge in a shore control centre the additional data could be presented on the bridge to complement what the crew can perceive for themselves. Although this would not be as remote as one envisages when discussing these systems. This would not be that different from current data reception and control from the bridge, the main differences would relate to the data received and the lack of manual tasks elsewhere on the ships (they would only work on the bridge). However, there could be the ability for them to retake control as part of a redundancy system. There is also the option to transfer between remote-control, and autonomous systems, and thus reduce their role further. This could be a way of phasing in the introduction of remote-control and autonomous systems.

Therefore, there are many roles that a crew can perform on-board a ship even with smart technology:

- Normal crewing assisted by smart technology,
- Remote-controllers from the bridge,
- Supervision when in autonomous mode,
- Authorising certain actions, while others are carried out autonomously,
- Pilotage (e.g. when near the coast), and retaking control when circumstances dictate.

2.3 REMOTE-CONTROLLERS

When remote-control is utilised on a manned smart ship it is clear that the remote-controllers would be considered as comprising the master and crew. However, with an unmanned smart ship this is not as clear. In order to ensure the applicability of maritime law, without having to make a lot of amendments, which is considered by many lawyers to be essential to the successful introduction of unmanned ships, remote-controllers need to be considered as the master and crew on-shore [2].

If they were not, this would involve the creation of new regulations to govern them, and although they could do so and borrow provisions from existing maritime law, this would take longer and may delay the introduction of unmanned smart ships. Arguably there are benefits from this delay: for example, the reliability of the technology can be demonstrated for longer on manned ships and other vehicles; and allow seafarers longer to adapt and retrain.

However, this writer is not convinced that these are substantial benefits, as training will begin when smart technology becomes more prominent on manned ships, and there will be willingness from seafarers to train to ensure employability. Additionally, the shipping industry is famously slow to develop in this field, thus the technology will be shown to be effective through other industries, and the shipping industry will not introduce smart shipping until it is satisfied with its ability (the risk cannot be too high, as manning has become less of a prevalent cost).

Therefore, the simplest solution is to interpret the master and crew as capable of being on shore. If necessary this could be reaffirmed through a convention that will state that for unmanned ships the remote-controllers will be the crew, with the manager (or chief controller) as the master. In turn these personnel (who are likely to be ex-seafarers) can be required to have the same training as a seafarer with the addition of remote-controller training.

3. LIABILITY IN MARITIME LAW

Maritime law is a specialist area of the law, and, as already been discussed, poses unique challenges. Maritime law is comprised of international and national law. International law comprises of conventions and customary international law. Primarily, this paper focuses on international maritime conventions, which often comprise of five defining features:

- Channelling liability to the shipowner,
- Strict liability,
- Limited liability,
- Compulsory insurance,
- Direct action against the insurer.

These features allow for third parties to receive compensation for damage as easily as possible. These conventions can be complemented by the provision of central funds (for instance, in relation to oil pollution there is the International Convention on Civil Liability for Oil Pollution Damage 1992 Protocol, as amended (CLC), and the International Convention on the Establishment of an International Fund for Compensation for Oil Pollution Damage 1992 Protocol, as amended (FUND)).

The conventions themselves place liability on the shipowner. Although through insurance, or compulsory insurance and direct action against the insurer, for larger ships, the insurer pays. Therefore, the shipowner has the responsibility to pay their premium or call, abide by the provisions of their insurance, and minimise the risk of incident. The master can be important for the channelling of liability to the shipowner, as the master represents the shipowner on-board. A master can also be personally liable.

International maritime law usually involves civil liability, but some conventions when violated also create criminal liability (e.g. International Convention for the Safety of Life at Sea 1974, as amended), for which the remedy is monetary.

However, this model of liability is not necessarily the best model for such a revolutionary change in shipping. Especially since smart shipping will eventually involve artificial intelligence and autonomy. A separate convention for smart ships will not be considered in this paper, but whether the law can be interpreted to impose liability differently than it does now.

4. LIABILITY OPTIONS

As Susan Lanoue said, "the potential for human injury due to a malfunctioning computer program is tremendous. No longer can computer-caused injuries be relegated to the realm of futuristic science-fiction novels" [3]. However, it is not just human injury that can be caused, there could be property damage or economic loss too [4]. Therefore, it is important that there is an effective liability system that imposes sufficient liability on the appropriate party.

Table 1: showing who could be held liable under each liability model. Numerous people could be held liable under each model or between models if an incident occurred.

Liability model	Who could be liable?
Perpetration-via-another	Programmer Manufacturer Owner Operator
Natural-probable-consequence	Programmer Manufacturer Owner Operator
Direct	Artificial intelligence entity

Gabriel Hallevy has been a strong proponent of exploring different liability models in relation to artificial intelligence. He explored three models of liability:

- The perpetration-via-another liability model;
- The natural-probable-consequence liability model;
- The direct liability model [5].

Hallevy has focused his discussions on criminal liability, but this paper will discuss the models in relation to civil liability as well. The first two models consider liability for the maker (programmer/manufacturer), or the user (owner/operator), whereas the third model imposes liability on the artificial intelligence entity itself (see table 1). If applied to the owner, the first two models could apply the law as it is, but justified differently. Hallevy considers these models as applicable to unmanned vehicles, which in future will include smart ships [6].

4.1 PERPETRATION-VIA-ANOTHER

Hallevy's perpetration-via-another model considers the artificial intelligence entity as an innocent agent, which although highly developed is still just a machine [5]. Therefore, the entity is not considered capable of the commission of the offence, and is the instrument of either the maker (as they design the entity to commit the act), or user (as they instruct them to commit the act) [5]. However, this model is not appropriate if the entity decides for itself to commit the offence [5].

4.2 NATURAL-PROBABLE-CONSEQUENCE

Hallevy's natural-probable-consequence model is also based on the involvement of the maker or user, but not them intending the offence (as under the perpetration-via-another model) [5]. This model is based on whether they could have foreseen the commission of the offence, and thus whether they were liable in not preventing it [5]. For instance, as an expert a programmer would be held to the standard of the reasonable programmer [7].

4.3 DIRECT LIABILITY

Hallevy's direct liability model is based on no human having involvement in the commission of the offence [5]. A human is not found to have a mental state for the completion of the offence [5]. Thus, the entity is considered as tantamount to a human (i.e. a legal person) [5]. As unmanned ships are developing to the point that they can perceive the environment themselves this model is appropriate. Having the ability to understand what they perceive, and interpret it, is equivalent to a human [6]. Therefore, it is important that by applying direct liability that a ship will be treated as having legal personhood [6].

David Vladeck considers legal personhood to be dependent on the development of full autonomy, and considered there to be no reason for them not to have legal personhood [8]. Vladeck summarises this as "they will not be tools *used* by humans; they will be machines *deployed* by humans…" [8]. It is thought that their ability to perceive and decide like humans will allow them to behave like humans, and perform to a higher standard without distractions [8].

However, there are many challenges posed by direct liability, which will be further explored later. These challenges include imposing remedies, insurance, and the nature of maritime law.

4.4 STRICT PRODUCT LIABILITY

Another option has been explored by other writers, including Lanoue, which is to apply strict product liability [3]. Lanoue recognises that a benefit of strict product liability lies in not having to prove negligence (or intent) [3]. It is based on a defect in the product, and although there has been discussion as to whether such control systems are products or services, it appears as though they are products; most doubt focuses on when they are custom-made, as they may be for larger ships [3].

Strict product liability is usually justified as providing reassurance for products that pose an additional risk [9]. This seems counter-intuitive when smart shipping supposedly represents safer shipping. However, new risks are considered to be riskier than risks that people are familiar with. Therefore, as a user the guarantee of compensation for a smart ship from the manufacturer may encourage the user to change from conventional ships to smart ships. In fact, the risk is smaller, yet the unknown creates the perception of greater risk. The same can be seen throughout the insurance industry, for instance when the flood risk is unpredictable it results in higher premiums [10]. Thus, strict product liability can be considered appropriate for smart ships which will increase safety. Additionally, this will be a small burden on manufacturers, because although they will be liable there will not be as many incidents for which they will have to pay.

Strict product liability is likely to place liability on the maker (manufacturer/programmer) instead of the owner, as it is based on the product being defective. Vladeck proposes that strict liability in relation to such smart machines/systems would not be based on the fact that vehicles are highly risky, but due to the reduced risk so that any incident is deserving of liability [8]. Strict product liability gives users confidence to accept the risk, even though it is a small risk [3],[9].

Whether strict product liability is justified by the greater perception of risk, or the reduced risk of incident it can be justified. However, the greater perception of risk is a justification that is more concurrent with current justifications.

There are many benefits to strict liability, including the ability to deter faults that could result in liability, and improve standards [3]. However, there is some doubt as to whether this can be achieved in machines that are controlled by computer systems, which are more intelligent and designed to be safer [9].

Unfortunately, intelligent systems such as computers develop 'bugs' easily, which could mean a lot of liability and could deter the industry developing. However, Lanoue notes that the common nature of faults has not been a problem for computer system manufacturers, so other companies may not be deterred [3]. Also, if it gets to the stage that these smart ships are safer than current ships, strict product liability will not cause as much of a financial burden.

Lanoue notes that it would not be suitable if the law allowed, in relation to cars, a strict product liability claim for a steering mechanism but did not if that steering mechanism was defective due to a computer program [3]. This argument focuses on ensuring equality between claimants in law. In maritime law, a lot of liability is fault based, but there is also limited strict liability, and all liability is channelled to the shipowner. Therefore, it could be argued that although strict product liability is an appealing option, it would introduce different liability systems based on how smart the ship is.

Strict product liability would also not preclude the maker, or their insurer, from recovering later from any persons that contribute [8].

4.5 COMBINING MODELS OF LIABILITY

It is possible that using one of Hallevy's models could find both the maker and user liable, or using a combination of the direct liability model and one of the others that the system could be held liable along with the maker and/or user (especially if there is complicity between the actors). Hallevy asserts that the direct liability model can apply independently, so the entity alone can be liable [5].

Additionally, the mode by which the manufacturer, user, or ship is found liable could be strict product liability. Although it is more likely to be the manufacturer who is held strictly liable. It would not be justified as it is under Hallevy's models. However, it would be the same result. Then they could try to pursue the user, or even the ship for their contribution (applying Hallevy's models). This would allow the channelling of liability that aids claimants, which is favoured in maritime law.

As Vladeck notes, if humans are still involved in decision making then the liability rules will be the same [8]. Therefore, if either the perpetration-via-another liability model, or the natural-probable-consequence liability model, the law will apply to humans and corporations as it does now.

It is also possible that all parties could be held liable, along with the ship itself (see figure 2). However, it seems redundant to hold the owner liable as well as the ship. Although the action of the ship may be utterly unforeseen and out of the control of the owner, if a ship is owned it will be effectively being holding the owner liable by two methods. The ship is not independent, and will depend on the resources of the owner (e.g. maintenance and finance) – there will still be a company for the ship, so holding the ship liable would be akin to holding the shipowner liable. Each ship will be independent, as ships are now through one ship companies, so this would not be remedied and increase sister ship arrest.

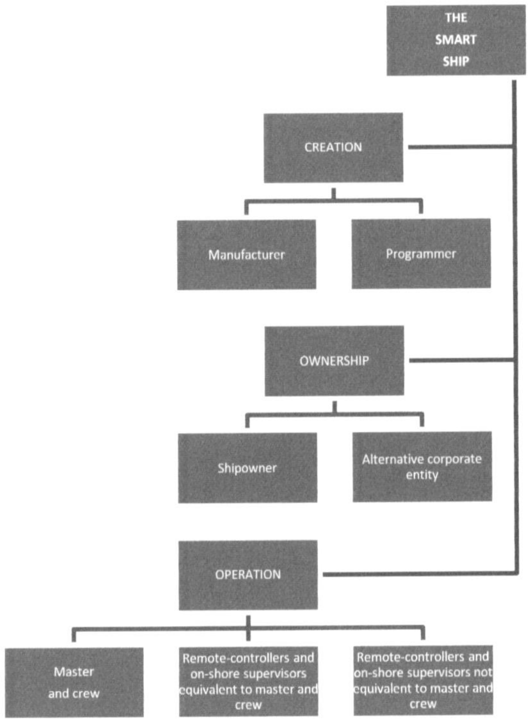

Figure 2: showing the parties in the aspects creation, ownership, and operation of a ship, and thus indicating that there are many different parties who could be held liable if there is an incident with a smart ship.

Therefore, it is possible that many could be held liable for an incident (most likely through a series of cases). This is why it is beneficial to channel liability. The issue then becomes who is it best to channel liability to: the manufacturer, the shipowner, or the ship?

4.6 WHY FIND THE SHIP LIABLE

As previously discussed, there are benefits to channelling and imposing liability on each of the parties, but this paper will now focus in more detail on the potential liability of the ship.

The arguments that favour holding the ship liable focus on the autonomy of the ship mean that the ship must be held to account and responsible by the law. For this to be effective, the ship would have to be programmed to understand the law, responsibility, and justice, and respect the law so that it complies with the law. This programming could mean that violations of the law are fewer than with seafarers, shipowners, and remote-controllers, who do not necessarily know the law as well (especially with the passage of time and forgetfulness); who could be more inclined to think that they are not breaking the law when acting under another motivation; or willing to disregard it. For example, a system can follow the rules of navigation in the Convention on the International regulations for Preventing Collisions at Seas 1972 and not incorrectly perceive a situation through a 'trick of the eye' and then make the wrong manoeuvre resulting in damage. Initially this programming could be ideal for ensuring the effectiveness of the law.

However, through time, the autonomous system will learn, and may determine that violating the law is more beneficial (especially if economical, or cost-effective, disregarding the cost of liability). This could be prevented by programming the ship with limitations that will not allow violations of the law, unless it is absolutely necessary and not for the selfish gain of the ship (as in *The Saint Jaques II* [2002] EWHC 2452, [2003] 1 LLR 203).

An autonomous ship will need to be programmed with these limitations to ensure that its learning is legally compliant. There will still be a role for the courts because if there is an incident, and the investigation finds that the decision of the ship was not clearly correct the data recorded will have to be presented to the court to determine liability (which will then require the legally compliant programming in that ship and others to be updated). These cases will also occur due to the fact that law is not always clear, and often intended to leave some flexibility, which leads to ambiguity.

5. HOLDING A SHIP LIABLE

As previously mentioned, there are many reasons for imposing liability on the smart ship. Most of these reasons depend on the autonomy of the smart ship. Thus, not all ships, and not all smart ships should be held liable: only autonomous smart ships. Other smart ships would require the application of current maritime law, or the law to be applied using one of the other models discussed in this paper (as in the section that discusses smart ships that utilise remote-control).

However, there are still practical problems with imposing liability on a smart ship, even if theoretically autonomy can justify liability and legal personhood can impose that liability.

5.1 DEFENCES

Hallevy, in relation to criminal law, asked whether a system that is malfunctioning, and its capabilities corrupted into making the wrong decision, can use the defence of insanity [5]. It can also be asked whether the system can suffer a loss of control [5].

Hallevy also asks whether the infancy defence would apply to systems, as they can develop an understanding of right and wrong [11]. This would be further supported when a system has not been programmed to understand the role of the law, and respect it. However, this programming would be there throughout its life, so this understanding would not be developed in the same way as in a human. Though it could still develop through the exposure of the system to the world and the knowledge it acquires. The problems with this are supported by the fact that corporations that have legal personhood, are unable to use a defence equivalent of infancy.

Hallevy also considers whether a virus infecting the operating system would be the equivalent to the influence of intoxication [11].

The important consideration to make here is that legal personhood is being suggested for the ship, which is more likely to be akin to the liability of a corporations and not a human. These defences are not available to corporations, and thus are unlikely to be available to a ship. Although they do raise interesting ethical and theoretical points making these defences available to the ship would not be practical.

5.2 REMEDIAL MEASURES

Punishments ranging from the death penalty, to imprisonment, to probation, and to fines are problematic when imposed on a system, so it may need to be asked whether other remedial measures will need to be developed instead [11]. For instance, corporations are held liable for any criminal offence that has a monetary penalty (though individuals in the corporation in certain circumstance will be personally liable).

5.2 (a) Capital punishment

Capital punishment is an effective measure for preventing an actor from committing further offences [5]. The same would apply in relation to a ship, if the whole ship or that control system was taken out of commission. However, this would not have one of the principle justifications of capital punishment, because it would deter another ship from committing an offence as it does not have a life to lose.

5.2 (b) Imprisonment

Hallevy considers imprisonment of an entity would prevent the entity from committing further offences for the specified period [5]. The inability to operate is tantamount to depriving the entity of freedom [5]. However, can a ship be considered as having freedom or liberty? Even with artificial intelligence, the ship would be made to work and would not have the freedom to do things other than work.

5.2 (c) Community service

Hallevy argues that the entity can be used for community service, as it will often be used for private benefit and it could be made to work for public benefit instead [5]. Doing work, other than what it is programmed to, is not a punishment: the reason for a voyage is not relevant, simply that it is on a voyage.

5.2 (d) Fines

Monetary penalties are also popular remedial measures, and of particular relevance to shipping. One method would be to take property, which could be the ship itself. Hallevy argues that since it is possible to fine a corporation that it is possible to fine a legal person (in criminal and civil law) [5]. As like a corporation it is still run by people, it represents an accumulation of people (wealth, knowledge, freedom to accept certain voyages), and becomes more than them. It develops its own wealth through voyages, as a corporation does through business [11].

However, currently, a ship is owned by a corporation – it is an asset of company, not the company itself. Thus, it would require a different understanding of a shipping company than is currently used. If this were possible, then the effect of the monetary penalty would be the same as it is now on the shipping company and its insurer.

5.3 MARITIME REMEDIAL MEASURES

5.3 (a) Arrest

Remedies and measures taken in maritime law specifically need to be considered in more detail now. Arrest of a ship is a form of obtaining security against the shipowner for a claim.

Arresting is considered to be problematic to Hallevy when there is not a physical body, but a ship does have a physical body to arrest [11]. However, there would be a physical ship to arrest, so this would solve Hallevy's problem.

The arrest of the ship could also serve to reinvigorate arguments regarding the fiction of personhood for *in rem* arrest of the ship in maritime law. It would make sense that the ship will be held to account, as the ship is its own owner, it is the shipowning corporation, it is responsible for the fault instead of the shipowner (as a separate entity) under the law currently (*in personam* arrest).

In American law, the fiction of personification still exists for *in rem* arrest, and Michal Chwedczuk recognises the impact that unmanned shipping can have on *in rem* arrest in America [12]. Chwedczuk hopes that despite the ability that unmanned shipping has to support the fiction that it will not and instead American arrest law will become as it is elsewhere (i.e. no longer apply the fiction of personification) [12].

However, the worrying aspect of unmanned or smart shipping is that it could reinforce it in America and resurrect the fiction elsewhere to the point that it is not an isolated fiction, but a greater fiction of maritime law. It could also lead to conferring legal personhood on the ship.

5.3 (b) Compensation from the shipowner

Another aspect of maritime law is obtaining compensation for any damage caused by a ship. The shipowner is the first source of compensation in maritime law, and liability is channelled to the shipowner in order to make it easier for claimants to get compensation. Although the money itself is likely to be from the insurer, it is still easier to make all claims to the shipowner and they can try to recover from other parties.

This could work for a ship as well, the ship could be insured as it is now and have the restrictions of the insurance contract programmed into it (meaning that it may be less likely to violate the terms, so it would increase the odds of the insurer paying and preventing incidents from such violations).

5.3 (c) Insurance

Insurance is the means by which claims under maritime law against a shipowner tend to be met. By imposing direct liability on ships the burden of compulsory insurance would be on the ship and not the shipowner. As Vladeck notes, imposing insurance requirements on the ship instead of strict liability on the maker would be another way of spreading the financial burden [8]. Maritime law conventions often include provisions for compulsory insurance, so the financial burden is spread as it is, so strict liability may not be necessary. However, the error in Vladeck's statement is that implies that it is a choice between compulsory insurance and strict liability. Yet maritime law will often combine the two for better claimant protection.

5.3 (d) Wilful misconduct defence in insurance

In the provisions for compulsory insurance, the insurer is allowed to utilise any defences that the shipowner is entitled to, and additionally when the claimant takes direct action against the insurer they utilise the defence that it was the wilful misconduct of the shipowner. This means that the shipowner will be liable from their finances, and the insurer will not pay. Importantly, the claimant will still be entitled to damages, though if taking direct action against the insurer, it will require the claimant to take the shipowner themselves to court.

The same could apply to ships – if the ship autonomously misconducted itself wilfully the insurer could avoid liability. Such misconduct may even be easier to establish, as the system could be prevented from fraudulently hiding that misconduct and the data recordings would be analysed automatically in the case. Then the finances of the ship (like a corporation) would be used to pay damages to claimants.

However, this defence may be worrying for claimants when autonomous systems and machines represent an unknown risk. It could be a concern that wilful misconduct will be become more prevalent, and the security that comes through an insurer will not be as available – thus defeating the purpose of compulsory insurance provisions.

5.3 (e) Compensation from a central fund

Central funds are a common component of numerous maritime conventions (e.g. FUND) in addition to compensation from the shipowner. Central funds are used to complement compensation from the shipowner (or their insurer). It is made of contributions from various parties. For instance, for the FUND for damage from oil pollution there is a levy on receivers of oil in State Parties (above a set amount), the cost of which is then passed onto the consumer. It is considered to be a pillar of how the conventions operate in order to ensure effective compensation.

However, the question becomes whether central funds would still be able to operate effectively if ships are unmanned and operate solely through smart technology. This question is whether these contributions will still be made.

This writer concludes that these contributions will still be made, as they are not made by the shipowner. For instance, under the FUND the contributions are made by receivers of crude or heavy fuel oil through a levy. Therefore, differences in the nature of the ship will not affect their roles as cargo receivers. Therefore, compensation will still be available under central funds.

5.5 WHICH REMEDIAL MEASURES ARE PRACTICAL?

Although the physical attributes of the ship allow for the imposition of greater remedial measures, this writer concludes that this is not viable, as it would introduce too much inconsistency between ships. The imposition of legal personhood is based on system developments, and thus in other areas of the law not as many remedies would be available for systems with similar capabilities (e.g. *in rem* arrest could not apply for systems without a physical presence). Therefore, the remedies should be those available for corporations as legal persons: fines in criminal law, and damages in civil law. This prevents maritime law being more severe on smart technology than other areas of the law, while also allowing for consistency between all ships and shipowners.

6. CONCLUSIONS

This paper assumed that the conventions and all other maritime law should still apply to a smart ship based on the purpose of the ship (e.g. dry-bulk cargo, oil tanker, passenger ship), and not involve a separate liability system for smart ships. This would represent to greater change in the law, and it is better to treat ships as equally as possible: fundamentally they are still ships, and ships before smart technology. The question considered in this paper, is whether they should mean the existing law is interpreted differently, so that liability is imposed on another party.

Although it is interesting to consider imposing liability on the manufacturer or the ship both are concluded to be unlikely. To encourage development and remain as similar as possible to current maritime law, liability should be channelled to the shipowner. Additionally, due to various types of smart ships (e.g. remote-controlled ships logically are the responsibility of the user, who is principally the shipowner) it is clearer to hold them all liable in the same way and the best way to do this is to hold the shipowner liable.

In relation to imposing liability on the ship, it is concluded that it is simply not practical. It raises a lot of confusion, despite the benefits that can it provide (e.g. justification for ship arrest, and providing more remedies than in relation to other forms of artificial intelligence). It would represent to greater change to the scope of liability for legal persons (in relation to autonomous smart ships). Therefore, liability should remain on the shipowner and not the smart ship.

7. ACKNOWLEDGEMENTS

I would like to thank my supervisors (Professor Mikis Tsimplis, Professor Andrew Serdy, and Dr Nicholas Townsend) for their continued input.

8. REFERENCES

1. MARITIME UNMANNED NAVIGATION THROUGH INTELLIGENCE IN NETWORKS, 'Impact on Short Sea Shipping', 2015. Available from: http://www.unmanned-ship.org/munin/wp-content/uploads/2015/10/MUNIN-D10-1-Impact-on-Short-Sea-Shipping-MRTK-final.pdf.

2. VAN HOOYDONK, E., 'The Law of Unmanned Merchant Shipping – an Exploration', *Journal of International Maritime Law*, 2014.

3. LANOUE, S., 'Computer Software and Strict Products Liability', *San Diego Law Review*, 1982-3.

4. LONGDIN, L., 'Liability for Defects in Bespoke Software: Are Lawyers and Information Scientists Speaking the same Language?', *International Journal of Law and Information Technology*, 2000.

5. HALLEVY, G., 'The Criminal Liability of Artificial Intelligence Entities – from Science Fiction to Legal Social Control', *Akron Intellectual Property Journal*, 2010.

6. HALLEVY, G., 'Unmanned Vehicles: Subordination to Criminal Law under the Modern Concept of Criminal Liability', *JLIS Special Edition: The Law of Unmanned Vehicles*, 2012.

7. NYCUM, S., 'Liability for Malfunction of a Computer Program', *Journal of Computers, Technology and Law*, 1979-80.

8. VLADECK, D. C., 'Machines Without Principals: Liability Rules and Artificial Intelligence', *Washington Law Review*, 2014.

9. DAHIYAT, E. A. R., 'Towards New Recognition of Liability in the Digital World: Should we be More Creative?', *International Journal of Law and Information Technology*, 2011.

10. UNION OF CONCERNED SCIENTISTS, 'Overwhelming Risk: Rethinking Flood Insurance in a World of Rising Seas' 2013. Available from: http://www.ucsusa.org/sites/default/files/legacy/assets/documents/global_warming/Overwhelming-Risk-Full-Report.pdf.

11. HALLEVY, G., 'The Criminal Liability of Artificial Intelligence Entities', *Social Science Research Network*, 2010.

12. CHWEDCZUK, M., 'Analysis of the Legal Status of Unmanned Commercial Vessel sin U.S. Admiralty and Maritime Law', *Journal of Maritime Law & Commerce*, 2016.

9. AUTHOR BIOGRAPHY

Hannah Stones holds the current position of PhD Research Student at the University of Southampton, and is based in the Institute of Maritime Law. Hannah is researching unmanned ships from an interdisciplinary perspective between law and engineering. Hannah's research is funded through the Vice Chancellor's

Scholarship, and the Leverhulme Trust through the Understanding Maritime Futures doctoral training scheme by the Southampton Marine and Maritime Institute.

EMBRACING A SMARTER FUTURE – HOW NAVIES COULD GAIN ADVANTAGE FROM SMARTER AND MORE AUTONOMOUS SHIPS SYSTEMS

J O Paul, QinetiQ, UK

SUMMARY

With the success of Unmanned Warrior 2016 demonstrating that unmanned and autonomous systems have a part to play in modern naval operations, QinetiQ are looking forward to see how autonomy technologies can be used in manned naval maritime platforms. As part of this, we have produced Concept Minerva, a Constabulary Vessel that investigates what equipment might be used in naval platforms in the not too distant future and what effect this will have on capabilities and manning. From this, we have concluded that adding automated and autonomous systems across the ship has the potential to profoundly influence manning, operator loading, maintenance & availability, information gathering, and the ability to deliver effects. However, realising these advantages will not necessarily be easy and there are many barriers to be overcome.

1. INTRODUCTION

This paper will introduce Concept Minerva, a constabulary vessel, which makes use of smart and autonomous systems to amplify and enhance the ships capabilities. It will discuss four main areas of interest:

- Open Architectures
- Offboard System Integration
- Highly Mechanised Handling Systems
- Smart Ship Management Systems

It will explain how the combination of these technologies will increase situational awareness, the ability to make timely tactical and operational decisions and deliver military effect, while also reducing operator loading. Through the use of automation and autonomy, a ship sensors and equipment can be used to gain the information edge and deliver effect in a more efficient manner.

The second half of the paper will look at the impact of these developments on design, procurement and the competency & training requirements of future crews. It will identify key areas in which a modern navy can adapt to take best advantage of the technology:

- Exploiting flexibility through open architecture and modular designs, making more use of modular equipment that is easier to upgrade through life and allows "plug and play" options to suit specific missions.
- Achieve responsiveness and agility through a split between procurement of the platform and the capability. Breaking dependency between platform build and equipment procurement.
- Identifying skills and training needed to deliver this new capability and question whether a paradigm shift in the way these vessels are operated is required.

2. CONCEPT MINERVA

Minerva (Figure 1) is a constabulary patrol vessel concept created for the purposes of exploring the benefits of autonomy on a naval platform. The technologies under investigation are all already available or at ever increasing technology readiness levels that will mature in the relatively near term through research and development projects. We will see how automation, autonomy, system integration and big data processing can be used to increase the effectiveness of a platform through reducing operator loading. This paper will investigate a few key areas where these can be applied; Command & control, Situational Awareness, Ship Management and the use of Offboard Systems to extend both data gathering and effect delivery operations.

Figure 1 - Concept Minerva

2.1 OPEN ARCHITECTURES

An open architecture approach, combined with modularisation and standardisation of both physical and electrical interfaces, allows for a flexible ship that is able to accept different modular mission packages that conform to the architecture's defined interfaces. This allows for a large amount of flexibility through the ability to swap in and out sensors and effectors with minimum impact to the system as a whole. This flexibility allows for easier upgrades of equipment and a ship to be modified in a timely manner to suit a specific mission.

The combat system's primary interface with the user comes from a Command & Control System (C2) which manages the flow of data and information into the system, displays it in the form of a tactical picture, and enables control of the platform's effectors. It gives the commander all the information necessary to make a tactical decision and the communication links necessary for him to command a response.

ViewFinder (Figure2) is a next generation C2 system which applies intelligence into the system to take loading away from the operator and carry out some of the 'mandrolic' and tedious tasks automatically or autonomously. Current command & control systems receive data from many sensory inputs with limited levels of supportive data processing; this requires operators to analyse the data or tracks from each of the sensors and then take a judgement on what is there and whether it is a threat. ViewFinder can do this autonomously through superior data fusion techniques and providing enhanced situational awareness by identifying abnormal behaviours.

Figure 2 - ViewFinder - Next Generation Command and Control and Tactical Picture Generation

ViewFinder has a data fusion engine that can overlay multiple sources of data of different types and combine them to gather all possible data on a given object. It is capable of fusing radars, sonar, ESM, datalink, Intelligence (Air Tasking Orders, Airspace Control Orders, Encyclopaedic Databases), IFF, AIS, ADSB and others. The system understands what data is attributable to which object and creates a very data rich picture of the operational environment. However this data then needs to be processed to identify and classify each of the objects in the picture; ViewFinder can analyse the data collected and also perform behavioural analysis of the object over time to provide a classification. ViewFinder then displays the tactical picture to the operators and monitors all tracks; it identifies well behaved tracks such as airliners in airlanes, and highlights suspicious behaviours. The user interface is highly configurable to display only the information each operator wants with different configurations for each console/role. It also allows users to interrogate the system to understand why units have been classified in a specific way and if necessary override the system.

This system has distinct advantages, by removing a number of monotonous tasks from the operators there can be a reduction in operator loading and possibly numbers of operators. While operator's attention can be diverted, ViewFinder's cannot, and it continuously and relentlessly reassesses every track. This delivers an accurate and constantly updating tactical picture which is not subject to human errors of poor communication, distraction or overloading.

2.1 (a) Command Aids & Intelligent Systems

As these systems become more intelligent they can start contributing to the decision making and command and control process. They can start considering responses to the events happening around them and evaluate those scenarios to come up with optimal responses to a given situation. These systems are at a relatively immature stage and not yet capable of the level of reasoning a human operator is but can still carry out a lot of the analysis quicker than a human. Thus the next level of autonomy on board a naval vessel is likely to be a command aid. Command aids can process the situation and provide suggestions of how to respond along with reasoning behind those choices, but critically the choice of which action to take remains with the human operator. These allow the operators to be assisted by autonomy while keeping overall control in human hands. This next section discusses potential for decision aids and handing over some control to autonomous systems.

In the future ViewFinder can be augmented further with Artificial Intelligence controlling the sensor package and tasking the sensors in the most effective manners. QinetiQ and it's partners are currently developing an intelligent surveillance system using the Sensing for Asset Protection with Integrated Electronic Networked Technology (SAPIENT) architecture. This system uses artificial intelligence both on the sensor and in the information fusion engine to allow the system to act autonomously, the result being that the system can autonomously cue sensors and control them in the best and most efficient manner against its prioritised objectives. As an example if the system has picked up an aircraft that might be a threat on radar but is having trouble classifying it, the system can redirect other sensors (electro optic and infra-red cameras for example) autonomously onto the track to gather more information rather than having to wait until the sensors have passively collected enough data. This technology could be plugged into the open architecture command and control system to provide sensor management and data fusion that gathers more data faster increasing the situational awareness of the platform and further reducing operator loading. SAPIENT is a fast maturing project and already had a number of successful trials on fixed infrastructure and is capable of transferring onto a moving platform.

So far we have reviewed the options for situational awareness, however the other side of the system, command & control, can also be augmented with levels of autonomy to help reduce operator loading.
Having gathered the tactical picture where suspicious or hostile tracks have been identified, further threat assessment can be carried out. This can be a mostly automatic process using an algorithm which assigns a threat level and priority to a track based on its classification, behaviours and other parameters, such as distance from the platform, speed, direction of travel, time before it is in range to attack. Prioritising the threats then enables decisions to be made on how to prosecute them and in what order. To take the idea further, an Artificial Intelligence engine could then decide on the most effective countermeasure or weapon system to use against each of the threats and in which order to prosecute them. There are a number of ways in which this could be achieved; using a "playbook" system that compares the current scenario to a set of pre-simulated scenarios and selects a response relevant to the matching scenario, using a tightly confined linear algorithm to assign a strict set of rules to how systems should respond, or a more non-linear solution taking advantage of AI to formulate a number of responses and consider their effectiveness and arrive at an optimised solution within a level of constraint.

Some vessels also rely on manually operated weapons systems, particularly smaller calibre machine guns. Executing command & control and communicating targets to the gunner can be a time consuming process with the possibility for misunderstandings. QinetiQ produce a module that can plug into the combat system and communicate targets and fire commands quickly to the gunner. POINTER uses an augmented reality gun site that gives the gunner an easy to understand heads up display. Targets can be broadcast into the display which can tell the gunner where to point the weapon and highlight the target once it is within the sites. It can also highlight friendlies with a different icon to help prevent friendly fire. The system also returns information on the weapon system back to the command and control system letting those in command know the arcs of fire, where the weapon is pointing and whether the threat is in the gun's blind arcs. This allows commanders to make better informed decisions on which weapons to task and the best heading for the ship to bring the threat into the weapons arcs of fire.

2.2 OFFBOARD SYSTEMS INTEGRATION

Concept Minerva proposes the use of mission modules that are easily replaceable. It is envisaged that these will be containerised units that can be placed on the aft deck of the vessel to provide mission specific equipment and payloads. This approach allows for the use of a selection of unmanned vehicles tailored to suit specific uses. For instance UAVs to increase Situational Awareness for missions like search and rescue and Exclusive Economic Zone (EEZ) security patrols or Unmanned Underwater Vehicles and Unmanned Surface Vehicles for mine countermeasure (MCM) and hydrography work (Figure 3).

Figure 3 - Unmanned Vehicles used for Mine Countermeasures Operations

Unmanned Vehicles can deliver many benefits to a platform, uppermost in mind is the ability to remove crew members from dangerous situations such as those encountered onmine countermeasure operations. However, these vehicles can put a strain on the operators with many different types of vehicle and control panels to operate them and their onboard payloads requires a large number of operators and soon it feels like the vehicles are far from unmanned.

QinetiQ and its partners have been collaborating on a research programme called Maritime Autonomous Platform Exploitation (MAPLE) to simplify the integration of unmanned vehicles into the command structure of a modern warship. It is an open architecture system that allows multiple vehicles of different types from different manufacturers to be controlled from one simple interface. The system has a top level goal based mission planner (GBMP) where the operator can command the squad of vehicles and set objectives. These commands and objectives are then flowed down and split between the individual vehicles to carry out individual tasks. The vehicle plans and progress are relayed back to the operator. Each vehicle type has a specific plugin to the system that can be removed and replaced from the system with the vehicles to allow mission specific configurations with minimum impact to the system during changeover. MAPLE also interfaces between the unmanned vehicles and the platform's command & control system so that details of unmanned operations are sent directly to the combat system as well as the data and information gathered by the unmanned vehicles.

Conversely the command & control system can send back commands and tasking to the unmanned systems.

MAPLE phase 3 culminated in a demonstrator unit called ACER that formed a core part of the Unmanned Warrior exercise. The unit performed above the DSTL client's expectations and demonstrated connection to over 25 different vehicles; it also performed a mine countermeasure vignette where it demonstrated squad behaviours and command & control of 10 vehicles including UUVs, USVs, and UAVs simultaneously in a credible exercise scenario.

As discussed above SAPIENT is an intelligent Operator/Command Aid tool which has the capability to assess a situation, identify whether it is threatening and propose (or execute) appropriate actions to manage the threat. Its use could be extended to take the information gathered by the various unmanned vehicles deployed by the command team and advise the command team of an emerging threat, providing Command with multiple sources of data to verify it's assessment, having tasked available sensors to collect sufficient information for it to form a conclusion. If required, SAPIENT can recommend actions to Command and, following endorsement, send the necessary command to other elements of the MAPLE system to deliver the required effect.

COMPACT is QinetiQ's rule based assurance system for autonomous vehicles. Originally developed for air vehicles it is a program that monitors the autonomy and prevents it from performing unwanted actions. These boundaries can include ensuring a vehicle does not try to manoeuvre outside its operating envelope, ensuring the vehicle follows the traffic rules of the environment it's in, and in a combat situation ensuring the system obeys the rules of engagement. This system makes verification and validation of autonomous systems a lot simpler, instead of verifying the autonomy algorithms you verify that COMPACT will prevent all unwanted behaviours and has the required control over the system to enforce the prevention. In MAPLE it can check the vehicle operations and plans to ensure they conform to the necessary rules and inform the operator of breaches and request corrective action or confirmation of intent by the operator.

2.2 (a) Conclusion

All these features offered by a new generation of systems and software offer a potent package, capable of allowing a small number of operators command and control a much greater number of unmanned system without overloading them. The autonomy in the system covers detailed planning and vehicle operation allowing a simplified interface with the operator who provides high level objectives and monitors the progress of the mission. Systems like COMPACT and SAPIENT will work as decision aids providing warning if the autonomy is trying to carry out undesirable actions and suggestions on how to react to evolving situations. An architecture similar to or based on MAPLE unlocks the full capability of unmanned systems and enables them to carry out complex missions in a variety of scenarios. The use of multiple UAVs can enhance a ships situational awareness by positioning sensors at range and monitoring a much larger area than the ship's sensors alone, enhancing its search and rescue and patrol capabilities. Using UUVs as search and prosecution vehicles and USVs and UAVs for data links you can create a mine countermeasures capability where the vessel stays remote from the minefield with only unmanned vehicles put at risk. A different mix of USVs and UUVs can be used for creating detailed hydrographic surveys.

2.3 HIGHLY MECHANISED HANDLING SYSTEMS

Handling systems are already used throughout industry with robotics and automation becoming more common place in warehouses and factory environments. Some examples are Volkswagen's automated spare parts warehouse and BMW Mini's automated production line in Oxford that demonstrates how mechanised systems can handle a wide variety of items of different sizes and weights. In the military too, mechanisation is starting to be adopted, the Queen Elizabeth Carrier's highly mechanised weapon handling system being a good example.

Highly mechanised handling systems reduce the manning need by directly replacing the man with a mechanical device to do the lifting. They have distinct advantages in the fact that they can lift any weight as long as they are designed for it, and can also be quicker. Personnel are removed from a task which can often require them to lift heavy weights in a challenging environment, especially in rough seas. The physical robotics reduces manning, removes crew from tasks with a higher risk of injury and can perform the task quicker. However more advantage can be sought from the control system by using a level of smart programming. Through having a smarter control system stores and magazines can be organised more efficiently to reduce wasted space and increase the capacity of the store room. It can even rearrange itself for certain missions making certain items more readily available depending on the mission and adapt based on the demand placed on it at a specific point in time. This allows magazines to be configured for higher capacities while ensuring the right item is available at the right time.

The main use in military vessels will be for weapons handling in the magazine through to loading of the weapon system. For instance, Minerva would have an automated magazine and transport system to the gun, thereby removing personnel from the dangerous task of munitions handling and loading. However there are other opportunities for handling other stores if there is a great enough requirement to justify the system. For instance on the future solid support ship there would be a potential for all store handling and replenishment at sea activities to be automated.

2.4 SMART SHIP MANAGEMENT SYSTEMS

The maritime industry is starting to adopt levels of smarter ship management, taking advantage of on-condition monitoring and analysis techniques to better understand how the ship is operating and manage it to operate in the most efficient way possible. The general theory is to use sensors to monitor enough properties of the ship to understand how it is operating and the health of the systems. This data then needs to be processed and then recommendations can be made; these can be on how to run the equipment to get the best efficiency or time between maintenance or detection of faults developing and a recommendation on preventative measures and required maintenance. This sort of thing is already being done by companies like ABB who can monitor ships equipment from shore based monitoring stations. They are able to detect faults developing before the crew and suggest remediative actions or organise corrective maintenance at the next port.

A good first step is for navies to adopt these systems on-board their ships and have personnel monitoring the systems so they can better understand the health of the vessel and plan maintenance accordingly. Done right there could be many advantages including higher availability of the vessels, removal of unnecessary preventative maintenance regimes and prevention of breakdowns through detecting issues early and cueing the required corrective actions. Eventually you can get confidence about the health of a ship and can reduce engineering onboard to lower levels, allowing the day to day running to be done by the management system.

However, there are issues with naval ships that make this system of working somewhat problematic, namely the production volume. The low numbers of naval ships present a series of problems; the cost of a land based analysis centre for commercial ships is paid off by reducing maintenance and manning costs on the ship, without many ships the savings are not as pronounced; these systems work on gathering and broadcasting large amounts of data on the health of the ship which could be intercepted; the systems also rely on pattern recognition and can cross reference faults against thousands or even millions of hours of operational data which does not exist for a bespoke naval design for a small number of ships.

One solution is to push down the route of commercial of the shelf (COTS) equipment that is being used widely and has the data that can be supplied by the manufacture on how the equipment runs, the faults and how they start to manifest themselves in the data. This will give you a library for pattern recognition and fault detection. Another solution is adding machine learning to the system. Machine learning can monitor the data and learn what is normal with a number of commissioning runs designed to specifically show the variety of operating conditions the ship will be working in. The management system can then see what normal looks like and identify patterns that are not normal. This allows a ship to self-analyse without broadcasting large amounts of data back to a large shore side data analysis centre and contact with shore could be limited to "phoning for help" when there is an issue. Furthermore the system will learn the common faults and the patterns, building a library so they can be recognised the next time the symptoms start to occur, and these libraries can be shared across the class of ship.

The solution is probably a mixture of moving towards more commercial products and using advanced analysis software and machine learning. This way the move to COTS, gives the economies associated with mass production as well as the supportability through a broad based logistics chain, but most importantly the data associated with the quantity of in service units and number of hours of running. Meanwhile machine learning and onboard processing can make the vessel and condition monitoring more efficient and help identify issues even in a bespoke configuration on a secure ship.

2.5 A HIGHLY CAPABLE CONSTABULARY VESSEL WITH A LOW COST FOOTPRINT

Today Concept Minerva utilises the ViewFinder software to unlock the full capability of its organic sensors through automated sensor fusion and track behavioural analysis. This gives the vessel a high level of situational awareness while reducing the manning and operator loading requirements. It ensures the vessel command personnel have the highest quality information with which to make decisions and enable the delivery of the commands to task assets in a timely and clear manner.

The use of modular offboard unmanned systems gives the vessel both flexibility and enhanced capability. A number of UAVs can be deployed with sensor packages to enhance and supplement the vessels situational awareness for search and rescue operations. A change of mission package can see the vessel deploy mine countermeasures equipment in a complete change of role. Through the adoption of a MAPLE-like or MAPLE architecture, information from the UXV sensors is easily integrated into the compiled tactical picture presented by ViewFinder to a simple UXV command interface is enabled by a system like SAPIENT. This way all the UXVs are a fully integrated extension of Minerva's installed equipment.

Mechanised weapon handling and automation will mean the magazine and gun are unmanned during operation with an operator controlling it remotely. This reduces manning and the risk of injury in what can be considered a particularly dangerous task. Meanwhile condition monitoring of the ships systems and manned analysis of them can allow early identification of issues and better management of maintenance. This leads to a ship with higher availability. Reduced manning also means less accommodation space required, therefore a smaller ship.

Overall these smart and autonomous systems allow a relatively small and cost effective ship to reach the capabilities and flexibility of a considerably larger ship while keeping costs and manning low by unloading operators and unlocking the information that can be gathered from the available sensory data.

In the future Concept Minerva can adopt more autonomy and see operations and command automated with humans overseeing and providing the man in the loop for weapons release. Much like senior personnel overseeing new recruits. Machinery spaces can also see reduced manning as autonomy is able to predict failures through processing the data of the on-condition monitoring systems and noticing the precursors, allowing for maintenance to happen in scheduled port stops and only when necessary. The question then becomes when should these operations be carried out autonomously rather than can they. How much control should an autonomous system have and how far can we trust them? What if they go wrong, will we be so reliant on them we can't operate without them?

3. HOW NAVIES COULD ADAPT TO TAKE ADVANTAGE OF THE TECHNOLOGY

The sorts of technology discussed open up new opportunities to reduce the through life costs of the vessels while providing a level of flexibility in the platform for upgrades and even changes in role. The following section will look at how these developments can lead to modular design, divorcing the capability and platform design cycles and how this can provide benefit. It will also consider the human factors impact of the technology and how it will shape a future navy in terms of manning.

3.1 OPEN ARCHITECTURE AND MODULARITY

To unlock the benefits of flexibility available from open architecture, the whole vessel must take on the ethos with naval architects adopting the same principles as the software designers. The design should allow for modularity, built on open architecture philosophies with clearly defined and published interfaces for all modules. This means the ship is designed for these modules to be replaced and/or upgraded and provides a known set of interfaces for any new equipment being installed.

A key area is the mission bay(s) which should be dedicated to provide a space for modular equipment loadouts that can be customised for each mission. These bays and deck should be designed to provide standardised physical interfaces for modules coming on board, as well as standardised connections to the command & control systems. This should allow very simple role changes for a vessel where a containerised capability can be removed and replaced with a different capability, or a special configuration for a specific mission, in a number of hours. An example could be Concept Minerva being loaded up with a number of UAV vehicles and associated launch, recovery and maintenance equipment in containers on the back deck; by docking in port these containers and vehicles can be unplugged, removed and replaced with a Mine Countermeasures capability of Unmanned Surface Vehicles and Unmanned Underwater Vehicles with associated maintenance containers over the course of a day. However these modules could contain any equipment and not necessarily just unmanned vehicles. This allows a vessel to flexibly perform multiple roles without needing major refit or even a vessel for each role.

The definition and standardisation of these modular interfaces, allow them to be delivered as "plug and play" modules. These modules can be developed away from the ship design or half way through the ships life and then delivered to add a new or increased capability. This enables flexible upgrade and capability development through life without the major impact of requiring a significant refit.

This philosophy can be pushed further beyond the mission bay and applied to the ships organic capabilities, such as sensors and even weapon systems. By providing standardised modular physical interfaces and connections into an open architecture command & control system, it allows for simpler upgrade and replacement of organic ship capability. Furthermore these interfaces can be standardised across the fleet enabling capability programmes to be run fleet wide, which allows for economies of scale through designing once for multiple classes rather than new designs for each. There will also be savings in procurement and manufacture through economies of scale and even opportunities to use more commercial or military of the shelf equipment.

3.2 DIVORCING THE SHIP DESIGN FROM THE CAPABILITY DESIGN

Obsolescence is a key issue facing military platforms. Currently the design of the platform and equipment is done together. This means that some equipment will be specified or have its design "frozen" midway through the ship design process and it will take 5 years or more before it is active on a commissioned vessel. Technology progression is now so fast that 5 years can see certain sensors and processing equipment a generation or two out of date meaning that the ship is already dealing with obsolescence before the first of class has been fully commissioned. It is now becoming accepted that certain classes will have an updated design midway through production to try and mitigate this issue, with the last few units of the class being produced to an updated version and the first few being updated to it in their first refit.

There needs to be a shift away from this design ethos to allow capability to be delivered while it is still working on state of the art technology. This can be done by dividing a platform into two distinct parts; the ship that is the core vehicle and equipment that can be sensibly designed and maintained for a long life, and the capability that is the sensors and effectors that deliver the combat edge and are required to be kept up to date throughout the platform's life to maintain the platform's superiority.

The ship is the aspect that takes time to design and build, but with thought it can be designed in a modular way so that it can incorporate the capability equipment. It should be designed in such a way that it has a high level of redundancy to support the capability through life; for example a generous power budget for the equipment and network capacity for equipment to communicate. This can then go into production before the capability design is frozen.

The capability design can then be progressed at the right pace during ship production so that it has current technology and is manufactured just in time for fitting out. This ensures the sensors and effectors are state of the art at the point the ship is commissioned rather than based on the technology of 5 or 10 years ago when the ship design was frozen for manufacture.

This split is enabled by modular design and open architecture creating a standardised structure in which to work. So the ship allows enough physical space and services for each piece of equipment with well-defined interfaces and the capability engineers design that equipment to fit those interfaces. The key is good interface management and system integration procedures to allow the two designs to progress separately from each other.

This benefit can be multiplied by expanding the modular interface standardisation and open architecture across the fleet. This allows capability design to be a fleet wide exercise meaning a capability (e.g. a new radar) being designed for a new class can also be a retrofit to other existing classes that share the architecture and interfaces. The capability development and integration can occur in isolation as a non-platform specific programme. This can enable quicker technology development programmes and upgrade cycles that can be rolled out fleet wide, leading to better and cheaper supportability of fleet wide capabilities.

3.3 WHAT EFFECTS WILL AUTONOMY HAVE ON THE FUTURE MANNING REQUIREMENTS AND THE SKILLS AND TRAINING REQUIRED TO OPERATE A VESSEL?

Autonomy is a disruptive technology and its adoption will disrupt the way in which navies crew their vessels and what skill sets the sailors will need.

While some systems will immediately produce savings in manning levels and reduce work loading on the operators, perversely, others may increase manning in the short term. Current unmanned vehicles and robotics can be fragile and somewhat unreliable and it has been seen with some UAV systems that they need more support staff than their manned equivalents. However, as the technologies matures the vehicles will become more robust and maintenance friendly, seeing reductions in personnel per system and with those personnel being able to support more than one system. Overall, autonomy has the potential to reduce manning and operator loading in a number of areas; by automating situational awareness processes in the operations room and allowing a single operator to take command of more assets; by providing assurance on the health of machinery allowing ships to sail with reduced engineering staff; and increasing the capability of a single ship.

The skill set of the operators will start to change as those in the operations room commanding offboard systems and reading situational awareness tactical pictures will require the skills to understand their displays and operate the equipment remotely. These skill sets may be more akin to those required for playing computer games rather than interpreting raw sensor data. This is a skill set change similar to those you see from a manned aircraft pilot to a UAV remote pilot. And while the amount of maintenance required at sea may reduce and require a smaller engineering team to carry it out. Each engineer will likely need to be trained in a much broader set of disciplines as there won't be enough engineers on board to maintain specialists. This all leads to a requirement for increasing computer literacy skills and increasing levels of intellect to deal with the information being given by the system and to fix it when it does have an issue.

While there is great potential for autonomy to deliver improved capability at a lower cost there are still many issues to be considered. How far will a crew member trust the autonomy? This trust will likely need to be developed or earned and autonomy may need to be introduced as a decision aid rather than as a fully automated system. This will allow crew members to start to trust the decisions the system makes, as well as allowing them to drill down into the reasoning of a decision and ask why it was made this way. Then there is the question of how much should autonomy be allowed to do? While information processing is considered relatively benign and a useful task for autonomy, one can raise many moral issues when the information it has processed is being used to direct a weapons system. The system may well be more efficient and less error prone in identifying a target, providing a firing solution, and directing the weapon system; however current Rules of Engagement dictate that there must be a man in the loop for weapons release and how many of the checks and balances required to fire a weapon should be handed over to an autonomous system? Should a man just make the ethical judgement and enable the system for weapons release and rely on the machine to have done everything else right or should personnel be more involved in the rest of the process?

There may even be a more fundamental issue to naval platforms; if you reduce manning significantly will the vessel still be able to operate and perform as expected. Reducing manning onboard can have unintended consequences and impacts on any vessel but it is most keenly felt on naval ships. Simple things like cleaning are usually done by junior rates, but if manning is reduced then it may become a struggle to keep the ship clean and disease free. More importantly the ship has an offensive role where it could be susceptible to damage and fire, which needs to be addressed by damage control parties. However if there is a very lean crew, will it be able to keep the ship operational and be able to carry out the damage control and firefighting duties. While some of these duties can also be made autonomous, for instance automatic carbon dioxide firefighting systems, not all damage control party tasks are easily automated.

So this leads to a fundamental question to the role of autonomy in naval ships is it to increase capability rather than reduce manning? Or do we need a complete paradigm shift in the way future autonomous naval ships are crewed and operated?

4. AUTHORS BIOGRAPHY

Justin Paul holds the current position of Technical Consultant at QinetiQ's Maritime Autonomy Campaign. He is responsible for developing business opportunities for QinetiQ in maritime autonomy through leveraging QinetiQ developed technologies and capabilities. His previous experience includes working for Babcock as a Senior Mechanical Engineer on maritime and defence related projects.

CAD TOOLS FOR DESIGNING SMART SHIPS IN THE WORLD OF THE INTERNET OF THINGS (IOT)

J A Muñoz and **R Perez,** SENER Ingeniería y Sistemas, S.A., SPAIN

SUMMARY

The Internet of Things (IoT) is a concept and a paradigm that considers pervasive presence in the environment of a variety of things/objects that through wireless and wired connections and unique addressing schemes are able to interact with each other and cooperate with other things/objects to create new applications/services and reach common goals. The smart vessels can integrate the developments on IoT to improve operation, usability and profitability for ship owners, users and workers. The different possibilities of introducing IoT in vessels need to start from the initial concept and the initial design. CAD systems need to be prepared and adapted in order to provide the management and the control of the elements connected to IoT in a vessel and the vessel itself. In this paper, the challenges of the IoT in vessels are discussed and a solution from the CAD system is presented focused in the shipbuilding concept.

NOMENCLATURE

A nomenclature is required for papers/technical notes using a large number of symbols, abbreviations and acronyms. SI units should be used wherever possible, and ordered alphabetically.

AI	Artificial Intelligence
AR	Augmented Reality
CAD	Computer Aided Design
DA	Data Analytics
DDOS	Distributed Denial of Service
IoT	Internet of Things
OS	Operating System
PLM	Product Lifecycle Management
PPL	Product Performance Lifecycle
VR	Virtual Reality

1. INTRODUCTION

As the latest literature Nobel Prize in literature Bob Dylan's song "Times are changing". However, in the latest times, the changes go as fast as the technology changes, and actually, it changes very fast. Naval industry is a very conservative sector where the changes enter very slowly. However, in order to evolve properly, it must to use technology tools incorporating current improvements of this year that become obsolete just a year later. Naval industry has long life cycles and needs to face with many technology changes along the entire lifecycle. The naval industry needs to harmonize these two opposite poles: modern technology and the long building lifecycle of the specific products.

If the marine industry wants to adapt to new technologies, it must start with the design process and prepare vessels for the inclusion of new technologies throughout their development.

2. SYSTEMS ENGINEERING

In this context, systems engineering emerges as the most adequate working methodology for a naval industry with a perspective of future.

If the naval industry wants to adapt itself to the new technologies it must to start with the entire design process of the product, in such a way that the product to be prepared for the new technology. The shipbuilding product has two well-defined and separated periods: the design phase and the operation life. These two elements can be sustained through working methodologies. Of particular note is systems engineering. Systems engineering appears to solve the problems that need to be addressed in sectors with high added value, such as products for defense or aerospace engineering, which are high-risk sectors that require a very good design and construction processes from very early stages to reduce costs and ensure planning and schedule within a budget. Systems engineering requires the use of more specific technologies in order to be applied effectively. In systems engineering, there are innumerable applications that can be used to provide solutions in the different stages of the life cycle of a ship.

CAD systems are traditionally used for the design and construction stage and in more recent times, the product lifecycle management (PLM) solutions are emerging to provide valuable tools for product management. However, these tools require an enormous amount of information that must be provided during the ship design and construction project stages. Therefore, it is necessary to find a way to join these two technologies: CAD and PLM so that all the value mutually contributed in a project not to be lost.

Nevertheless, this is only a side of the problem. There are different technological trends appearing as the engines of the great changes that are affecting us. Among them, the

IoT (internet of Things) stands out because of the impact it has directly on the world we see and touch. How it affects or can affect this trend to the world of shipbuilding?. What can the software industry bring to enter the IoT to the shipbuilding industry?. First, we need to analyze what is actually the Internet of Things nowadays.

3. THE REALITY OF IOT NOWDAYS

A very recent survey made by the SAS Institute Inc. on the impact of IoT on companies [1] showed that the penetration of related initiatives in several fields of the industry is growing and that most of those fields could be of direct application in shipbuilding industry, see figure 1.

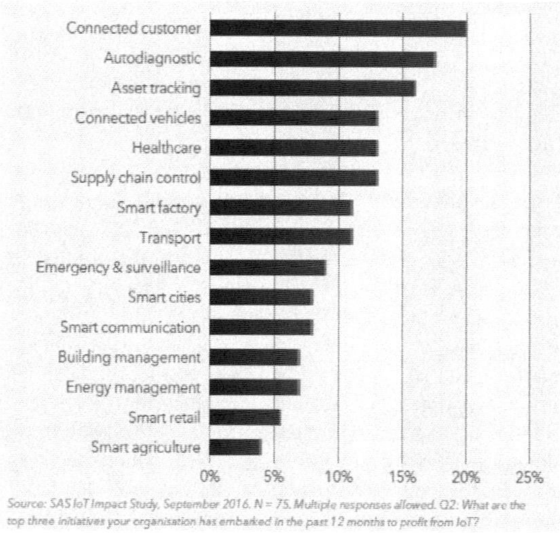

Figure 1: SAS IoT Impact study

Although the greatest penetration was in connectivity with customers, fields like auto diagnostic, asset tracking, connected vehicles, supply chain control, smart factory, transport or smart cities were among the most active in the IoT. Implementing initiatives in these fields within a ship is converting a ship into a smart ship. However, ships live in the sea and this circumstance greatly conditions the application of advances such as IoT in the elements that are pieces of a ship.

Ships vessels are small pieces of humanity in the sea; they are like medium sized cities, medium industries, leisure places or whatever anyone imagines. All the necessities that a society can have, shall be necessities in the ship. If the society progresses in some knowledge or technological area, sooner or later, the ship vessels will have to address with that progress. It is obvious that IoT can provide many advantages to ships. However, life in the ship has specific risks that have to be taken into account when anyone thinks in the IoT for ships. Apart from the environmental circumstances of ships, there are a number of challenges to be faced and problems to solve if we want to implement valid IoT solutions in smart ships.

3.1 CHALLENGES OF THE IOT

We can recognize several areas where IoT presents problems that need a common solution: security, standardization, scale challenge and business orientation

We are still very aware of the recent DDoS attack by thousands of malicious software infected devices connected to the IoT. What could happen with a ship awaiting responses from internet in the middle of a storm with a load of millions of dollars on board or with thousands of people inside?. Security is the most important problem to be solved in a connected IoT ship. There are several issues related with this question: identification of devices, protection against attack, control of updates, redundancy, etc. There are still no guaranteed solutions to this issue.

The rapid growth of IoT has led to an uncontrolled growth of devices connected to the Internet through particular solutions of the manufacturers. Each device manufacturer has built their own IoT solution by connecting their hardware to their cloud server to service their requests. According to Ahmed Banafa wrote in [2], "As the industry evolves, the need for a standard model to perform common IoT backend tasks is becoming more relevant". Moreover, it becomes necessary to have a greater standardization if it is desired that this connectivity is a pole of growth of current society.

Another issue can be known as scale challenge. The large variety of devices that have relationships with each other, but that have different workflows and of course different update times. If updating them freezes the operation of the system, the working time of the system would be significantly reduced. It is necessary to control the update cycles or perhaps to directly prevent them until it is decided that it is time to do it globally or selectively.

Figure 2: The future in ship design

The success of IoT will not go hand in hand with the early adopters, but with the companies and this will only be possible if the initiatives have a clear component oriented to the business, that is to say to add a value. It is necessary to identify which of the initiatives provide a clear value to the business, but taking into account that this value can be in various terms and not all directly economic. In the field of ships there are many possibilities as shown in figure 2. The clearest ones have to do with the optimization of energy costs, fuel consumption, choice of routes, safety at sea, but also in the work itself inside the ship.

One of the conclusions of all the analysis about the consequences of the IoT is that objects must be re-thought from the early design.

The fact is that ships are subject to harsh environmental conditions makes that any technological advance to be applied inside a ship have to take into account from the very beginning. Currently all the designs in the modern society are made by information technology tools, computer aided designs, supported with adequate databases and with lifecycle management of the information. To address the challenge of the IOT for ships, it is necessary the CAD-CAM-PLM tools to adapt in different manners.

4. ADAPTATION OF THE NAVAL INDUSTRY TO THE NEW TECHNOLOGICAL CHALLENGES

As Stuart Taylor declared [3]: "The platform is the key to success. The “things” will get increasingly inexpensive, applications will multiply, and connectivity will cost pennies. The real value will be created in the horizontal platform that ties it all together, i.e. the new OS."

In an article published in the Harvard Business Review, Michel E. Porter and James E. Heppelmann, [4] described the new technology stack: “Smart, connected products require companies to build and support an entirely new technology infrastructure. This “technology stack” is made up of multiple layers, including new product hardware, embedded software, connectivity, a product cloud consisting of software running on remote servers, a suite of security tools, a gateway for external information sources, and integration with enterprise business systems”. If a company wants enter in this business must have products that connects with the products of other companies. So, the necessity of having an opened product is increasing.

From our point of view, the mentioned platform it is not exactly an OS, but a set of interconnected applications, which share the same data about the product. Data that have been appearing since the early design phases and feed that single database, not only with the design attributes but also with the desirable values for optimal performance. Around this database will appear the applications that control the life cycle of the components, the sensors that collect the information of operation or operational life of the same objects and devices.

The key is that all devices, sensors, components are not connected directly to the IoT, but connected to a management core that controls the data to interchange to IoT for those devices allowed to enter in that network. In this way, it is possible to take advance of the fact that all the existing components in the ship have its respective model created, incorporated, and defined from the CAD system. We could say that CAD is the birthplace of the elements of the ship. Therefore, it is the CAD system that knows them and from where they can be assigned the mission and the level of participation in that world of IoT.

5. CONTRIBUTION OF A CAD SYSTEM TO THE IOT

CAD tools are at the beginning of the product life cycle, but they also have a strong relationship with the production cycle, providing the information needed for construction in all aspects. Those CADs with a compact and homogeneous database can extend their contribution to both the life cycle management tools and the IoT connectivity management application of the ship's elements.

To make possible all of this, it is necessary that CAD applications have certain characteristics. A CAD system like FORAN has to evolve to become a global solution. In our vision, the CAD system will not only be the heart of the design but also the vehicle of communication between the products, their manufacturers and their operation. Having this idea in mind, we wonder: what should be those characteristics that turn a CAD system into a global design solution? We think the answer is the integration of systems. It is impossible to grow enough quickly just with organic grew. So, the companies have to look for the future of their products thinking in terms of integration. The question now is which is the way of getting the integration. To answer this question, it is necessary to look into the future and to identify the technological trends that will be developed in this area.

At the recent WebSummit technology conference in Lisbon past November 2016, Mike Schroepfer, chief technology officer at Facebook, declared: "Connectivity, artificial intelligence and virtual reality" are the keys to designing the future in the next ten years. We share the same vision and we think about what will be the application of these trends in the naval field. Any future technology for software solutions, the world of IoT, smart ships and any other technological developments of the products shall be supported by some of these three trends. They are not disjointed roads, they are trends that converge, and support each other.

5.1 CONNECTIVITY

We can see the connectivity from different points of view:

- Connectivity of design systems
- Connectivity of design and construction systems
- Connectivity of ship systems to each other and to the world of IoT.

5.1 (a) Connectivity of design systems

The design of a ship needs very different design applications. It seeks to optimize the product from different points of view, which requires design systems that, to a greater or lesser extent, need to be connected. They must be able to talk to one another, exchange data, models or forms. It is necessary to have data communication agreements, through either APIs or common interfaces. CAD exchange formats have extended their scope. The connection of CAD systems with the product life cycle management systems is indispensable to bring value to the product itself. The PLM will not only be part of the design and construction stage but also extend to the operation of the ship. The PLM stands as the controller of the product, the repository of the unique truth, where the information is the only valid one. The different objects and equipment of the ship should not be autonomous entities that decide by themselves, but they should be supervised by a system that takes into account the product as a whole. A ship is a product with a high value by itself, but also has a lot of value for what it contains, so it cannot be left to the edge of technological autonomy. It is necessary to have a higher responsibility and this responsibility must be of a human intelligence.

The past 10/21 IoT DDoS attacks have changed the rules. It has been shown just how vulnerable the internet, which is now an integral part of the critical infrastructure of the world, by persons whose identity is not immediately ascertainable. Until this vulnerability is addressed, it will cast a serious shadow over the future of connected technology, a future in which much hope and massive resources have already been invested [5]. This is a reason to maintain all the access of the elements of a ship and the ship itself under control, as it is shown in figure 3.

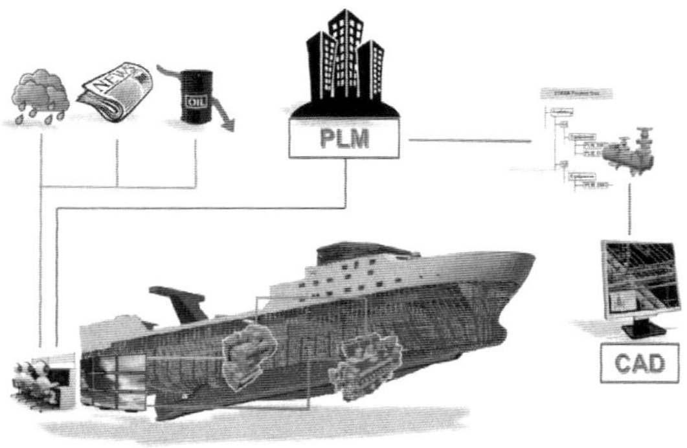

Figure 3: All the ship information under control

5.1 (b) Connectivity of design and construction systems

Along the production process, the communication between the shipyard and the design office is constant and both must have updated information about the product and its building phase. At this stage, very interesting and increasingly achievable new technologies appear. Each of the assembly plans must be consultable from the workshop. The assemblies will have indelibly printed QR codes that uniquely identify each of the elements, but can also incorporate identification or access to the assembly drawings. Even more, video sequences could be made to show how this assembly should be built, and mobile devices shall read the QR code and they will access to the building video. CAD systems that have virtual reality solutions can perform these assembly sequences, record them and store them as part of the assembly information as it was done with the drawings. Workers may consult simultaneously or alternatively this set of information. This information circulates in the own network of the shipyard that is definitely a network of its own. However, it is necessary to ensure that the information has the appropriate level of confidentiality. For this, it will be necessary to provide security certificates to devices so that only those who have the appropriate level of security can access the corresponding information. This is of the most importance in military building, but it is also interesting to preserve the shipyard's own knowledge in its area of privacy.

The different parts to be mounted, not only can incorporate their QR codes with the information described above, but also can incorporate RFID or other location devices. These well-placed and secured devices will allow them to be located in the assembly line, which will allow the controllers to know their finishing condition and therefore the progress of the construction. Obviously, these devices must have characteristics different from the simple RFID tags of the products existing in the stores. They must be sufficient robust and have enough scope so that they can be detected in the workshop area of the assembly line. This information serves not only to know the progress of the construction but also to support the management of materials and the necessary purchases

5.1 (c) Connectivity of ship systems to each other and to the world of IoT

Finally, there is the connectivity of the devices to the IoT. We understand that objects, products of a ship may be interesting its connectivity with the world of IoT, but such connectivity must be controlled and governed from within the vessel system as a whole and with human supervision. Such control can be made more easily if the management applications of such a global solution have the necessary information and are capable of centralizing connectivity.

For each of the elements that must participate in this ecosystem, the design system can define the information that is necessary to share, i.e., which sensors must be equipped in those connectable elements. The application will manage and centralize the information of those elements allowed to be connected in network and will be in charge of managing direct or indirect connectivity with the rest of the world. The solution proposed by FORAN is the incorporation of a set of server applications that allow linking the different levels of work, all around a unique and powerful relational database. Within the work levels are the access of client applications of design, client applications of visualization and control, but also it can incorporate client applications of the objects. These client applications will have the functionality designed for the purpose that designer or the project have decided. They may transmit operational information, diagnosis, results or what would have been considered appropriate in the design. This information has to be handled by the server application that will determine the actions to be taken with it and execute them. You can transmit diagnostic information to request maintenance actions or manage the results to take action on other elements or simply inform about the state of the devices and objects to the control systems or the computers on board.

This solution is based on the following key components: design database, application server and design applications oriented to facilitate the integration of all kinds of information of the designed objects.

The global design solution is configured as a set of applications deployed on a Tomcat application server:

- Design client applications, which communicate directly with the BBDD. It would be what forms the formerly known as the CAD system
- Application server
- Client applications of users
- Client applications of devices, sensors, etc. ...

The architecture of FORAN system can be seen in figure 4 and it consists of the following:

1. **FSERVER**. It is the heart of the solution. It communicates with all the servlets and with the different clients. It communicates with database in order to manage all the different projects. It can work simultaneously with different projects. Each project is a ship. The FSERVER can activate different servlets to have access to all the logic of the different design applications. These servlets are enabled to get information, create models, make calculation and in general all the things that can be done within a ship design. We can observe this servlet as the heart of the architecture because it mobilizes all the rest of the servlets in the application server.
2. **FAUTH**. A servlet that controls the access to users. It sets the security level to users and to devices. When a user or a device request some kind of access either to the DB or to the Internet, FAUTH authorizes the access or not in accordance with the security policy in the internal directory of users and devices. Moreover, the FAUTH control the access to the different concepts stored in database accordingly with the level of security of each user. Something similar shall be done for the devices.
3. **FCLOUD**. Is a servlet that acts with the clients, they are either users or applications and devices (objects in the ship). When it receives a request, depending of the kind of request it activates the correct application to provide the service. This server application is known with the generic name of PSERVER. The FSERVER communicates with the PSERVER in order to request the functionality designed for the given PSERVER. For instance, this functionality might be to generate reports, to create CAD interchange files, to make specific calculations or to connect the device with internet to get updates of software, to transmit diagnostic information or whatever you can imagine to do in the IoT. The PSERVER can have a large variety of functionality. We can imagine a device that controls the state of an internal and watertight door or a valve. The connection between the device is made with specific PSERVER and the client application through FCLOUD could order to close or to open.
4. **FVAULT**. Actually, it shall be an intermediate repository under security control and document controlled, that contains any kind of generated files, i.e. CAD files, report files, etc.
5. **FPLM.** A servlet that communicates with the PLM system of the organization.

The internal communication between the servlets is made by using CORBA. The communication between the client applications and the FCLOUD is made by passing and receiving serialized objects of type JSON.

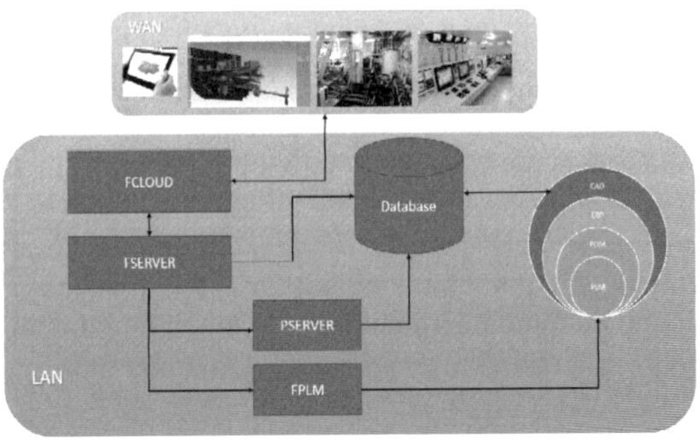

Figure 4: Architecture of the FORAN solution

5.2 VIRTUAL REALITY AND AUGMENTED REALITY

The CAD tools are in the core of the systems that provide virtual reality (VR) and augmented reality (AR). But these systems also provide the feedback to the CAD tools and to the engineering process of designing the ship, because they can track the building and mounting process. The integration of the CAD tools with the VR systems shall be through the connection of VR systems to the unique model database. The VR shall have access to the model even in real time. The VR system is able to project the model over the current situation. Design teams shall interact collaboratively with the functional 3D virtual prototype in an immersive 3D virtual reality environment, so the design teams will have a real-time information of how the model is built and the degree of advance of the building. This is what it is known augmented reality. There is a very wide range of possibilities, that we will see appear in the very near future.

SENER is starting its collaboration with ESI-Group to increase the integration of the FORAN system with the ESI-Group product ICIDO for the use of virtual models in the factory environment for the validation of the constructions. But after the ship is built, the CAD tools, together with the VR system will allow to create and to maintain a virtual model that could be consider as the twin model of the ship. This twin virtual model can be used for predictive maintenance and for predictive performance evaluation. But, in order this can be of utility it is necessary to feed the model with current and real performance data of the real model. In this point is where the artificial intelligence (AI) and data analytics (DA) enter in action.

5.3 ARTIFICIAL INTELLIGENCE AND DATA ANALYTICS

Devices connected to the network not only will request updates but they also shall provide useful information about their normal activity and performance. This information will be gathered by the server and it could be used for different purposes. The information related to the performance could be analyzed to design improvements for future designs or to be used in the virtual model for getting improvements in the operations.

Additionally, it could be possible to evaluate the behavior of the devices in the virtual environment in order to predict how the model will act with determined parameters. The virtual model, also known as twin model will learn with the real data provided with the real design. To support this feature, it is necessary to continuously monitor the different elements of the ship. The information so recovered has to be adequately analyzed within the scope of the cyber-physical realm systems. In this way, the different integrated systems extend the edge of the PLM to a new concept known as Product Performance Lifecycle. This concept has been recently presented by [6] at the Needham Emerging Technology Conference in NY past May, 2016.

6. CONCLUSIONS. A NEW CONCEPT, THE INTERNET OF SHIPS

Future smart ships must be connected to IoT or they will not be smart. The connection of smart devices within a smart ship must be human controlled. The control should start from the design tools because they control the shipbuilding process from the early stages of the design up to the final production. The set of design tools, product lifecycle management and device must be inter connected among them and will be the platform for the smart ships connected to IoT. The information shared in the scope of the IoT must be managed by the human along the whole lifecycle of the ship, starting from the beginning of the initial design. This need requires the CAD tools to be prepared with specific characteristics to handle that information. This new ecosystem, opened, incorporating the new trend of technology but adapted to the specific environment of shipbuilding will be the Internet of Ships.

7. ACKNOWLEDGEMENTS

The authors would like to express sincere gratitude to SENER for providing the support and information for this paper.

8. REFERENCES

1. SAS Institute, «Internet of the Things, Visualize the impact» 2016.

2. A. Banafa, «IoT Standardization and Implementation Challenges» Newsletter IEEE Internet of Things, July 2016.

3. S. Taylor, «10 Predictions for the Future of IoT» Big Data Quarterly, March 2016.

4. M. E. Porter y J. E. Heppelmann, «How Smart, Connected Products Are Transforming Companies» Harvard Business Review, October 2015.

5. S. Cobb, «10 things to know about the October 21 IoT DDoS attacks» Welivesecurity.Com, October 2016.

6. C. Tanasescu, «From Virtual Prototyping to Product Performance Lifecycle» 2016. [On line]. Available: http://www.esi-group.com.

9. AUTHORS BIOGRAPHY

Jesus A. Muñoz holds the current position of FORAN System Product Manager at SENER Ingeniería y Sistemas S.A. He is responsible for the technical development guidance of the FORAN System. He is also professor of Applied Systems Engineering at the Marine Engineering School of the Technical University of Madrid.

Rodrigo Perez is the head of Naval Shipbuilding Business Development in SENER. As well as Jesus, he is also professor in several subjects at the Marine Engineering School of the Technical University of Madrid.

INFLUENCE OF LOSS OF SYNCHRONISATION BETWEEN SIGNALS FROM VARIOUS MARINE SYSTEMS ON ROBUSTNESS OF PREDICTIVE MODELS ALGORITHMS

A Swider, Rolls-Royce Marine AS, Norwegian University of Science and Technology, Norway
E Pedersen, Norwegian University of Science and Technology, Norway

SUMMARY

The marine industry is now entering the challenging phase of smarter shipping, including on-board monitoring systems, and advisory tools. Measurements from sensors in various equipment and installations play a crucial role. The quality of available signals and the data collection system is then a critical aspect to achieve robust results. This paper focuses on the quality of the data collection system, specifically the loss of synchronisation between various signals. Real system examples will be presented, showing the introduced delays. The impact of loss of synchronisation on predictive models algorithms used in marine applications will demonstrate the importance of data synchronisation. Additionally, a method of evaluating the synchronisation between measurements is proposed.

NOMENCLATURE

a, b	Regression coefficients (-)
a_{opt}, b_{opt}	Optimal regression coefficients (-)
corr	Correlation (-)
DAS	Data Acquisition System
E	Expectation (-)
F	Frequency (Hz)
I	Current (A)
$I_{measured}$	Measured current (A)
MSE	Mean Square Error
R	Resistance (Ω)
t	Time (s)
U	Voltage (V)
U_{est}	Voltage estimation (V)
$U_{measured}$	Measured voltage (V)
var	Variance (-)
ρ	Correlation coefficient (-)

1. INTRODUCTION

The marine industry is now entering the challenging phase of smarter shipping, including on-board monitoring systems, and advisory tools. Smart ships will revolutionise ship operations, and are the future of the marine industry [1]. Modern vessels will be equipped with various on-line data collection and advanced monitoring systems. The on-board measurements from sensors of many installations play a crucial role, and their availability expands the functionality of marine products. The aim of data analysis in the marine application is developing on-shore and on-board advisory tools using prediction of propulsion power [2] or ship performance monitoring [3], as well as enhancing knowledge about specific systems and components, and the relationship between systems. The example of use of predictive modelling can be statistical models, which lets us predict the fuel consumption of a vessel based on speed and sea state. Developed models can be utilised to voyage optimisation.

The importance of the installed sensors and data processing is increasing as the ship industry enters the phase of digitalisation and Shipping 4.0 [4]. A crucial aspect, which needs to be considered during data analysis, is the influence of uncertainty [5].

Machine learning algorithms and statistical modelling [6] become widely used tools in equipment monitoring and advisory systems. However they are sensitive to data quality, and in particular to relationships between subsystems retained as correlations in the data. Various Data Acquisition Systems (DAS) [7] introduce disturbances that can influence the output of the algorithms. The importance of DAS and effect of the data collection is often omitted in the process of data analysis, and some of researchers tend to assume that DAS does not introduce distortions. However, processing of data from several marine systems creates many challenges. One of them is lack of data synchronisation, which can be introduced by specific systems setups and different time intervals. In the analysis of tremendous amounts of data (Big Data), ignoring systematic errors like significant time delay to the signal during data collection can have serious consequences, and can lead to improper output response while doing off-line analysis. A delay can also affect the reliability of on-board monitoring systems. Additionally, correlations between specific signals can be corrupted and influence results produced.

This paper will focus on real examples from marine systems, to visualise loss of synchronisation between various signals, and their impact effect on the predictive models algorithms used in marine applications. We thereby demonstrate the importance of data synchronisation. The simple example of Ohm's law and influence of the loss of synchronisation will be presented, and then extended to the marine system instance. We will propose a methodology to detect the delay from various sensors, both with and without time vector, including hidden delay which is hard to detect if there is no knowledge about the DAS.

2. THEORETICAL EXAMPLE AND MOTIVATION

Synchronisation relates to the phenomena, which occurs at the same time, in Greek syn-"together" + khronos "time" [8]. The impact of loss of synchronisation will be shown via a simple example below, to pinpoint the undesired effects loss of synchronisation has on the output of a prediction model. It is assumed that the model analysed contains two systems: registering voltage and current at the resistor (Figure 1).

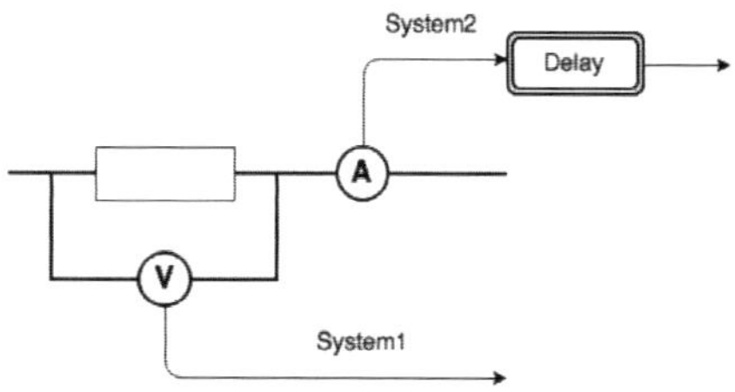

1: Conceptual example of two measuring systems: voltage (System 1) and current (System 2) registration.

According to Ohm's law, the relation between the current and voltage can be expressed by:

$$U = I \cdot R \tag{1}$$

where I – current [A], R – resistance [Ω] and U – voltage [V]. Based on this trivial example we can show degradation of the predictive model when synchronisation is lost.

We employ the model $U=aI + b$, that is a linear regression (a simple prediction model), which aims to find the relation between input and output variables assuming the mean square error [9] optimisation criteria. Thus, the error defined as

$$\begin{aligned} \mathrm{MSE}(a,b) &= \mathrm{E}[(U - U_{\mathrm{est}})^2]|_{U_{\mathrm{est}}=\mathrm{aI+b}} \\ &= \mathrm{E}[(U - aI - b)^2] \end{aligned} \tag{2}$$

should be as small as possible. The aim of the prediction model is to find the values of a and b, such as MSE will be minimised. Calculating a and b is based on following derivatives:

$$\frac{\partial \mathrm{MSE(a,b)}}{\partial \mathrm{a}} = 0 \tag{3}$$

$$\frac{\partial \mathrm{MSE(a,b)}}{\partial \mathrm{b}} = 0 \tag{4}$$

where all partial derivatives should be equal 0, which ensures a minimum for the mean square error. Substituting into the formula (3) and (4) equation (2), we have:

$$\frac{\partial \mathrm{MSE(a,b)}}{\partial \mathrm{a}} = -2\mathrm{E}[UI] + 2a\mathrm{E}[I^2] + 2b\mathrm{E}[I] \tag{5}$$

$$\frac{\partial \mathrm{MSE(a,b)}}{\partial \mathrm{b}} = -2\mathrm{E}[U] + 2a\mathrm{E}[I] + 2b \tag{6}$$

where the solution becomes:

$$a_{\mathrm{opt}} = \frac{\mathrm{E}[UI]-\mathrm{E}(U)\mathrm{E}(I)}{\mathrm{E}(I^2)-\mathrm{E}^2(I)} = \frac{\mathrm{corr(U,I)}}{\mathrm{var(I)}} \tag{7}$$

$$b_{\mathrm{opt}} = \frac{\mathrm{E}(U)-\mathrm{corr}(U,I)\mathrm{E}(I)}{\mathrm{var}(I)} \tag{8}$$

Therefore the prediction model is given by equation:

$$U_{\mathrm{est}} = a_{\mathrm{opt}}I + b_{\mathrm{opt}} \tag{9}$$

The predictive model presented in the example above is very often used in practice, when the measurements of one variable is very expensive or impossible. In marine systems, it can for instance be the shaft power and shaft torque. Based on Ohm's law, the optimal results are $a_{\mathrm{opt}}=R$ and $b_{\mathrm{opt}}=0$, so further analysis will show the negative influence of the loss of synchronisation. It can be presented that minimal mean square error (MSE) of the optimal linear prediction is given by:

$$\begin{aligned} \mathrm{MSE_{min}} &= \mathrm{E}[(U - U_{\mathrm{est}})^2] = \\ \mathrm{E}[(U - a_{\mathrm{opt}}\, I - b_{\mathrm{opt}})^2 &= \mathrm{var}(U)(1-\rho^2) \end{aligned} \tag{10}$$

If signals are synchronised, the correlation coefficient ρ in the model is equal to one (because the relation between current and voltage is linear according to (1)), and the MSE is equal to:

$$\mathrm{MSE_{min}} = \mathrm{var}(U)(1-1) = 0 \tag{11}$$

For further analysis, let us assume that signals from the system in Figure 1 are given by equation:

$$R = 1; I = \sin(2\pi Ft)\,; U = IR = \sin(2\pi Ft) \tag{12}$$

where F=0.1Hz. We assume that the voltage measurement system is errorless so that $U_{\mathrm{measured}}=U$, but that the system measuring the current introduces a delay dt. This causes a lack of synchronisation, according to

$$I_{\mathrm{measured}} = I(t-dt) = \sin\big(2\pi F(t-dt)\big) \tag{13}$$

However, it should be noticed that in real circuits there is always an inductive and capacitive load, which also introduces a lead/lag of voltage versus current.

It can be demonstrated that the correlation coefficient ρ which appears in equation (10) depends on the delay dt according to the following equation:

$$\rho(dt) = \cos(2\pi F dt) \tag{14}$$

The relationship between the delay and correlation coefficient and the MSE error is presented at the following figures (Figure 2-3).

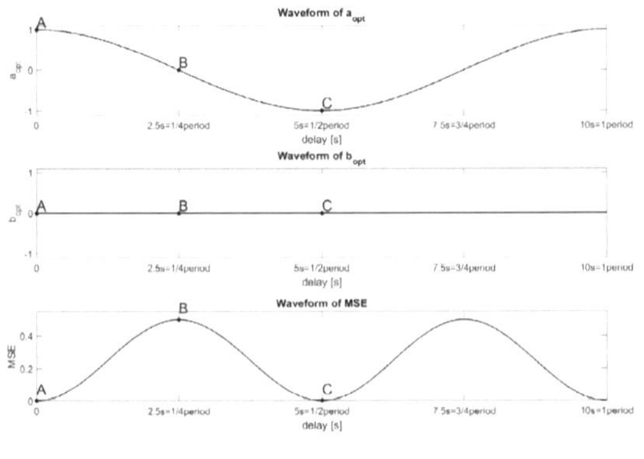

Figure 2: Depicts waveforms of a_{opt}, b_{opt} and MSE for the prediction model from equation (9).

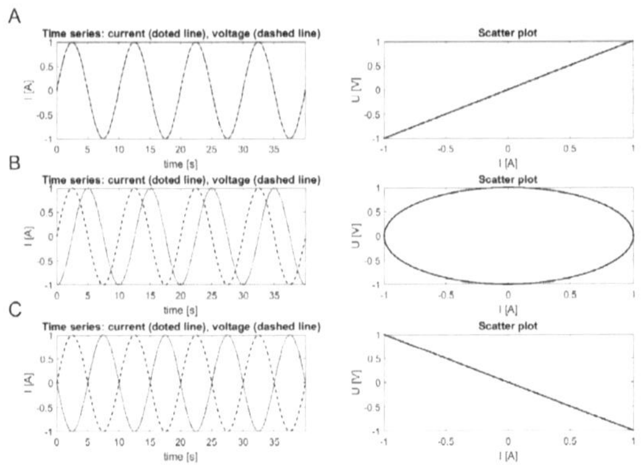

Figure 3: Depicts waveforms (I: continuous line, U: dashed line) and scatter plots for three cases: A: full synchronisation ($dt = 0$), B: loss of synchronisation ($dt = 2.5$s), C: loss of synchronisation ($dt = 5$s).

We can see that a_{opt} is a periodic function with the same period as the analysed current and voltage signals. It is easy to show that $a_{opt}=\cos(2\pi F dt)$, $b_{opt}=0$ and that the latter is thus independent of dt. Mean Square Error is given by the equation

$$\mathrm{MSE}(dt) = \mathrm{var}(U)(1-\rho^2) = 0.5(1-\cos^2(2\pi F dt)) \quad (15)$$

Three reference cases, A, B, and C are shown in Figures 2 and 3 for the three different time delays. For point A, $dt = 0$, which means that signals are synchronised and the measurement system (Figure 1) does not introduce any delay. In this example $a_{opt}=1$, $b_{opt}=0$ and MSE=0, i.e. $U_{est}=1 \cdot I_{measured} + 0$, which is consistent with Ohm's law.

For point B dt is equal to 2.5s, which means there is a loss of synchronisation. Coefficients of the regression model take the following values: $a_{opt}=0$, $b_{opt}=0$ and MSE=0.5, so the model equation is: $U_{est}=0 \cdot I_{measured} + 0$. Hence, with the delay of 0.25 of the signal period (phase shift = $\pi/2$), the prediction model is not consistent with physical laws. In this case, the measured signals are orthogonal, so the correlation coefficient $\rho=0$ and MSE has the highest possible value, MSE=var(U). Results obtained based on the analysis of the measurement system with a resistor represent the relationship between voltage and current for induction.

Point C in Figures 2 and 3 depicts the situation when the delay dt is equal to 5s, which is equivalent to 0.5 of the signal period (phase shift = π). Derived coefficients of the prediction model takes following values: $a_{opt}=-1$, $b_{opt}=0$ and MSE=0, so the model equation is: $U_{est}= -1 \cdot I_{measured} + 0$. The obtained model has the opposite sign to the correct results. In the practical applications, such an error is possible to detect based on engineering knowledge.

These results which are based on a theoretical example clearly illustrate that the delay between signals from analysed systems (loss of synchronisation) has crucial influence on the output of the prediction model. The worst case is the delay equal to 0.25 of the signal period, which leads to gross misinterpretation. For the achieved model, it would be concluded that there is no linear relation between measured variables and the correlation coefficient $\rho=0$. Figure 3 presents waveforms in the time domain and scatter plots for three different cases A, B and C when one signal is delayed. For case B, the scatter plot shows that there is no linear relation between voltage and current, which defies the laws of physics. The example was simulated based on periodic signals, which are common in marine systems. Based on these results, it can be concluded that if the delay is inconsiderable in relation to the signal period (dt<<0.12 of the signal period), then loss of synchronisation is not significant for the prediction model's output. In practice, this means that a delay between systems can be acceptable for slowly varying signals. Examples of the slowly varying signals include trim of the vessel, mean power consumption, and vessel speed. High frequency data (short period signal) will be more challenging, and even insignificant delays will be very harmful. Examples of such signals include vibrations and other signals used for condition and equipment monitoring. Other groups of signals will be those related to the motion of the vessel, like roll and pitch, which include wave frequency. If information about the wave is crucial, then the delay of 0.25 of the signal period i.e. 1.2-2.5s (wave period is 5-10s) will be significant for the output of the model.

It can be concluded that models that rely on high frequency data are more sensitive to loss of synchronisation, meaning that small time delays can distort the results.

3. TOOLS TO ACHIEVE DATA SYNCHRONISATION IN OFF-LINE ANALYSIS

Based on the theoretical example presented in the previous paragraph, loss of synchronisation ($dt \neq 0$) can introduce errors to a prediction model. Based on the conclusions above, one of the standard phases in the data pre-processing should be data synchronisation.

In practice, synchronisation can be done in two ways:

- Based on the TIME VECTORS, which have to be recorded in each subsystem/component separately. It should be known if the recorded time comes from the time when the measurement is taken or when data enter the logger. This is discussed in Section 4.
- Based on the CROSS-CORRELATION FUNCTION, if the time vector is not registered or influenced by distortions for some systems. This approach requires data redundancy and availability of the same signal from several sources, see Section 5.

4. SYNCHRONISATION BASED ON TIME VECTOR AND INFLUENCE ON PREDICTIVE MODELS

The data collected on-board the vessel come from different sub-systems. To avoid the phenomenon presented in Section 2 there is a need for data synchronisation. Synchronisation can be achieved based on the time vectors. The time vector is defined as the vector, which contains the time instants of measurements. Examples of delays between various systems are presented based on data from the offshore vessel, see Table 1.

Table 1: Example of delays

System	Time of the measurement	Delay
System 1 (S1)	26-Feb-2016 19:26:47	
System 2 (S2)	26-Feb-2016 19:32:19	S1-S2 ~6m
System 3 (S3)	26-Feb-2016 19:15:22	S2-S3 ~17m
System 4 (S4)	27-Feb-2016 19:15:23	S3-S4 ~24h

In this paper, we focused on the extreme cases of 24h, 1200s and 600s delays which were found in various real systems. Large delays are probably the effect of lack of clock's synchronisation from individual subsystems. The motivation is to visualise loss of synchronisation between signals and their impact on predictive models. We thereby demonstrate the importance of data synchronisation.

In Figure 4 two time series are presented (duration 33h): power of the engine and the current, both in %. Signals come from two independent systems and are registered in the same data logger. The upper subplot of Figure 4 depicts that there is a significant delay (dt=24h) between signals. The lower subplot depicts time series after synchronisation, based on the time vectors. The results obviously shows the desired effect of signal synchronisation (overlapping).

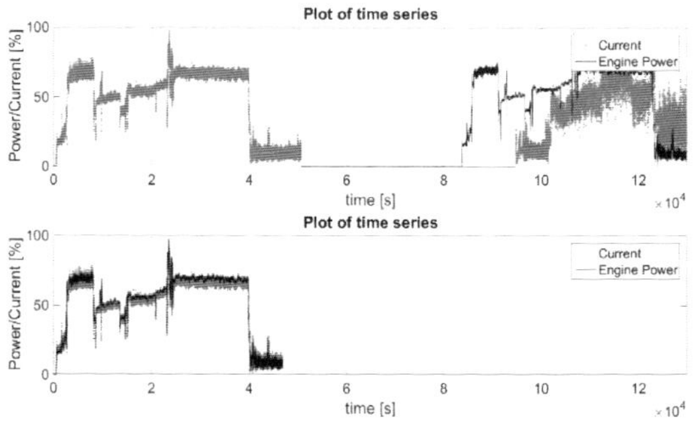

Figure 4: Time series: Power of the Engine in % (black) and Current in % (grey) for two scenarios: before synchronisation (upper subplot), after synchronisation (lower subplot).

The scatter plots of the data and two associated predictive models (built based on data from example presented in Figure 4) are shown in Figure 5. Models are prepared before and after synchronisation. There is a physical relation between power, current and voltage, given by equation:

$$P = UI \tag{16}$$

where P is the power [W], U - the voltage [V] and I is the current [A]. In the analysed case of the vessel switchboard, the voltage is constant, current and power are expressed in %, it means that based on the equation (16), there is a linear relation between power and current. The upper scatter plot of Figure 5 (prepared before synchronisation) does not clearly show the physical relation between power and current. The scatter plot is misleading, resulting in wrong conclusions, since it appears as separated "clouds" of data. The parameters describing model quality [6], take the values:

- Root Mean Squared Error: 28.2%,
- Adjusted R-Squared 0.00839
- F-statistic vs. constant model: 1.1e+03, p-value = 1.53e-240

The model describes only 8% of variability in the response: this occurred because of the loss of synchronisation. The beneficial effect of synchronisation is shown on the lower scatter plot of Figure 5. The linear relationship between current and power is evident and consistent with equation (16). Parameters describing the predictive model take the following values:

- Root Mean Squared Error: 1.83%
- Adjusted R-Squared 0.993
- F-statistic vs. constant model: 6.89e+06, p-value = 0

which confirms that the model is better that the one before data synchronisation (model explains 99% of variability in the response). We can conclude that synchronisation has important influence on the prediction model output and correct interpretation. While building a data-driven model without synchronisation, we cannot expect that results will be repetitive. Analysed signals are not purely periodic signals but conclusions from Section 2 are still applicable. This example showed the influence of the extreme delay of 24h.

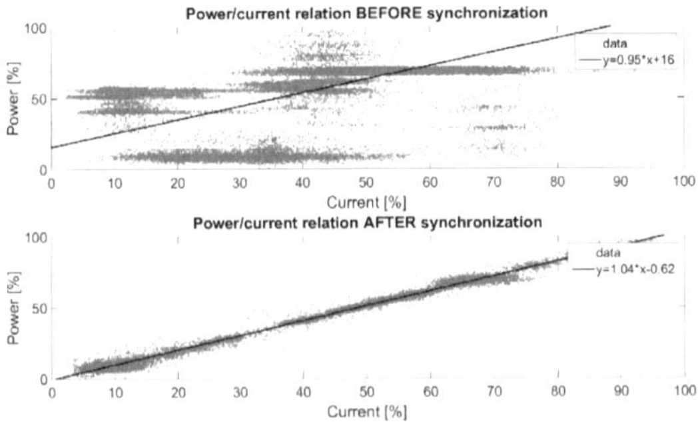

Figure 5: Scatter plots of the Power of the Engine and the Current, both in %, for two scenarios: before synchronisation (upper subplot, prediction model: y=0.95x+16), after synchronisation (lower subplot, prediction model: y=1.04x-0.62).

The example presented in Figure 6 shows the case when the delay dt=1200s for two signals: the main engines power and main propulsion power, both in % (duration 33h). On the upper plot of Figure 6, we can see the time series before synchronisation. The time delay from this example is not so visible in this scale, however it effects the relation between consumed and produced power. That relation is very often plotted on-line on the vessel bridge to show the status of the available power. The delay between consumed and produced power can mislead the crew of vessel. The lower subplot of Figure 6 depicts data after synchronisation, which are evidently overlapping.

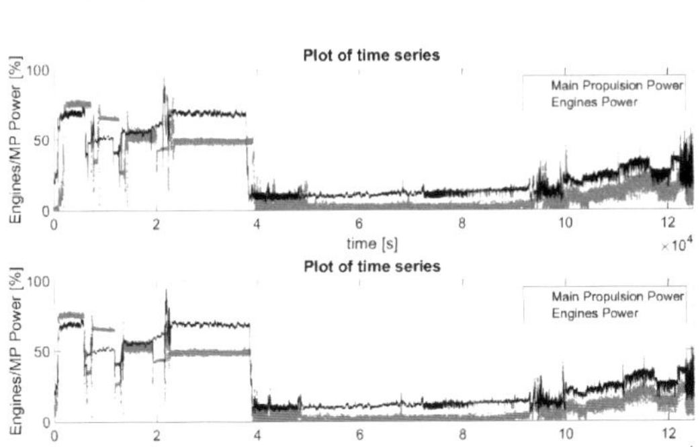

Figure 6: Time series: Power of the Engines in % (black) and Power of Main Propulsion in % (grey) for two scenarios: before synchronisation (upper subplot), after synchronisation (lower subplot).

The scatter plots of the data and two associated models from produced power of the engines and consumed power of main propulsion units are shown in Figure 7. The relationship between signals can be described by the equation:

$$P_P = P_c + P_O \tag{17}$$

where P_P is power produced by engines in %, P_C is power consumed by propulsion [%] and P_O is the residual power (other thrusters, hotel loads, losses). A roughly linear relationship is expected (based on the proper pre-processing) saying that the consumed power depends on the produced power. Unfortunately, the upper plot of Figure 7, where the relation between consumed and produced power is plotted, shows no clear dependency, that again motivates the need for synchronisation. The scatter plot depicts "clouds" and the linear relation between output and input variables is disturbed. Parameters describing the predictive model take following values:

- Root Mean Squared Error: 10.7%
- Adjusted R-Squared 0.774
- F-statistic vs. constant model: 4.28e+05, p-value = 0

The model describes 77% of variability in the response and this occurred because the loss of synchronisation. The lower scatterplot of Figure 7 shows the clear relation between produced and consumed power after signal synchronisation. The parameter describing model quality takes the values:

- Root Mean Squared Error: 7.48%
- Adjusted R-Squared 0.89
- F-statistic vs. constant model: 3.09e+05, p-value = 0

These statistics indicate that the model is better fitted to the data. The model explains 89% of variability in the response. The synchronisation based on time vector again helped to achieve consistency with engineering knowledge results.

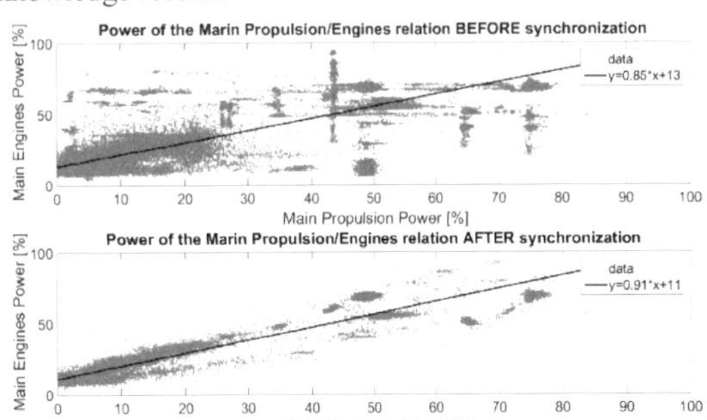

Figure 7: Scatter plots of the Power of the Engines and the Power of Main Propulsion, both in %, for two scenarios: before synchronisation (upper subplot, prediction model: y=0.85x+13), after synchronisation (lower subplot, prediction model: y=0.91x+11).

The last example of unsynchronised data is presented in the Figure 8 and 9, with the delay $dt \approx 600$s. The time series of speed of the vessel together with power consumed by main propulsion are shown in Figure 8. Time series before synchronisation (upper subplot) depicts that the vessel accelerates without raising the main propulsion power - an unrealistic situation. The physical relation between analysed signals is correct for the data after synchronisation, as shown in Figure 8 (lower subplot).

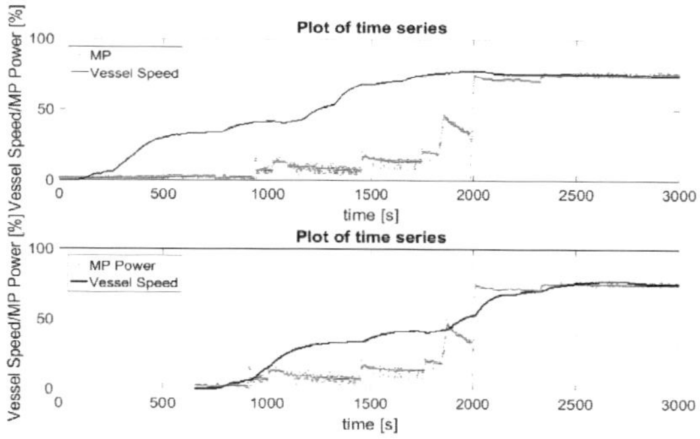

Figure 8: Time series: Vessel Speed % (black) and Power of Main Propulsion in % (grey) for two scenarios: before synchronisation (upper subplot), after synchronisation (lower subplot).

The scatter plot of the data and two associated models (before and after synchronisation) are shown in Figure 9. Fitted model on the upper plot of Figure 9 is the 4th order polynomial and express the relationship between Vessel Speed and Main Propulsion Power, both in %. The polynomial order was chosen to reflect the good fit to the data and minimise the RMSE. The general relationship is retained but from physical interpretation it is not realistic to achieve 60% of the vessel speed with only a few percent of the main propulsion power. That suggests the model is unsuitable. The parameters describing model quality take the values:

- Root Mean Squared Error: 15.9%
- Adjusted R-Squared 0.749
- F-statistic vs. constant model: 2.24e+03, p-value = 0

The model describes 75% of variability in the response, this occurred due to the loss of synchronisation. The scatter plot from the data after synchronisation is described sufficiently by the model of the 2nd order polynomial, which is depicted by the lower subplot of Figure 9. In that case, the resistance curve for the vessel is achieved so we can conclude that the model describes the physical law. Parameters describing the predictive model take the following values:

- Root Mean Squared Error: 7.92%
- Adjusted R-Squared 0.936
- F-statistic vs. constant model: 1.23e+04, p-value = 0

and indicate that the model has better quality and lower complexity compared to the model fitted to the data before synchronisation. The model explains 94% of the variability in response.

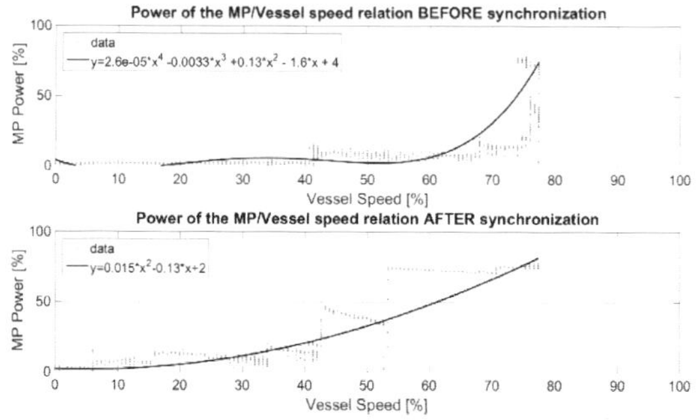

Figure 9: Scatter plots of the Vessel Speed and the Power of Main Propulsion, both in %, for two scenarios: before synchronisation (upper subplot, prediction model: $y=2.6\cdot10^{-5}x^4-0.0033x^3+0.13x^2-1.6x+4$), after synchronisation (lower subplot, prediction model: $y=0.015x^2-0.13x+2$).

All examples show the need of synchronisation and influence of the delay between signals from various marine systems. It has been shown that simple predictive models are not robust on the lack of synchronisation and physical relations between signals can be distorted resulting in improper model fitting and wrong conclusions. In this paper it was decided to show simple prediction models to visualise clearly the phenomenon of loss of synchronisation, which are applicable in all machine learning algorithms. It can be expected that results from more complex models will hide the problem with synchronisation, so that is why before analysing Big Data from marine systems there should be a clear understanding of data set and knowledge about the Data Acquisition System. The physical relationships need to be clearly described by the model and this is only possible when we have control and overview of the data. The interpretation of results based on fitted model should be linked with physical interpretation. Instead of only improving the achieved statistical model is it recommended to look at the data set without trusting their quality and relationships. This paragraph is fundamental to proper data analysis for on-line learning and advisory systems. The off-line analyses are less demanding since the delay can be investigated in archived data sets.

5. SYNCHRONISATION BASED ON CROSS-CORELATION FUNCTION

In the case when data synchronisation based on time vectors from various systems is not feasible, it is possible to synchronise data based on cross-correlation function. The function is given by the following equation (when it is assumed wide sense stationarity of both signals):

$$R_{xy}(\tau) = \mathrm{E}[x(t)y(t+\tau)] \tag{18}$$

where τ is time lag, E is expected value operator and R_{xy} is the cross-correlation function. [10]
Assuming, that signal *y(t)*:

$$y(t) = x(t - dt) \tag{19}$$

is a delayed signal *x(t)* by the delay *dt*, the cross-correlation function has its maximum for $\tau=dt$ associated with the delay value. This is shown in Figure 10.

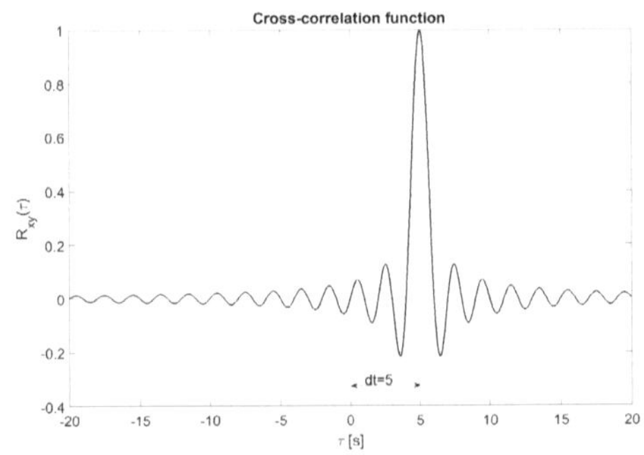

Figure 10: Example of a cross-correlation function between two delayed signals *x*(t) and *y*(t), where y(t)=x(t-5).

Based on the property of the cross-correlation function it is possible to estimate the delay between the same signals registered by different subsystems [11]. This approach is widely used in hydroacoustic systems [12], telecommunication [13] and identification systems. The delay, between two signals can be estimated by the following equation:

$$dt = \operatorname{argmax}\left(R_{xy}(\tau)\right) \tag{20}$$

where: *dt* is the estimated delay, $R_{xy}(\tau)$ is the cross-correlation function. Based on the presented equation, there is a need of registering the same signal twice and data redundancy by each system. By cross-correlation function the estimated time delay will be the average delay for the analysed data set. In practice the delay can change during the time (e.g. from 1s- to some extreme values when there are gaps in the data logging system). To get a more accurate estimation of the changing time delay it is recommended that data be split into smaller time intervals (time windows) and the time delay calculated. The trade-off between length of the time window and variability of the delay in the system needs to be applied. The example of the varying time delay between two similar signals is presented at Figure 11. The upper plot of the figure shows overlapping time series of 1 month of data. The lower plot of Figure 11 depicts varying estimated time delay, which motivated the block processing to get a clear overview of the delay.

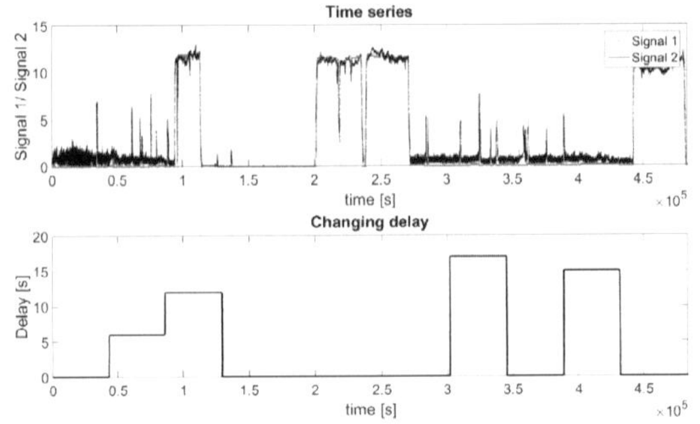

Figure 11: Time series of the two signals (upper subplot) and estimated changing time delay (lower subplot) for the time window 12h.

This is a crucial stage and from the authors' experience, the delay can vary in time, so the time windowing approach gives better results. We would like to emphasise that the variable delay throughout time is an additional dimension of data synchronisation difficulty, which should be compensated, while data is being analysed.

6. HIDDEN TIME DELAY

In previous paragraphs, it was described the loss of synchronisation and influences of the delay on the output of simple machine learning algorithms (predictive models). The proposed solution was based on the time vector or cross-correlation function. In this section, we describe additional problem with the synchronisation, which can be crucial for the results of on-line machine learning algorithms, off-line analysis but it is very often omitted by the Data Scientists.

To describe the problem, the simple example will be given, see Figure 12.

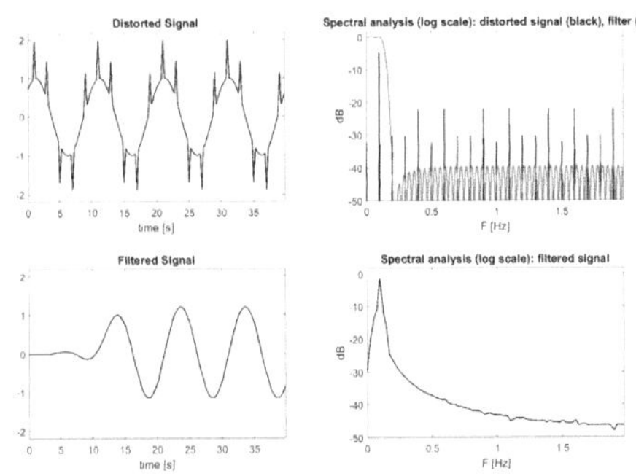

Figure 12: Influence of signal filtration on the time delay. Time and spectral analysis for distorted signals (upper subplot) and for filtered signal (lower subplot).

Figure 12 depicts the influence of filtration (low pass filtration) on the time delay. The input signal to the filter consists of two components: desired sine waveform and spike noise. The aim of the filtration is to remove the noise without the distortion of the desired signals.

Unfortunately, in reality digital filters distort the signal by the introduced delay (lower subplot in Figure 12). More details about this phenomenon can be found in [12]. This type of delay is very often omitted in real systems, but can have serious consequences on the quality of the data-driven model. In practice it is difficult to detect such a delay (time vectors will not help to find the delay introduced by the system filtration), which can be included in one or more registered signals. In aiming to detect such a problems, it can be recommended to carefully analyse the architecture of the Data Acquisition System. A helpful tool to detect the filtration included in the system can be spectral analysis. Figure 13, shows an example of the spectrum of the filtered and non-filtered signal from the same system.

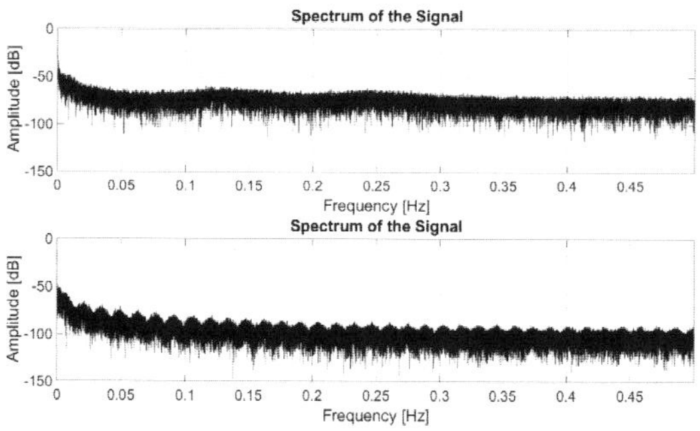

Figure 13: Signal Spectrum for filtered (lower subplot) and not filtered signal (upper subplot)

Regular shape of the spectrum (like the one presented on the lower plot of the Figure 13) can indicate that the pre-processing stage was applied before data were stored. It is recommended for investigation to distinguish if the spectrum of the signal comes from filtration. It is important to estimate the delay introduced by the filtration and give consideration in further analysis. We can recommend applying spectral analysis and DAS investigation, which can be challenging.

7. SUMMARY

In the marine industry there is increasing emphasis on data analysis, therefore measurements, their quality, and sensors and Data Acquisition Systems play an important role. Crucial is the systems approach and influence of the DAS on data analysis. Therefore, we cannot look at these topics separately, assuming sufficient data quality and building data-driven models for advisory systems without a critical scientific approach. It is recommended to look at the data analysis area containing three phases: sensor and DAS investigation, proper data pre-processing algorithms and data-driven model building.

Based on the results presented we can conclude that the loss of synchronisation has significant influence on data-driven models. The delay between signals needs to be checked and compensated for reliable results. To summarise, simple prediction models presented visualise clearly the phenomenon and the effect of loss of synchronisation. Three examples given in this paper confirm that statistical models are not robust given a significant delay between signals. We have to be very careful while analysing signals which come from different subsystems and components. The relation between the delay and period of the signal can be used to roughly estimate the influence of the delay. However synchronisation is recommended in all cases. For high frequency data (short period signal) even insignificant (short) delays will be very harmful. It can be concluded that models that rely on high frequency data are more sensitive to loss of synchronisation, meaning the small time delays (even few seconds) can distort the results.

The key to solving the synchronisation problem is the time vector, registered in all subsystems. The time vector when the signal is measured should be monitored and synchronised, for example with the GPS. The cross-correlation function can be used to solve a synchronisation problem, when the time vector is distorted. For using the cross-correlation function, the same signals should be registered from different subsystems. In practice the delay can change during the time, so it is recommended to split the data into time intervals where length is a trade off between time intervals and variability of the delay.

In the paper additional problems related to a hidden time delay were described shortly. It is recommended, specially for on-line monitoring systems, that the filtration delay is compensated.

8. ACKNOWLEDGEMENTS

The authors gratefully acknowledge advice and support from Professor Helge Langseth from Norwegian University of Science and Technology together with colleagues from Rolls-Royce Marine AS and anonymous reviewers.
The research for this paper was financially supported by Rolls-Royce Marine AS and the Norwegian Research Council as a part of the Industrial PhD program.

9. REFERENCES

1. JOKIOINEN, E., 'Remote and Autonomous Ship – The next steps, *AAWA project* 2016.

2. PEDERSEN, B.P. and LARSEN, J., 'Prediction of Full Scale Propulsion Power Using Artifical Neural Networks', *COMPIT*, 2009

3. HANSEN, S.V., 'Performance Monitoring of Ships', *Ph.D. thesis, DTU*, 2011

4. Rødseth, Ø. J., Perera, L. P., Mo B., "Big Data in Shipping – Challenges and Opportunities", *COMPIT*, 2016

5. ALDOUS, L., SMITH, T., BUCKNALL, R., THOMPSON, P., 'Uncertainty analysis in ship performance monitoring', *Ocean Engineering*, 2015

6. HASTIE, T., TIBSHIRANI, R. and FRIEDMAN, J., 'The Elements of Statistical Learning, Data Mining, Inference, and Prediction', *Springer-Verlag*, 2009.

7. BENDAT, J. and PIERSOL, A., 'Random Data Analysis and Measurement Procedure', *Wiley*, 2010.

8. http://www.dictionary.com

9. KAY, S., 'Intuitive Probability and Random Processes using MATLAB', *Springer*, 2005

10. KAY, S., 'Fundamentals of Statistical Signal Processing, Volume III: Practical Algorithm Development', *Prentice Hall*, 2013

11. SHIN, K., HAMMOND J., 'Fundamentals of Signal Processing for Sound and Vibration Engineers', *Wiley*, 2008

12. SALAMON, R., 'The Hydrolocation Systems', (in Polish) *Gdansk Scientific Society*, 2006

13. HAYKIN, S., MOHER, M., 'An Introduction to Analog and Digital Communications', *Wiley*, 2009

14. MITRA, S., 'Digital Signal Processing: A Computer-Based Approach', *McGraw-Hill*, 2010.

10. AUTHORS BIOGRAPHY

Anna Swider holds the current position of Project Engineer/Industrial PhD Candidate at Rolls-Royce Marine AS, Norway. She is responsible for research within data analysis for vessel power system design support. She graduated in 2009 – MSc. Eng. from the field of Electronics and Telecommunications, Real - Time Systems and in 2011 - Eng. Ocean Engineering and Ship Technology, both from Gdansk University of Technology. Her previous experience includes an engineering job and research for data analysis and pre-processing of data in various applications.

Eilif Pedersen received the M.Sc. degree in marine engineering from the Norwegian Institute of Technology, Norway, in 1983. He has been with the Norwegian Marine Technology Research Institute, as a Senior Research Engineer until 1999 when he joined The Norwegian University of Science and Technology, as an Associate Professor. His areas of expertise are in the field of modeling methodology and simulation of dynamic multidisciplinary and mechatronic systems focusing on machinery system dynamics, internal combustion engines, vibrations, thermal- and hydraulic machines, fuel-cell system dynamics, and hybrid power plants for marine applications. He has held multiple positions within the university such as Vice Dean of Education at the Faculty of Engineering Science and Technology, Head of Master Programs in Marine Technology, Leader of the Research Group of Marine Systems and Head of Machinery Laboratory at the Department of Marine Technology. Currently he is the Leader of the Power Systems and Fuels work package at the Smart Maritime Center for Research-Based Innovation.

AN OPEN FRAMEWORK FOR SMART SHIP APPLICATIONS DEVELOPMENT

S. Koroneos & **P. Bennett,** Soldecom, Greece

SUMMARY

This paper is focused on the implementation of an open framework for Smart Ship applications development & deployment based on Hybrid Cloud architecture, & the IoT Fog paradigm. Our approach, contrary to the current concept that involves collection of on-ship sensor data & their transition for processing and consolidation in shore-based clouds, focuses on creating an onboard private cloud & moving the processing closer to the data generation sources. As a result, we create an entity that can operate autonomously, with facilities to capture, store, process & serve, ship, cargo and other sensor data, to local and remote actors via suitable communication networks. Additionally the ship may consume data services from third parties to augment its operations on an autonomous basis. We conclude by highlighting some practical issues regarding onboard accommodation of the sorts of computer and communications facilities we describe, and point to useful emerging industry frameworks & guidelines.

1. INTRODUCTION

Some 50,000 merchant ships carry 90% of world trade in goods, and are operated by 1 million ship's crew [1]. The value of this trade in 2015 was USD 16.5 Trillion [2]. Merchant shipping is an integral part of a global supply chain that includes a diversity of actors.

A transportation industry hot topic addresses application of contemporary and emerging Information and Communications Technology (ICT) to improve the efficiency of goods transportation in a transformational way. The Internet of Things (IoT), Cloud, and Big Data Analytics are often cited as enabling technologies.

1.1 DEFINITIONS

Smart Ship is a term often used to encapsulate the discourse in Merchant Shipping circles, and can mean different things to different people. We prefer the following definition:

A Smart Ship is defined as an entity that can operate autonomously, with onboard facilities to capture, store, process and serve, ship and cargo sensor data, to local and remote actors via suitable communication networks. In addition it may autonomously consume data services from third parties to augment its operations.

In contrast an entity where the sensor data is simply collected and forwarded to on-shore facilities for processing may be referred to as a "Connected Ship".

1.2 WHY AUTONOMY IS IMPORTANT

Pooling and processing of ship data on the ship itself, and at fleet levels provides new opportunities for analysis to provide increased data utility and attendant efficiencies regarding ship, fleet and enterprise operations.

Also the ship's data may be connected, in a broader sense, to actors in the global supply chain on a selective basis to achieve information sharing and coordination efficiencies.

Another example of a beneficial higher order type of autonomy is the participation in "community" information schemes. For example exchanging weather information. In this instance, each ship in the community becomes effectively mobile weather station providing and consuming weather information. Enhanced autonomy opens the door to any number of such possibilities.

1.3 WHY INTEROPERABILITY IS IMPORTANT

For our purpose we define interoperability as follows:

The ability of computer systems and their associated application software to exchange and make use of information freely and easily in way that is independent of the specifics of the underlying technologies. Use of specified and mutually understood data formats and communications protocols provides for syntactic interoperability. Beyond this, semantic interoperability is required to interpret automatically the exchanged information in a precise manner. To achieve semantic interoperability, parties to the exchange need to refer to a common reference model that specifies the content of the exchanged information unambiguously.

Interoperability is important for several reasons:

- It maximises the ability to use and share information from operational and software development standpoints
- It leads to faster, less expensive applications software development
- Products from different vendors work together so increasing the ability to purchase underlying

technologies from different vendors at competitive prices

- Underlying technologies can be changed and adapted with minimal change to applications software. Given the 20 year or more lifetime of a merchant vessel this is a key concern

An example where this 'open systems' approach is working to great effect is the fast development and ready availability of smartphone applications.

In contrast most ship automation software tends to be implemented by specialist suppliers in proprietary way that deals with a particular aspect of operations such as propulsion, navigation, or electricity generation and distribution.

Proprietary ship instrumentation and automation software has knowledge of sensor characteristics and communications methods in its domain, but this is hard coded and hidden in particular implementations.

Thus ship sensor data is currently generated in silos and is hard to get at. Crucially, this limits sensor data utility.

1.4 THE TECHNOLOGIES

The IoT paradigm is defined in many ways. The Oxford English dictionary provides a starting point:

The interconnection via the Internet of computing devices embedded in everyday objects, enabling them to send and receive data

A deeper search will reveal myriad definitions depending on the perspective of the agency providing the definition. Indeed, an IEEE document devotes 80 pages to the subject [3]. We have selected one of the many definitions as being sufficient for our purpose [4]:

IoT is a global network infrastructure, linking physical and virtual objects through the exploitation of data capture and communication capabilities. This infrastructure includes existing and evolving Internet and network developments. It will offer specific object identification, sensor and connection capability as the basis for the development of independent cooperative services and applications. These will be characterized by a high degree of autonomous data capture, event transfer, network connectivity and interoperability.

Sensor data captured via the Internet can be pooled and put to work in Cloud facilities. These provide storage, processing and Internet connectivity, on demand, in a highly extensible fashion. This is accomplished through orchestration of multiple servers, each containing a plurality of multi-core processor chips.

Cloud-based big data software utilises the multitude of processing resources by parallelising workloads to achieve high performance.

Analytics software, that includes descriptive, diagnostic, discovery, prescriptive and predictive (machine learning) techniques, provides the means to process pooled sensor data to produce Business Intelligence (BI) and predictive capabilities for use by the diversity of actors in the Smart Ship Use Case.

Sensor Fusion is powerful technique that combines sensor data, or data derived from sensor data, so that the resulting information has more value.

Finally Open Systems Frameworks provide the means to achieve interoperability.

In the following sections we explore how these technologies can be applied to Merchant Shipping to enable better and more efficient transportation of goods. We commence by considering the Smart Ship Use Case.

2. THE USE SMART SHIP CASE

Figure 1 provides a Unified Modelling Language (UML) diagram that provides a big picture view of the key Actors and functions involved in Merchant Shipping.

2.1 UML DIAGRAM

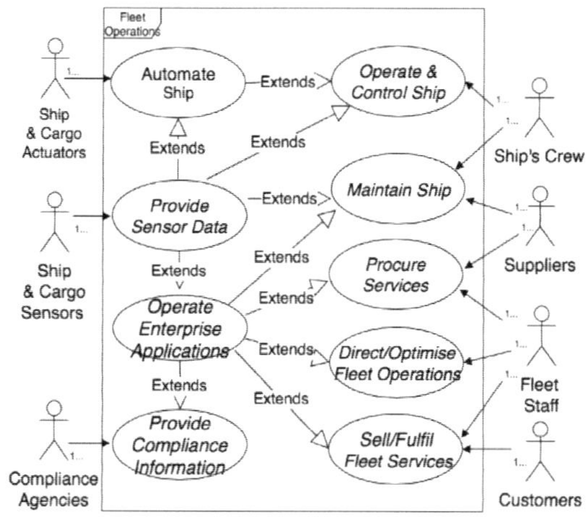

Figure 1: Smart Ship Use Case

2.2 KEY ACTORS & THEIR ROLES

2.2 (a) As directed by Fleet Management, the *Ship's Crew* are responsible for hands-on operation and maintenance of their ship in a safe, compliant and efficient manner.

2.2 (b) As directed by the Fleet Board, the *Fleet Staff* are the onshore people who are responsible for fleet and business operations, planning, business development and optimisation.

2.2 (c) The *Customers* or cargo owners require expeditious, coordinated, and reliable cargo shipment by sea at competitive costs. End-to-end visibility of cargo delivery and assurance of cargo health are also key requirements.

2.2 (d) The *Suppliers* represent a diverse and vast global ecosystem of providers of relevant goods and services that need to coordinate their activities with ship and fleet operational requirements.

2.2 (e) *Compliance Agencies* include the International Governments and Non-government agencies (principally the Classification Societies) that ensure compliance with International Maritime Organisation (IMO) regulations for the safety and security of shipping, and the prevention of marine pollution by ships. The primary international agreements are the International Convention for the Safety of Life at Sea (SOLAS), the International Convention for the Prevention of Pollution from Ships (MARPOL), and the International Convention on Standards of Training, Certification and Watchkeeping for Seafarers (STCW).

2.2 (f) *Ship & Cargo Sensors* are the diversity of idiosyncratic devices that generate instrumentation and control data on a ship. Data is generated continuously, and often frequently, as ordered time series data. The data can measure every aspect of ship operations and systems, its environment, and its cargo. (This can even extend to the crew with the latest developments in wearable sensors). Some sensor data, closely associated with actuators, is used to inform closed-loop automation systems such as the autopilot, propulsion control and ship stabilisation. Other data may give rise to alerts, or provide guidance to the crew to enable them to operate and control the ship on an open loop basis. Sensor data may also include Radio Frequency Identification (RFID) data, bar codes and other forms of identification typically associated with cargo tracking.

2.3 (g) *Ship & Cargo Actuators* are the diversity of idiosyncratic devices involved in closed loop control of ship and cargo machinery.

2.3 KEY FUNCTIONS

Regarding the functions shown in Figure 1 that are required to fulfil key actor roles and goals, it will be noticed that *Provide Sensor Data* touches and impacts all of the other Use Case activities. In formal Use Case terms, it *extends* these activities. The implication is that optimal provisioning of ship sensor data can improve the key functions to the benefit of all of the Use Case Actors.

A key goal is to move from current practice, where sensor data is collected and used in a stove-piped fashion, to a situation where it is pooled and made widely accessible by all of the actors in the Use Case. This enables transformation from the current situation where sensor data is used for anomaly detection and selective control on a ship, to a situation where it can be used for optimisation, prediction and coordination information that can be used to improve ship, fleet, enterprise, and supply chain efficiencies.

Regarding the supply chain, examples of actors are provided in Figure 2.

Goods Producer	Port Service
Goods Consumer	Port Authority
Shipping Agent	Customs Authority
Shipping Broker	Compliance Agent
Shipping Pool	Insurer
Trading Firm	Bank (Trade Finance)
Road Transport	Air/Sea Rescue
Rail Transport	Coastguard Service
Air Transport	Weather Service
Logistics Firm	Mapping Service
Warehouse Operator	Navigation Service...

Figure 2: Examples of Supply Chain Participants

2.4 INDUSTRY KEY CHALLENGES

The key challenges faced by the merchant maritime industry can be summarised as follows:

2.4 (a) Ship Management - reduction of maintenance and other operating costs, maximising ship availability, preserving ship value

2.4 (b) Navigation Management - fuel usage efficiency, schedule optimisation, reduction of emissions and spills, ensuring a safe passage

2.4 (c) Port Operations - Reduction of time in port, berthing plan optimisation, ensuring seamless transhipment of goods

2.4 (d) Inland operations - ensuring end-to-end cargo visibility, optimising the container repositioning process, coordinating with hinterland operations

The expectation is that all of the above can be impacted in a beneficial way through Smart Ship developments.

3. CURRENT SITUATION

3.1 SHIP SENSORS

A large merchant ship may employ thousands of sensors.

Maersk is a leader in the application of Smart Ship technology. In 2014, they published facts and figures about their new breed of Triple-E Class Container Vessels. One interesting fact is they were collecting data from some 3,000 sensors on each vessel [5]. Early in 2016, they announced a Remote Container Management System that provides centralised tracking of refrigerated containers [6]. Given the Triple-E Class can carry up to 18,000 containers, each of which carries a smart sensor measuring temperature, humidity and container integrity, we can begin to see the trajectory of merchant ship sensor deployment.

Other container carriers are implementing similar schemes, and we can expect this trend of instrumentation of cargos as well the ship itself, to develop to include, for example, tankers and bulk carriers.

Another perspective regarding sensor growth can be obtained from information published by the big ship builder Hyundai Heavy Industries (HHI), Figure 3.

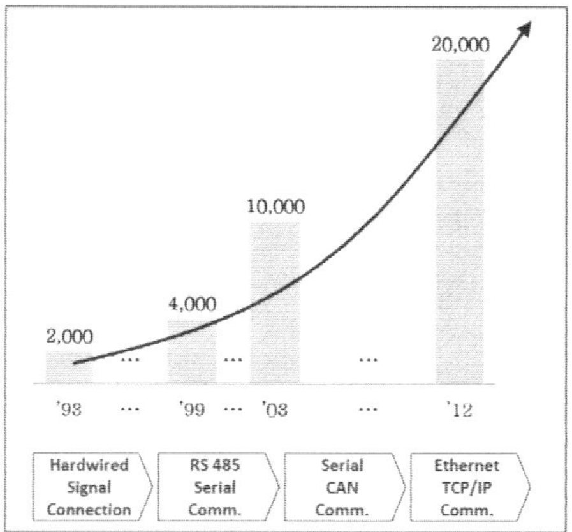

Figure 3: Plot of ship sensor growth - HHI Research 2013

This is a rapidly evolving situation where we are seeing virtuous circle dynamics in sensor development and deployment. Cheaper and better sensors allied with better sensor connectivity and data gathering leads to more sensor data in the open cloud. Better applications increase the utility of the gathered data, thus completing the circle.

3.2 SHIP AUTOMATION

As noted, current ship automation solutions tend to be implemented by specialist suppliers in a proprietary fashion. Thus ship sensor data is generated in silos and is hard to get at, limiting sensor data utility.

Integrated bridge and control systems do bring some of the information together, but tend to be implemented as end-to-end and closed proprietary systems.

Numerous specialist suppliers provide sensors. Each sensor may have its own operating characteristics. They connect to ship systems via a diversity of connection technologies largely adopted from Industrial Automation. These include:

Wire and fieldbus technologies:

- NMEA 2000
- Controller Area Network (CAN)
- Common Industrial Protocol (CIP)
- PROFIBUS
- HART

Ethernet technologies:

- Ethernet/IP - Ethernet implementation of CIP
- PROFINET - Ethernet implementation of PROFIBUS
- Regular Ethernet operating with TCP/IP

Also a variety of wireless technologies are used, for example, to gather data from cargo containers.

As is the case in Industrial Automation, adoption of mainstream TCP/IP enabled Ethernet is increasing on ships, as large scale Ethernet adoption has led to radically increased performance and lower costs, and this potentially allows access to sensor data at a single service gateway. In this context some smart sensors are also becoming Ethernet enabled.

The largest Asian shipbuilders, HHI and Daewoo are integrating Ethernet LANs into new ship builds.

The Japan Ship Machinery & Equipment Association (JSMEA) is orchestrating an industry research project called Smart Ship Application Platform.

The objective is to investigate the best means of bridging the ship machinery and Internet worlds. The research model features a Ship LAN that is connected to the key ship systems to enable gathering of field data. The model also features an onboard database. Importantly the results are to be made available in two proposed ISO standards:

- ISO WD 19847 – Shipboard Data Servers to Share Field Data on the Sea
- ISO WD 19848 – Standard Data for Shipboard Machinery & Equipment

4. SMART SHIP ENABLEMENT PATTERN

4.1 THE BASIC IoT PATTERN

In this pattern, Figure 4, a diversity of dispersed Internet enabled Things may contain one or more sensors that can connect to a central cloud using ubiquitous and standard Internet protocols. The sensor data is ingested and pooled in the central cloud. Cloud-based applications use big data analytics, machine learning, advanced visualisation technologies, and dashboards to provide and share Business Intelligence and predictive capabilities.

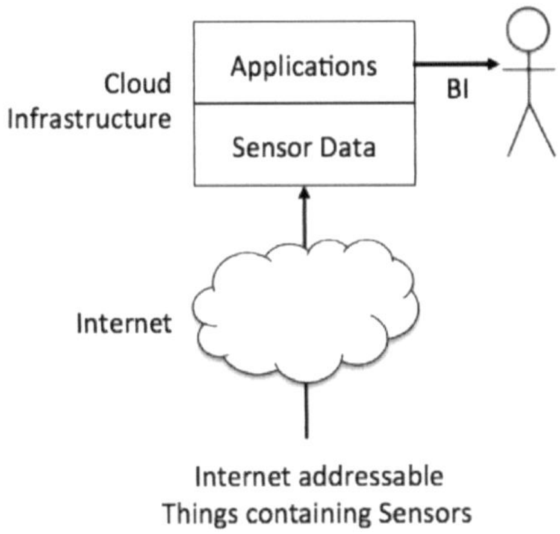

Figure 4: The Basic IoT Pattern

4.2 THE FOG PATTERN

The Fog Pattern, Figure 5, is an elaboration that has been developed to deal with cases where a Thing may contain a large number of sensors that produce a lot of high frequency data. A typical example is a factory floor.

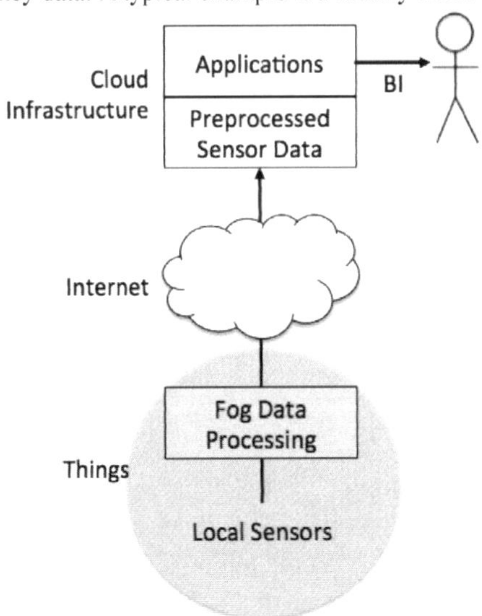

Figure 5: The Fog Pattern

In this situation, it does not make economic sense to serialise the data, to transmit it and then unpack it for processing in real-time in a remote cloud, particularly as the raw and processed data may be used locally by factory floor workers.

The economic balance may indicate that it is better to process the raw data close to where it is generated using distributed servers. In this scheme, pre-processed summary data is uploaded to the central cloud for consolidation with data received from other field operations.

A further reason for adopting this pattern is if communications bandwidth is constrained and/or unusually expensive.

The Ship Use Case satisfies all of the above criteria.

First, as we have seen, a large merchant ship can be expected to host thousands of sensors of all types that, on aggregate, can produce a lot of streaming data that increases in proportion to sensor sampling rates which can by up to 50 Hz or more. For example Figure 6 shows the outputs, in terms of Euler angles over time, from a ship motion sensor.

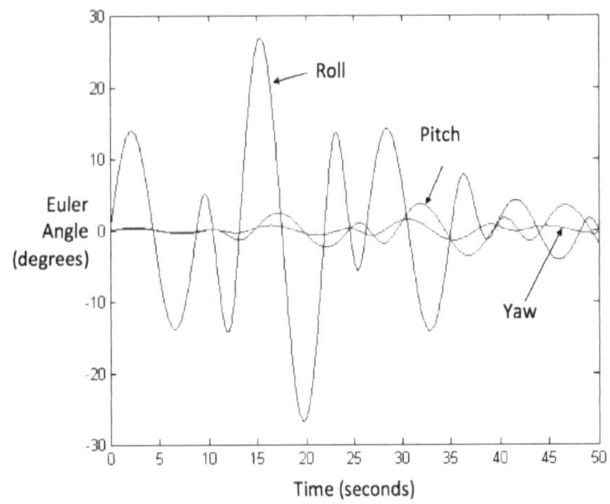

Figure 6: Outputs over time for a ship motion sensor

It can be surmised that, to capture the changes in Euler angles faithfully, a sample rate of up to 50 Hz will be required. Rotating machinery will require much higher sampling rates.

For the ship's cargo, sampling of container sensors, for example, may be at a lower rate. One sample every 10 seconds is sufficient to check sensor health, and to obtain an early warning of undesirable trends. Given potentially thousands of containers, each equipped with a smart sensor gathering data on temperature, humidity and container integrity this can, nevertheless, add up to a lot of additional sensor data.

Second, there is as strong operational requirement for the ship's crew to use the results of processed sensor data locally onboard their ship.

Third, the wireless satellite technologies needed for ship-to-shore broadband communications are inherently constrained and expensive compared with equivalent shore-based technologies.

Regarding the latter, the benefits of the latest generation of High Throughput Satellite (HTS) VSAT services are promoted as a Smart Ship enabler.

Indeed, they do promise a step up from the popular Inmarsat Fleet Broadband service that is used by the ship's crew for ad hoc Internet access and IP Telephony.

Inmarsat Fleet Xpress is representative of the new breed of HTS services. The Fleet Xpress maximum uplink speed (ship to shore) is quoted as 5 Mbps, and the maximum downlink speed (shore to ship) is 50 Mbps [7].

Regarding available bandwidth in practice, account needs to be taken of other issues inherent in the geostationary satellite services that are the mainstay for marine data communications, namely:

- Significant Bit Error Rates (BER) caused by atmospheric, third-party radio interference, multipath effects, and other conditions requiring retransmission of data
- Degradation or loss of service in adverse weather conditions
- Shared channel contention and congestion which means that maximum quoted information rate may not be obtained in practice
- TCP/IP overheads – This Internet protocol was not designed for satellite usage. High inherent latency (0.5 seconds round-trip delay for Geostationary Satellites), and other factors can result in impaired data transfer efficiency

Figure 7 shows estimated ship sensor data rates (Mbps) at increasing average sensor sample rates in the range 1 to 10 Hz for a typical large merchant ship. To produce this we have assumed 10,000 sensors generating 14 bytes of raw data per sensor made up of timestamp, sensor ID, and sensed value. We have overlaid the plot with the nominal uplink capacity for a single Fleet Xpress terminal.

It can be seen that the link capacity can be quickly exceeded as the average sensor sampling rate is increased.

This is a fairly rough and ready analysis, but we think it is sufficient to indicate that that sensor data needs to be processed and stored on the ship where it is produced and used.

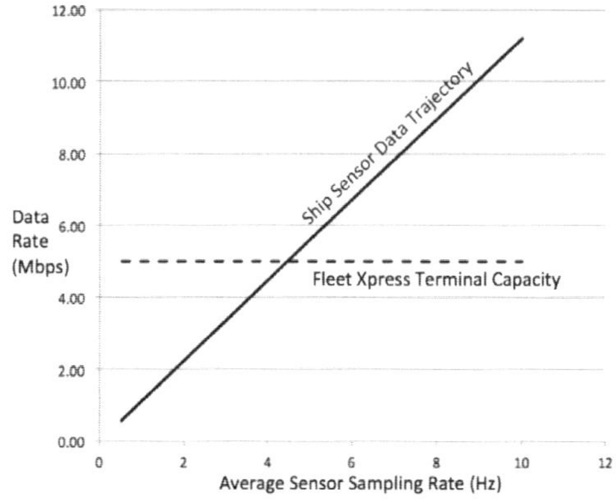

Figure 7: Ship sensor data at increasing sampling rates

Pre-processed summary data can be uploaded to a central cloud for consolidation and use at the fleet and enterprise levels. There is nothing in this arrangement that would prevent Fleet Level Staff, or indeed third parties, given the appropriate permissions, from making ad hoc requests to the ship database, or subscribing to particular sensor data streams on an ad hoc basis.

5. ON-SHIP FACILITIES

5.1 A HEYDAY OF BRITISH ENGINEERING

As a short diversion, consider this:

In the 1960's, Ferranti Wythenshawe produced compact process control computers, the Argus 400 and 500 series. They were applied to challenging applications in the Navy and elsewhere. In fact, the Queen Elizabeth 2 (QE 2) Steamship, commissioned in 1969, was equipped with one.

The Cunard Director of Engineering at the time recalls, in a contemporary article, how the Argus 500 was used [8]:

"Operating with 16K words of memory, the Ferranti computer was used for continuous monitoring of 700 sensors, performing real-time systems control, and generating logs and alarms automatically. It was also used for hotel stock control and weather routing. Another application performed prediction of freshwater requirements. Yet another application optimized fuel usage. This alone was sufficient to pay back the investment".

5.2 A SHIP CLOUD

Since 1969 the relative performance of processors has improved by leaps and bounds, for example:

Argus 500	1
Smartphone chip	100
Multi-core Server chip	1,000,000

If the Argus 500 could accomplish what has been described on the QE 2, imagine the transformational outcome achievable by the application of modern, low cost technologies on a merchant ship.

Today's midrange computer servers provide a huge amount of processing power and storage in a small footprint at ever decreasing costs. They come with abundant commercially supported open source software that can create an on-board private cloud that is a microcosm of the infrastructure used by large-scale Internet service companies and enterprises.

The servers can be obtained from a diversity of vendors in a highly competitive market, and can be employed on a mix and match basis.

We think use of such off-the-shelf solutions is important to enable easy upgrade of the onboard systems over the life of the vessel. Moreover, this type of solution provides a platform that facilitates development, deployment, and operation of applications in a way that is familiar to mainstream ICT developers and operational staff.

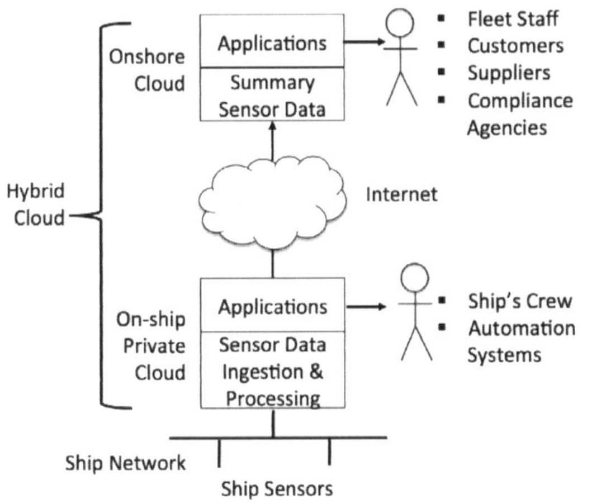

Figure 8: Proposed Hybrid Cloud Arrangements

In effect, we are proposing a hybrid cloud, Figure 8. This pattern mixes, on-premises, private cloud and third-party public cloud services with orchestration between the two platforms. The arrangement provides a highly flexible and scalable means for workloads to move between private and public clouds as needed.

5.3 GENERAL ARRANGEMENTS

Figure 9 outlines the general onboard arrangements we envisage.

The ship servers will require accommodation in suitable Unattended Machinery Spaces (UMS). They will be connected to thousands of sensors located throughout the ship, including machinery, automation and navigation systems, the ship structure and cargo.

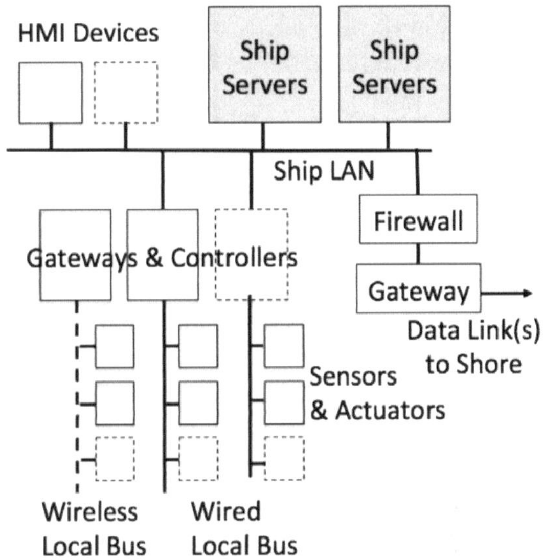

Figure 9: Onboard equipment arrangements

An Ethernet Ship LAN will connect the ship servers directly to Ethernet enabled sensors & actuators where these exist. Alternatively connection will be via Gateways and Controllers that connect to sensors and actuators via legacy fieldbus and wire networks. Also connection to wireless networks will be required via suitable hubs and gateways to sensors that might be attached to movable objects. Examples are containers and crew, for that matter, who may be equipped with wearable sensors.

Connection will also be required to Human Machine Interface (HMI) screens for the crew.

Finally the Ship LAN will connect, via a Firewall, to the ship broadband gateway(s).

As we have noted, new ship builds are tending to include ship LANs for connection to sensors, typically based on fault tolerant, high-speed industrial Ethernet. Ideally, they will provide open gateways to facilitate ingestion of ship sensor data by third parties. Access to ship equipment actuators via such open gateways will also be beneficial to achieving a higher degree of automation in due course.

For older builds, there is an opportunity for retro fitting specialists to provide the necessary connectivity.

To obtain the desirable benefits of interoperability, connection not only requires bridging of arcane ship protocols to standardised Internet communications protocols, but also information frameworks and codified metadata to provide context for use of observed sensor data. Ideally, this should provide application developers with familiar and widely used web application development constructs.

This is an area where existing frameworks developed for other sectors may have application, as we elaborate later.

6. SENSOR DATA TREATMENT AND USE

The essential processes are data ingestion, raw data stream processing, real-time analytics, storage, logging, ad hoc analysis of stored data, and finally usage/display of results.

6.1 OPEN LOOP CONTROL

Regarding usage, the simplest case is non-intrusive one-way data collection for open loop control. This is focused on more granular and comprehensive monitoring of the ship and its environment, and:

- Displaying this information to the ship's crew, and fleet staff in real-time via integrated instrumentation services
- Using this information to feed onboard and onshore applications that process the data to generate advisory information

6.2 CLOSED LOOP CONTROL

A more ambitious use of sensor data will be to direct signals back, on a selective basis, to closed loop control systems actuators from applications running on the ship cloud. This will take advantage of pooled sensor data and sensor fusion techniques, and possibly contextual information delivered from onshore. This will require configuration of the Ship LAN to ensure high speed and low latency connection to the relevant controllers and actuators. The highest level of fault tolerant computing and communications systems performance will also be required. The payoffs will be better and less expensive ship automation.

6.3 DATA INGESTION

Data from sensors will be ingested as streaming and time-ordered information. Fundamentally, it will consist of packets of data containing a Sensor ID, Sensor Data Value, and a Timestamp.

6.4 ONBOARD DATA PROCESSING

The data will be processed, initially, by stream processing software. Functions will include cleaning, filtering, alert generation, real-time instrumentation, and local logging of data received from individual sensor streams.
Also higher order processing using complex event processing (CEP) on multiple streams may be used to generate correlated event data. This is where consolidation of ship sensor data really begins to pay off, providing for advanced techniques such as fusion of sensor data to feed, for example, precise situational awareness and dynamic positioning applications.

6.5 USAGE OF ONBOARD DATA

6.5 (a) Open Loop Control

Processed streaming data will drive comprehensive and highly granular information in real-time to ship instrumentation. Storage and consolidation of key event data in an onboard database will be performed as a basis for onboard information systems used by ship's crew for local ad-hoc analysis and reporting. The information may also be sent to shore on a selective basis to implement, for example, a cargo monitoring service.

The data will also be available to onboard applications that employ advanced analytics of the type normally available in onshore clouds to provide onboard intelligence and diagnostics advisory information. The analytics available onboard will include descriptive, diagnostic, discovery, prescriptive and predictive (machine learning) techniques. Sensor data fusion techniques also have considerable potential, as will use of advisory and contextual information received from shore.

Here are some illustrative examples of onboard applications:

- The fusion of sensor data from high definition cameras, sonar, thermal imaging, conventional Radar and LIDAR combined with high performance analytics could provide better situational awareness, and collision avoidance information
- Continuous hull & superstructure stress & strain measurement linked to motion, navigation and other sensors allied with predictive capabilities could provide advisory information regarding heading/course changes allowing the Master to make immediate course corrections for severe situations
- Advisory information for optimising fuel usage could be generated through real-time monitoring of shaft power & fuel flow combined with information from offshore as well as local environmental information
- Provision of information for optimising the quantity and disposition of ballast & trim for lowest hull resistance could be provided through analysis of cargo and fuel consumption, and other dynamic factors such as on-route weather and sea conditions, the approach to shallow waters
- Continuous health monitoring of ship machinery & equipment (including IT systems and infrastructure), cargo and crew

- Cyber security monitoring, for example scanning for anomalies in incoming data

Output and querying facilities for ship's crew will be provided by large screen display panels, workstations and mobile devices connected to the ship LAN.

Crew configurable visualisation techniques, live mimic diagrams, and instrumentation dashboards will be used to provide big picture and drill-down capabilities.

6.5 (b) Closed Loop Control

In this case, ingested data will be fed back to actuators in real-time to provide advanced ship control applications. The applications may also use contextual information received from offshore such as speed and navigation instructions, or third-party provided information services.

Examples of these applications are:

- Integrated dynamic positioning using all of the ship's propulsion and guidance actuators along with environmental information. This approach may be based on forecasting the vessel's motion, rather than acting on present conditions
- Use of predictive models to inform the autopilot to provide enhanced & adaptive course control
- Optimisation of energy use by monitoring and controlling energy consumption on a ship-wide basis
- Automated monitoring and shutdown of equipment when a critical state is sensed

6.6 ONBOARD DATA STORAGE

The data stored onboard may become a large file that can be uploaded periodically to an onshore history file for batch analytics processing when the ship is in port.

Additionally, it is possible that onboard database could be queried on an ad-hoc basis by shore-based information systems, applications, and Fleet Staff, given appropriate access controls.

The amount of information stored on the ship will be a policy matter that balances storage costs with on-ship applications enablement.

6.7 ONSHORE DATA UPLOAD

Selected data will be uploaded to the onshore cloud in real-time or a periodic basis via the ship's broadband service, again according to a policy that balances communications cost with on-shore applications enablement.

6.8 USE OF ONSHORE DATA

Data will be pooled from all ships in the fleet and used onshore for the following purposes:

6.8 (a) To feed enterprise analytics and applications that provide insights into ship and fleet operations. Fleet staff will use these applications to fulfil their roles namely, business operations, planning, business development and optimisation. The applications may also be shared, on a controlled basis, with Third Parties. For example:

- Suppliers who need to coordinate their activities with ship and fleet operational requirements, and
- To fulfil Compliance Agency reporting requirements

6.8 (b) To inform Fleet Staff and Third Parties about selected ship operational data on an ad-hoc query and reporting basis.

6.8 (c) To enable Enterprise scale applications using the variety of big data analytics techniques we have previously described.

Examples of application areas include:

- High resolution Fleet status, disposition & tracking
- Navigation & speed advisory services
- Cargo tracking & scheduling
- Cargo status & health monitoring
- Cargo utilisation efficiency
- Crew health and welfare
- Fleet supply chain integration
- Ship, machinery & equipment predictive maintenance
- Supplier coordination
- Automated regulatory compliance
- Ship, fleet & enterprise cyber security

Configurable and advanced visualisation techniques and dashboards will be used to provide big picture and drill-down capabilities. Connection to existing enterprise applications such as SAP will also be provided.

6.9 APPLICATION DEVELOPMENT

The onshore cloud will also be a repository for ship configuration data and provide the environment for ship applications software development and deployment using contemporary DevOps (Development & Operations) technologies. This refers is a widely used set of practices that emphasises the collaboration and communication of both software developers and other information technology professionals while automating the process of software delivery and infrastructure change.

Employing constructs that are familiar to Internet Developers will facilitate application development. These include:

- Representational State Transfer or REST. This is an architectural style for designing networked applications. REST applications use simple HTTP requests to create, read, update and delete data
- JavaScript Object Notation (JSON), which is a lightweight data-interchange format
- XML technologies for coding information such as sensor metadata in machine-readable form
- Domain focused Software Development Kits (SDKs) and reusable software libraries

There is also a need for a framework to enable interoperability. This will provide a means for sensor data discovery and the context for sensor data use.

Use of a framework will release developers to focus on delivering value rather than worrying about daunting details of sensor protocols and data formats. This is calculated to accelerate and simplify application development. The framework will also minimize software changes required to accommodate changes in infrastructure and equipment.

7. SENSOR CONNECTION FRAMEWORKS

There are many initiatives that are key to enabling software development in an open and connected world. Our assessment is that two stand out as holding promise for bridging the ship and Internet worlds.

7.1 OPC UNIFIED ARCHITECTURE (UA) [9]

This standard is widely adopted in Industrial Automation. It is embodied in International Electro-technical Commission (IEC) 62541 standards. The OPC Foundation makes the following claims:

OPC UA is an extensible framework that provides for:

- Platform independence: from an embedded micro-controller to cloud-based infrastructure
- Security: encryption, authentication, and auditing
- Extensibility: ability to add new features without affecting existing applications
- Comprehensive information modelling

7.2 OGC SENSOR THINGS API & ASSOCIATED STANDARDS [10]

OGC has developed a mature and comprehensive framework for Sensor Web Enablement (SWE). The SensorThings API is a fresh initiative that is grounded in the SWE standards. It provides an open and unified framework to interconnect IoT sensing devices, data, and applications over the Web. It builds on web protocols such as MQTT and HTTP, Figure 10, and applies the easy-to-use REST style that is familiar to Internet Application Developers. The result provides a uniform way to expose the full potential of the IoT and to achieve interoperability.

The API allows easy integration into existing Spatial Data Infrastructures or Geographic Information Systems that are highly relevant to the maritime industry. It appears to be an elegant and modern framework that is web and programmer friendly. It is also associated with other useful standards.

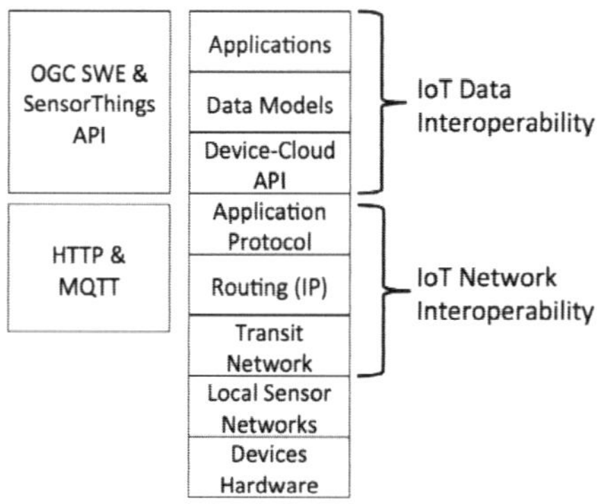

Figure 10: OGC SensorThings API Applicability

8. OTHER SMART SHIP ENABLERS

8.1 AN INDUSTRY DATABASE

In addition to connectivity framework used to bridge the Internet and Ship worlds, there is a need to catalogue metadata for the diversity of sensors used in merchant shipping to ensure interoperability. This is a dynamic situation. What we think is needed is a shared industry database (probably using XML) provided as a Web Service. This may best be accomplished through a consortium.

A Fleet Operator will draw on the database and other sources, for example CAD ship drawings and specifications, to build a manifest (or configuration file) of sensors for each ship in the fleet. Suppliers will update the database as and when they introduce new or enhanced products or the ship configuration is changed.

8.2 SHIP DESIGN AND CONSTRUCTION

Regarding new ship design and construction, we see the following as practical enablers:

Ideally there should be suitable Unattended Machinery Spaces to accommodate the industrial rack-mounted servers we are proposing. Preferably there should be two spaces that are separated to mitigate the risk of loss or denial of access through fire or flood.The spaces will need to conform to Classification Society UMS requirements, and also reflect onshore Data Centre best practices. In particular, modern servers generate a lot of heat that will have to be extracted. They also require humidity control and filtered clean air. Also clean power and uninterruptible power supplies (UPS) will be needed.

Also accommodation of thousands of sensors located throughout the ship and extending to the cargo will be required. The Ship LAN that will need to act like an spinal cord connecting sensors to the equipment located in the server rooms. Ideally open gateways to facilitate ingestion of ship sensor data by third parties will be provided. Access to ship equipment actuators via such and open gateway will also be beneficial to achieve a higher degree of automation in due course.

8.3 CYBER ENABLED SHIP SECURITY

A key requirement is adoption of best practices regarding unauthorised access and intrusion prevention from Internet Agents intent on causing harm and disruption via malware and cyber attacks is a key requirement

The Classification Societies are providing marine industry focussed guidance regarding this [11], [12], [13]. The last reference also provides useful guidance on Cyber Enabled Ship classification.

9. NEXT STEPS

Soldecom is building a proof of concept solution based on the approaches and technologies we have described. This will be a learning experience that leads to evaluation of solution components and framework elaboration. Following this we will be seeking to work with an early adopter Shipping Fleet Customer to develop an on-ship test bed to confirm and refine operational requirements, and to prove the framework in practice. We will also be contributing to key industry cooperatives focused on Smart Ship enablement.

10. CONCLUSIONS

Smart Ship development and deployment is at a nascent stage, but is gathering momentum. New ship builds need to be specified to accommodate the sorts of computer and communications facilities we have described. For existing ships there appears to be an opportunity for retrofitting specialists.

There is also a need for an industry-wide initiative to classify and encode sensor metadata if the promise of interoperability and its attendant benefits are to be realised. Also, in this context, adoption of an open framework such as OPC-UA or OGC SensorThings should be investigated. Last but not least the shipping industry has to move from the notion of "Connected ship" to the "Smart ship" which will operate as an integral part of the unified logistic chain.

11. REFERENCES

1. International Chamber of Shipping Website
2. World Trade Organisation Website
3. 'Towards a definition of the Internet of things', *IEEE*, 27 May 2015
4. The EU CASAGRAS Project
5. MAERSK LINE, 'Taking the Lead in World of Change', *Maersk Annual Group Magazine,* 2014
6. CHURCHILL J., 'Containers Listen and Talk', February 2016
7. Inmarsat Website
8. KAMEEN T., 'QE2: A Computer Aboard', *British Marine Progress, 1969*
9. OPC Foundation Website
10. Open Geospatial Organisation Website
11. BIMCO, 'The Guidance on Cyber Security on-board Ships', February 2016
12. LLOYDS REGISTER, 'Cyber-enabled ships - Deploying information and communications technology in shipping', February 2016
13. LLOYDS REGISTER, 'Cyber-enabled ships, ShipRight procedure-autonomous ships', July 2016

12. AUTHORS BIOGRAPHY

S. Koroneos is a cofounder and CEO at Soldecom. He is responsible for the development of the Soldecom Unified Communications Platform that provides best in class VOIP and other services for large shipping fleets.

P Bennett is a Chartered Engineer and advisor to Soldecom. He has deep experience in streaming data information systems and analytics gained principally in Financial Services. This commenced in 1972 with the development of a market price collection and display system at the London Stock exchange that used a Ferranti Argus 400 process control computer.

REQUIRED FLOODING SENSOR ARRANGEMENT FOR RELIABLE AUTOMATIC DAMAGE DETECTION

E Takkinen, **P Ruponen** and **P Pennanen**, NAPA, Finland

SUMMARY

Rapid detection of flooding, and reliable damage estimation are required for an efficient decision support system, especially on passenger ships that can carry thousands of people onboard. For this purpose, properly located flooding detection sensors are needed. This paper presents a study on the required sensor arrangement on a large passenger ship. Three alternative sensor arrangements are compared with several different damage scenarios. For this, a state-of-the-art decision support system with time-domain flooding prediction is used. The results are analysed with a recommendation for a minimum sensor arrangement for similar ship designs.

1. INTRODUCTION

In severe flooding accidents the captain of the ship is put under heavy pressure in making the decision of whether to evacuate the people or not. In the cases of the *Costa Concordia* and the *Sewol,* the order for abandoning the ship was given far too late. The basis for this important decision is a quick understanding of the extent, nature and severity of the flooding case. Reliable flooding detection by automatic sensors is the only viable method for acquiring information about the flooding without loss of valuable time and jeopardizing the personnel of the ship. Floodwater must always be detected before the stability of the ship is compromised.

Fast and accurate detection of flooding is thus a requirement for correct decision making onboard a damaged ship. Therefore, the International Maritime Organization (IMO) has set out minimum requirements for flooding detection systems in IMO MSC.1/Circ.1291 [1]. These requirements are sufficient for identifying the flooded watertight (WT) compartments only, because of the fact that also on/off flooding detectors are recognized as sensors. In order to get information about the amount of floodwater and the flow rates needed for predicting the severity, reliably working level sensors are needed.

However, even when using good quality level sensors, a more extensive sensor coverage is needed for an accurate assessment of the breach size and location. In addition, the increased number of sensors provides some redundancy if some sensors are not working, e.g. due to the impact of the damage. Simple monitoring of the floating position of the ship alone is not enough for a correct assessment of the actual damage case, although this information may support the primary data from the level sensors. This has already been noted in IMO SDC2/INF.6 [2], where updated recommendations for flooding detection were presented based on the results of the EU FP7 project FLOODSTAND.

Recently, a novel decision support system (DSS) for flooding emergencies has been presented in Pennanen et al. [3]. This system utilizes time-domain prediction of progressive flooding [4] based on accurate flood level sensor data. The results, including damage stability assessment and estimated available time for an orderly evacuation and abandonment are presented to the user following the Vessel TRIAGE method, [5]. The present study uses this DSS for assessment of the sufficient level sensor arrangement for a fast and reliable flooding detection and prediction of progressive flooding.

2. REGULATIONS

As mentioned, IMO requires all passenger ships build after 2010 to be equipped with a flooding detection system, [1]. The IMO MSC.1/Circ.1291 only states that the system needs to monitor the dry spaces since all tanks normally have level gauges. Measurement of the floodwater levels is not mandatory since simple on/off switches are also acceptable. However, in order to do a rapid and reliable assessment of the damage, both the current level and level rate are needed. Therefore, switches should only be used in rooms without connections to other rooms, such as double bottom voids.

The IMO MSC.1/Circ.1291 [1] states that: *"sensors should generally be installed at the centreline of the space (or alternatively at both the port and starboard sides). In addition, any watertight space that extends the full breadth of the ship or with arrangements that would seriously restrict the transverse flow of water should be provided with sensors at both the port and starboard sides."* Based on the full-scale tests [6] the A-class fire rated bulkheads will seriously restrict the flooding if the fire doors are closed. This has already been noted in the proposal for improved guidelines for flooding detection systems [2], developed within the FLOODSTAND project.

The scope of the present study is to compare different level sensor arrangements and their impact on both flooding detection and time-domain analysis of progressive flooding within a decision support system framework.

3. ANALYSES FOR DECISION SUPPORT

3.1 BREACH ASSESSMENT METHOD

The breach in the hull of a damaged ship can be assessed on the basis of the measurement data from the flood level sensors. The details of the developed algorithm are published separately [7], but for convenience a short description is provided here.

The sensor data provides only the current level and the level rate (time derivative of level) in the monitored room. Based on this data and the numerical 3D ship model, the net volumetric flow for each room with a sensor can be calculated with a central difference method:

$$\frac{dV}{dt} \approx \frac{V\left(H(t) + \frac{\delta t}{2}\frac{dH(t)}{dt}\right) - V\left(H(t) - \frac{\delta t}{2}\frac{dH(t)}{dt}\right)}{\delta t} \tag{1}$$

where δt is the time step, typically about 2.0 s is applied, and dH/dt is the measured level rate in the room.

The flow rate through an opening at time t can be calculated by applying Bernoulli's equation:

$$Q(t) = \int_A C_d \sqrt{2g \cdot \Delta H(t)} dA \tag{2}$$

where C_d is the discharge coefficient, g is acceleration due to gravity, $\Delta H(t)$ is the effective pressure head of water and A is the area of the opening. In practice air compression inside a flooded room can be notable, [8], but for simplicity it is sufficient to consider only the hydrostatic pressures.

The equations (1) and (2) provide the fundamentals for the breach assessment. However, there is one equation with many unknown factors. Thus the problem needs to be simplified by assuming a constant discharge coefficient of 0.6 and a fixed location for the breach. In addition, any progressive flooding from the breached room is ignored. Consequently, the following equation for the area of the breach can be obtained:

$$A \approx \frac{\frac{dV_i(t)}{dt}}{C_d \sqrt{2g \cdot \Delta H(t)}} \tag{3}$$

The breach location is set to the side of the room, extending from the bottom to the top, if the room is limited to the hull surface or it is within the $B/5$ limit from the surface at design draft. Here B is the breadth of the ship. For the rooms that are inside this limit surface, the breach location is set to the bottom of the room.

When flooding is detected in a new room, the source of floodwater needs to be checked. If there are possible flooding paths from previously flooded rooms the newly detected water is likely progressive flooding. If no other source is found the room is defined to be breached.

3.2 TIME-DOMAIN FLOODING PREDICTION

When the breaches in the hull of the damaged ship are known, the progressive flooding to undamaged rooms can be calculated with time-domain flooding simulation. The applied method is presented in detail in [9] and [10]. The flow rates in openings are calculated from Bernoulli's equation, (2). In order to achieve a fast computation, a relatively long time step of 30 s is used. The numerical method is based on implicit scheme, which ensures numerical stability of results also with a long time step. In practice a 3 h prediction can be done within about 3 min.

The system will automatically repeat the breach assessment every 5 min. If new flooding is detected and its source cannot originate from an already flooded room, an additional breach opening is defined. The starting point for the next prediction is also updated based on the latest measurement data for water levels in the rooms.

4. CASE STUDY

4.1 LARGE PASSENGER SHIP DESIGN

The study of the required sensor arrangement was done with a slightly modified version of the FLOODSTAND sample ship [11].

The ship is a Post Panamax, 125 000 GT, large cruise ship with a total capacity of 5 600 persons on-board. Main dimensions of the ship are presented in table 1. The ship is divided into 22 watertight compartments.

Table 1: Main dimensions of the study ship. [11]

Gross Tonnage	125 000 GT
Length overall	327 m
Length at design waterline	300.7 m
Design draught	8.5 m
Breadth at design waterline	37.4 m
Max number of passengers	4200
Number of crew	1400
Total persons on-board	5600
Number of WT-compartments	22

4.2 SENSOR ARRANGEMETS

The study was done with three different sensor arrangements, versions A, B and C. Sensor amounts are presented in Figure 1 and Table 2. Sensor coverage is defined as the percentage of total dry space volume with level sensors below the bulkhead deck.

Version A fulfils the absolute minimum requirement of the IMO MSC.1/Circ.1291 [1] for flooding detection system. This means that some of the WT compartments are equipped with only one sensor at the centre line. Version A has 70 sensors in total.

Version B is based on the revised recommendation of the FLOODSTAND project, [2]. This means that all the WT compartments are equipped with two sensors on each deck. One on each side of the ship. In the bow of the ship, small WT compartments are equipped with only one sensor. The bulkhead deck (BhD) has only two sensors, in the aft steering gear room. In total, version B has 89 sensors.

Sensor arrangement C covers all the larger rooms in the ship. This means that some WT compartments have more than two sensors on each deck. This version also has sensors on the bulkhead deck. The total amount of sensors in this version is 134.

Table 2: Studied sensor arrangements (BhD=bulkhead deck).

Ver.	Info	Sensor count below BhD	Sensor count on the BhD	Sensor coverage
A	Minimal requirement	68	2	73.7 %
B	FLOODSTAND recommendation	87	2	88.1 %
C	All the large rooms equipped with sensors	95	39	93.2 %

4.3 STUDIED DAMAGE SCENARIOS

A total of six different realistic damage cases were studied. These are illustrated in Figure 2, and the details are presented in Table 3. All damage cases were at least two compartment damages, so that they would cause significant flooding in the ship. Some of the damages are very similar to the real life damage scenarios, like *Costa Concordia* and *Monarch of the Seas*. In all the cases, all the watertight doors were considered to be closed. Also all the fire doors to staircases and escape trunks in the lover decks were considered to be closed. In addition, the doors to the cold rooms and stores were closed. It was assumed that the real status of all doors was known. It is worth mentioning that the monitoring of the fire door statuses is crucial, since the effects on the flooding process by fire doors can be significant [12].

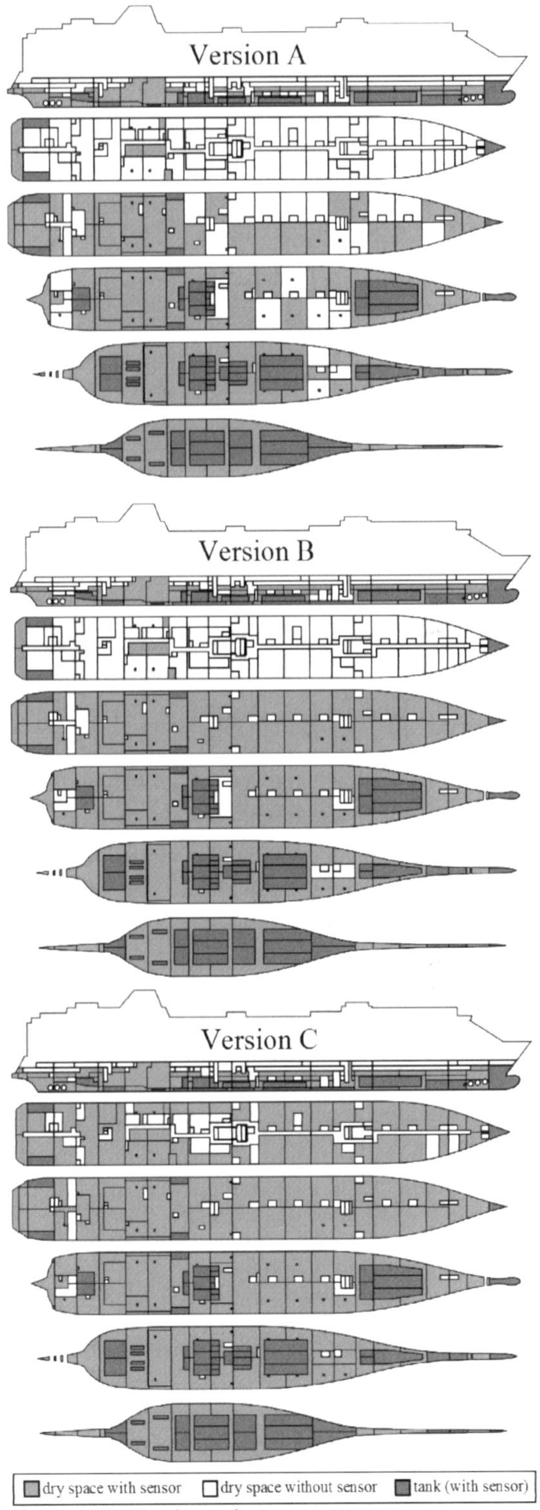

Figure 1: Comparison of sensor arrangements

Table 3: Overview of the studied damage cases.

Damage case	Info	Number of damaged WT compartments	Equilibrium or capsize time
Case 1	Side damage to aft part of the ship.	2	43 min 44 s to equilibrium
Case 2	Side damage to forward part of the ship.	2	5 min 44 s to equilibrium
Case 3	Side damage to aft part of the ship. Both engine rooms are breached.	5	Capsize after 7 hours and 42 min
Case 4	Long bottom damage with small breaches to tank top.	6	Capsize after about 60 min
Case 5	Long and narrow side damage to middle part of the ship just under water line.	4	Flooding takes about 60 min, stable equilibrium with large heel angle
Case 6	Long side scrape damage to fore part of the ship, damaging also rooms above the bulkhead deck.	5	Capsize after 1 hour and 30 min

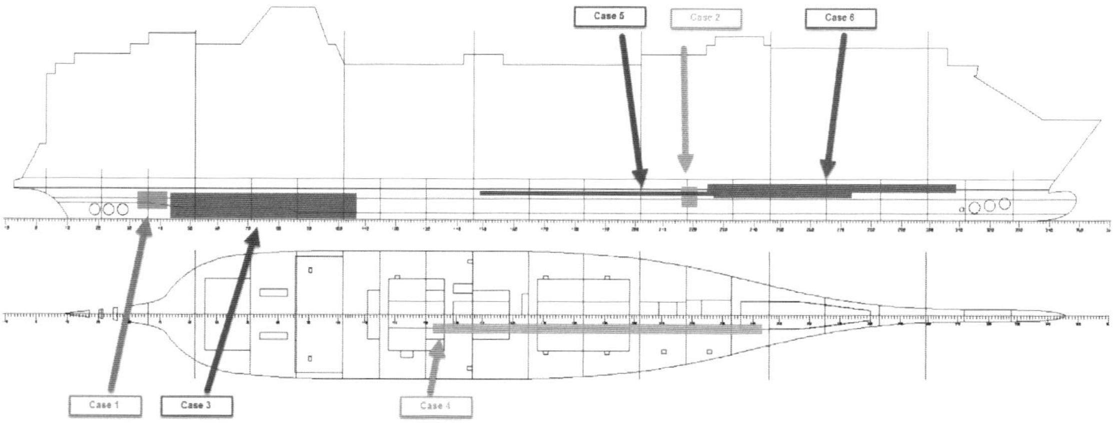

Figure 2: Studied damage cases.

These damage scenarios were first simulated with the NAPA software, using a short time step of 2 seconds. From the simulation results, artificial measurement signal data was created for each sensor. This sensor data was then used to perform automatic damage and progressive flooding predictions. These results are analysed in the next chapter. All simulations were done in calm water, and the applied signal data had no noise, e.g. due to sloshing inside the flooded rooms.

In the simulations, the same door statuses are used as in the reference calculation. Also the initial loading condition is the same.

5. RESULTS

Results were analysed by comparing the automatically created breach sizes and predicted development of flooding to the reference results.

Case 5 is presented here in more detail as an example. The rest of the results and study methods are described in the original study by Takkinen [13]. An overview of the results for all the cases are presented in the breach assessment and in the progressive flooding calculation in sections 5.2 and 5.3.

5.1 EXAMPLE CASE 5

Case 5 is a long raking damage to the starboard side of the ship, just under the intact waterline. It extends over 5 WT compartments. The breach is only 10 cm wide. The total breach area is 6.0 m^2. The flooding takes 60 minutes, and the ship finds a stable equilibrium with a heel of 10 degrees to port side.

Immediately after the damage occurs, the ship heels very strongly to the starboard side. After a while the flooding progresses to undamaged rooms, and the heeling starts to equalize. Due to the asymmetry of the staircases, the flooding progresses to the port side compartments on the lower decks. This makes the ship heel to the port side in the later phases of the flooding. This is a good example due to the large changes in the heeling angle and immersion of the breach. These phenomena bring out a lot of problems if the breaches have been modelled incorrectly.

The real breach is very close to the water line on the starboard side, and when the ship heels to port side, the breach rises from the water. As a result the flooding stops. This now creates a discrepancy between the modelled breach and the simulated breach. The modelled breach is always defined from the bottom of the room to the top of the room [7]. In this case the real breach is located close to the top of the room and is vertically very narrow. The difference between the reference breach and predicted breach are visualized in figure 3. This causes the ship to equalize to a about 4 deg smaller heel angle in version B and C compared to the reference calculation.

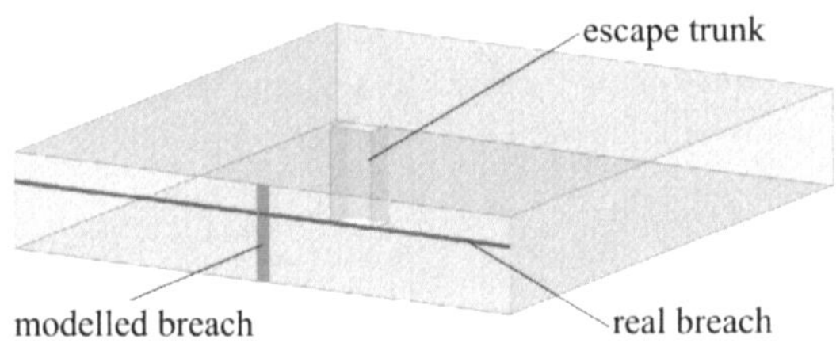

Figure 3: Comparison of the real breach (red) and estimated breach (green) in a single room in Case 5.

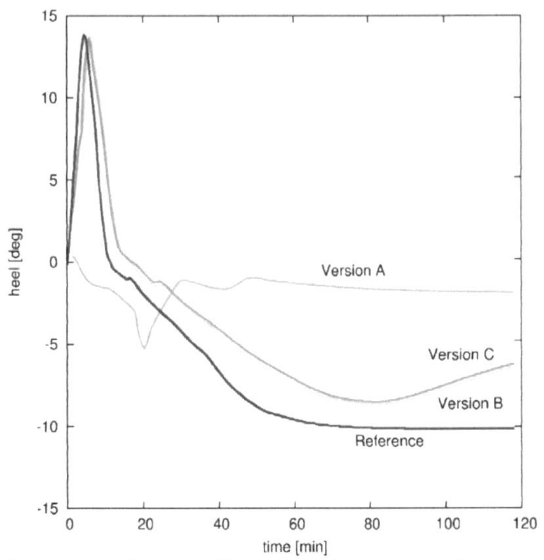

Figure 4: First prediction results compared to the reference heel angle in Case 5.

The reason why the prediction with the sensor arrangement version A is not following the reference simulation is that some of the modelled breaches are on the wrong side of the ship. This is due to the flooding sensors being located too close to the centre line of the ship.

In the later phases of flooding, version A predicts the ship to heel to the starboard side. When the DSS clearly gives a wrong prediction, the credibility of the system has been lost, and even if the system starts to predict correctly in later phases, the user may not believe the system anymore.

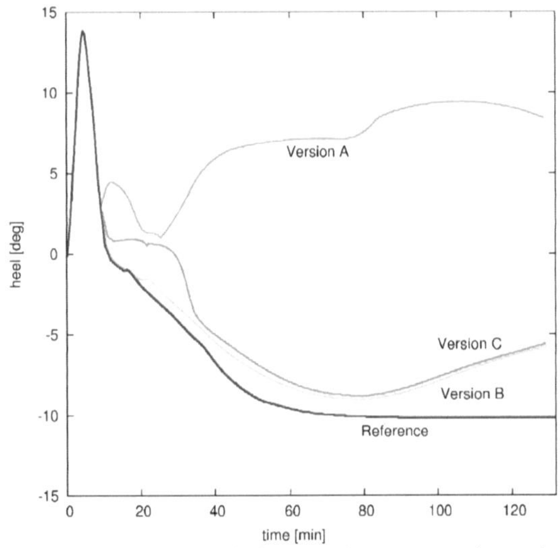

Figure 5: Third prediction results compared to the reference heel angle in Case 5 Prediction starts 12 minutes after the damage.

5.2 BREACH DETECTION OVERVIEW

Overview of the results are presented in Table 4 with scale *"Poor"*, *"Moderate"*, *"Good"* and *"Exellent"*. Here *"Poor"* means that either most of the breaches are not detected, they are on the wrong side, or that they are in the wrong location. For *"Moderate"* most of the breaches are detected well, but there is an error in at least one important breach, which affects the prediction calculation significantly. *"Good"* means that all the breaches are detected and only small differences occur in the sizes of the predicted breaches compared to the actual breaches. *"Excellent"* means that all the breaches and their sizes are predicted correctly. Some larger errors in double bottom voids are still accepted.

In most of the cases with the arrangement version A there was not enough data to detect all the breaches with sufficient accuracy. However, large areas, like the engine rooms, are well covered by the sensors also in this version, and the results for these areas are the same as in the other sensor arrangement versions. This can be seen in cases 3 and 4. In cases 5 and 6, where the ship suffers long raking side damages, version A performs poorly. These damages are in areas where there are crew cabins, and there is only one sensor at the centre line of the ship for these compartments. Due to this, some damages are generated on the wrong side of the ship. Thus the version A gets the overall grade *"Moderate"*.

The differeces between the versions B and C were very small, even tough version C has much more sensors. The benefit of these additional sensors is not very significant since they are providing only little additional information. The only notable difference is for Case 6, where the rooms above the bulkhead deck are also breached. Both versions receive the grade *"good"* for the automatic breach detection even tough verison C has 51 % more sensors.

5.3 PROGRESSIVE FLOODING PREDICTION

Results from the progressive flooding calculations are presented in Table 5 with the same scale as in the breach detections. Here *"Poor"* means that the flooding prediction is clearly wrong for a longer time, and gives the crew incorrect information. *"Moderate"* means that the flooding prediction follows the reference results, but gives overly optimistic results. *"Good"* means that the results are basically right but there is a difference between the reference results and the prediction results. *"Excellent"* means that prediction results follow the reference results closely.

Due to the shortages in breach detection, the version A performs only moderately in the progressive flooding calculations. Especially for cases 5 and 6, where the breaches are in compartments that have only one sensor at the centre line of the ship in version A Consequently, version A receives an overall grade of *"Moderate"* in the flooding prediction.

The differences between the versions B and C are small. In cases 1 and 3 version B, that has less sensors, performs better than version C. This happens because the additional sensors in verison C change the calculation according to the flooding level that they receive. The only case where version C actually performed better than version B was case 6. This is due to the sensors on bulkhead deck. Because of the total number of sensors in relation to the prediction performance, version B get overall grade *"Excellent"* and version C gets *"Good"*.

Table 4: Overview of the breach assesment accuracy.

Case	Version A	Version B	Version C
1	**Moderate**, one of the most important breaches was not detected.	**Excellent**, all the breaches detected with satisfying accuracy.	**Excellent**, all the breaches detected with satisfying accuracy.
2	**Moderate**, some breaches were not found and large size differences occurred.	**Moderate**, some breaches were not found and large size differences occurred.	**Moderate**, some breaches were not found and large size differences occurred.
3	**Good**, all the breaches detected with good accuracy.	**Good**, all the breaches detected with good accuracy.	**Good**, all the breaches detect with good accuracy.
4	**Good**, most of the breaches detected.	**Good**, most of the breaches detected.	**Good**, most of the breaches detected.
5	**Poor**, only few breaches detected in the first round, and some breaches detected on the wrong side of the ship.	**Good**, all the breaches correctly detected.	**Good**, all the breaches correctly detected.
6	**Poor**, most breaches were not found and large size differences occurred.	**Moderate**, some breaches not found and large size differences.	**Good**, most of the breaches detected, also on the bulkhead deck.
Overall grade	**Moderate**, big problems in some cases to detect the breaches. Only damages to the engine room and tank top were detected well.	**Good**, in most of the cases the breaches were detected with good accuracy. Experienced some difficulties, especially if the damage is above the bulkhead deck.	**Good**, the accuracy is good in all the cases. However the benefit compared to version B is not significant, especially when considering the amount of extra sensors used.

Table 5: Overview of the flooding prediction accuracy.

Case	Version A	Version B	Version C
1	**Moderate**, heel and floodwater prediction follow the reference.	**Excellent**, heel and floodwater prediction follow the reference very well.	**Good**, heel and floodwater prediction follow the reference well.
2	**Good**, heel and floodwater predictions follow the reference.	**Excellent**, heel and floodwater predictions follow the reference well.	**Excellent**, heel and floodwater predictions follow the reference well.
3	**Good**, DSS predicts the ship to capsize about right.	**Excellent**, DSS predicts the ship to capsize exactly right over one hour advance.	**Good**, DSS predicts the ship to capsize about right, but earlier than it actually happens.
4	**Good**, predicts correctly the ship to capsize. First a little later and then earlier than it actually happens.	**Good**, predicts correctly the ship to capsize. First a little later and then earlier than it actually happens.	**Good**, predicts correctly the ship to capsize. First a little later and then earlier than it actually happens.
5	**Poor**, breaches in wrong sides and too late causes the calculations to be very wrong.	**Good**, predictions follow the reference nicely, however predicts the heeling to decrease more than it actually does.	**Good**, predictions follow the reference nicely, however predicts the heeling to decrease more than it actually does.
6	**Poor**, flooding prediction gives too optimistic results.	**Moderate**, the missed breaches on the bulkhead deck make predictions somewhat inaccurate. DSS predicts the ship to capsize later than she actually does.	**Good**, detected breaches on the bulkhead deck make sure that the prediction is fairly accurate.
Overall grade	**Moderate**, does not guarantee the DSS to work properly.	**Excellent**, guarantees the DSS to work in all situations. Damages above the bulkhead deck are however problematic	**Good**, Not much better than version B, but a more expensive solution

6. CONCLUSIONS

The only realistic and reliable way for automatic breach detection in a cruise ship, with complex internal structures, is to use flooding level sensors. Automatic and reliable breach detection, combined with progressive flooding calculation, enables accurate and realistic assessment of damage stability and available time for orderly evacuation and abandonment.

It is worth noticing that even in optimal conditions, automatic damage detection is still an estimate. It is impossible to model the exact shape and size of a breach based on only the available sensor data. However, if the location and the size are estimated with reasonable accuracy, also progressive flooding can be predicted with good accuracy. The time to detect all the damages is also very important so that the crew gets the information about the situation rapidly. If the first predictions are giving clearly wrong information, the credibility of the whole decision support system is compromised, and even if the system is giving correct estimates later on, the crew may no longer believe in them.

The performed study on a large cruise ship design with three different sensor arrangements confirmed that the revised recommendations for flooding detection systems in SDC2/INF.6 [2] are enough for a fast and reliable flooding detection within the framework of an advanced decision support system for flooding emergencies. On the other hand, the absolute minimum amount of sensors as required by IMO MSC.1/Circ.1291 [1] is not enough in many cases, since flooding can progress for a long time before it is even detected by the sensors.

7. REFERENCES

1. IMO MSC.1/Circ.1291 "Guidelines for Flooding Detection Systems on Passenger Ships", 9 December 2008

2. IMO 2014. Guidelines for Flood Sensor Placement and Technical Requirements, SDC2/INF.6, submitted by Finland.

3. PENNANEN, P., RUPONEN, P., RAMM-SCHMIDT, H. 2015. Integrated Decision Support System for Increased Passenger Ship

Safety, *RINA Damaged Ship III*, 25-25 March 2015, London, UK.

4. RUPONEN, P., PENNANEN, P., LINDROTH, D. 2015. Prediction of Survivability for Decision Support in Ship Flooding Emergency, *Proceedings of the 12th International Conference on the Stability of Ships and Ocean Vehicles, STAB2015*, Glasgow, UK, pp. 987-997.

5. NORDSTRÖM, J., GOERLANDT, F., SARSAMA, J., LEPPÄNEN, P., NISSILÄ, M., RUPONEN, P., LÜBCKE, T. & SONNINEN, S. 2016, "Vessel TRIAGE: A method for assessing and communicating the safety status of vessels in maritime distress situations", *Safety Science*, vol. 85, pp. 117-129.

6. JALONEN, R., RUPONEN, P., WERYK, M., NAAR, H., VAHER, S. 2017. A Study on Leakage and Collapse of Non-Watertight Ship Doors under Floodwater Pressure, *Marine Structures*, Vol 51, pp. 188-201.

7. RUPONEN, P., PULKKINEN, A., LAAKSONEN, J. 2017. A method for breach assessment onboard a damaged passenger ship, *Applied Ocean Research*.

8. RUPONEN, P., KURVINEN, P., SAISTO, I., HARRAS, J. 2013. Air compression in a flooded tank of a damaged ship, *Ocean Engineering*, Vol. 57, pp. 64-71.

9. RUPONEN, P. Progressive Flooding of a Damaged Passenger Ship, *TKK Dissertations* 94, 2007.

10. RUPONEN, P. 2014. Adaptive time step in simulation of progressive flooding, *Ocean Engineering*, Vol. 78, pp. 35-44.

11. KUJANPÄÄ, J., ROUTI, A-L. 2009 "Concept Ship Design A" FLOODSTAND Deliverable D1.1a.

12. RUPONEN, P. 2017. On the effects of non-watertight doors on progressive flooding in a damaged passenger ship, *Ocean Engineering*, Vol. 130, pp. 115-125.

13. TAKKINEN, E. 2016 Effect of flooding sensor arrangement on automatic breach assessment for damaged passenger ship, M.Sc. Thesis, Aalto University, Finland https://aaltodoc.aalto.fi/handle/123456789/21624

8. AUTHORS BIOGRAPHY

Mr. Esa Takkinen holds the current position of Technical Consultant at Napa Ltd. He is responsible for loading computer and decision support system deliveries. He wrote his master thesis about the effects of flooding sensor arrangement on damage assessments for NAPA. He graduated 2016 as M.Sc. Naval Architect.

Dr. Pekka Ruponen holds the current position of Lead Research Engineer at Napa Ltd. He is responsible for ship stability and safety related research and development. His previous experience includes development and validation of a time-domain flooding simulation method. He has over ten years of experience in damage stability calculations and he has authored several journal articles and conference papers in the field of ship stability and safety.

Mr. Petri Pennanen holds the current position of Senior Adviser at Napa Ltd. He is responsible for regulatory relations, concentrating on rule development work at IMO and close co-operation with other regulatory bodies like classification societies. His previous experience includes passenger ship design and software applications used on board ships. During his 19-year career at Napa, he has been heading loading computer deliveries and customer support, moving later on to software product development and research activities in ship stability and decision support, being finally responsible for the business related to loading computers and decision support as a senior product manager.

THE UNGOVERNED SPACE OF MARINE FIRE SAFETY ENGINEERING

Dr Carl Stephen Patrick Hunter BA(Dunelm) Hon DSc FRINA FIMarEST MRAeS, Coltraco Ultrasonics, UK

SUMMARY

This paper explores continuous monitoring on smart ships to increase safety, regarding the "Ungoverned Space" of marine fire extinguishing systems and Protected Space integrity. Going above and beyond regulatory compliance to secure additional benefits is vital to increasingly high value and safety critical vessels. Traditionally, the industry cares little about this ungoverned space. However, with the impact of fire being more widely recognised, new research and development explores ultrasonic technology to monitor the extinguishing systems through inspection and integrated, continuous monitoring. The latter gives eyes to shore-based operators at a time of fewer crew and hopes for autonomous shipping. New smart ship technology counters the environment in which a "safety first" culture remains un-pursued and unrewarded. There is a real cause for concern – now there is an impetus to change. Ultrasound is now being put to more advanced, innovative uses in shipping for fire safety solutions.

1. INTRODUCTION

Gaseous extinguishing/suppression systems are installed to protect against special hazards in critical infrastructure at sea. They deliver the infrastructure resilience our advanced maritime society requires. If the hazard is special and the vessel is critical, then this is the case for the constant monitoring of the fire systems that aim to deliver their protection. In fire safety engineering, there are four key areas which must be considered to best analyse the efficacy of the system: contents of agent in the cylinders, weight of agent in the cylinders, pressure, and protected space integrity. This paper explores these areas, demonstrating that there exists an "Ungoverned Space" in fire safety engineering in the marine sector which must be addressed, so that the industry can work together to improve safety of life, asset and vessel at sea.

1.1 WHAT IS THE "UNGOVERNED SPACE"

Who would build a ship or offshore platform and install a power generating or auxiliary machinery without installing emergency power systems or monitoring their condition states? Who today would consider installing an alarm system without monitoring its overall status, not only its actuation, and integrating the whole system into the ship safety management system, with central monitoring being an essential part of it? These are basic engineering principles: building in redundancy and constantly monitoring critical systems.

Yet when it comes to marine fire maintenance, never mind constantly monitoring, of gaseous fire extinguishing systems, there is a lack of knowledge among the majority of the industry, around the potential risks. Awareness should exist about the huge expenses incurred by fire, both in terms of costs upfront from the damage and long-term due to reputation and unknown losses or damage to seafarers, vessel integrity, and cost of downtime and recovery. Above all is the risk to human life presented by fire. Poor maintenance of suppression systems risk accidental fatalities due to lack of training about the lethal properties of CO_2: when released it reduces oxygen levels to extinguish fire. Limited appreciation of the need for room integrity testing of protected spaces leads to minimal regulation compliance which could mean a failure of the fire system suppressing a fire because the room is unable to hold the discharged gas due to leaks of the space into which it actuates.

1.2 ARE STAKEHOLDERS WORKING TOGETHER TO IMPROVE FIRE SAFETY?

Following a fire, 40% of facilities experiencing business dis-continuity do not survive another 12-18 months afterwards. FM Global say that a third of businesses affected by fire go out of business within 3 years [1]. The damage of fire cannot be over-stated. It is necessary for stakeholders to work together to improve fire safety. The stakeholders comprise the ship owners and its leadership team. Critically it also involves the mariners who occupy its spaces, the insurers who safeguard their lives, their consequential loss or injury and the damage to their families, the asset value itself and the fire company who design, install, commission and maintain the marine clean agent system.

An engineered marine fire system typically lasts 20 years, longer than some vessels! More automatic fire protection systems mean fewer fires which must be contained by the ship in its isolation. But we understand the FRS far more than the fire engineering solutions embedded in the vessels themselves. Information about fire systems is not usually publicly available. One day customers will question the fire safety systems of the assets they depend on. Customers expect 100% availability of the asset and when using it reasonably, they expect 100% safety in it. Is it not time that assets such as vessels should be monitored 100% of the time?

1.3 ADDRESSING THE UNGOVERNED SPACE: EXPERIENCE & EDUCATION

There exists much "ungoverned space" in the marine industry. There exist few Engineers in the industry who fully understand the design, installation, testing, maintenance and safety of gaseous firefighting systems in vessels or offshore marine structures. There is a lack of knowledge of the characteristics of the various extinguishants and the types of fire for which they are suitable. For too many years the industry has been left to too few brilliant – largely British and American - experts to determine safe outcomes. But as the world changes, so must the industry integrate technological solutions to provide a bulwark against wider industry misinterpretation, minimal, occasional and even flagrant disregard of the application of standards and good global engineering practise. Education is key to creating standards which all can understand and apply.

1.4 FIRE SAFETY AT SEA

Shipping, Naval and Oil and Gas sectors are all safety critical, with possible catastrophic and expensive results in the event of fire. Incidences of fire on ships have been increasingly reported in the press [written in November 2016]. For example, the International Maritime Risk Rating Agency (IMRRA) have declared fire safety to be the most common deficiency on tankers in October 2016, with 126 incidences [2]. A white paper has been published by the International Union of Marine Insurance (IUMI) suggesting methods for improving fire safety standards on container ships. IUMI's Loss Prevention Committee Vice Chairman Uwe-Peter Schieder advocated in *IHS Safety at Sea* that "Major fires on container vessels count among the worst hazards in global shipping". This was published in article arguing "Fire risk grows as ships get bigger" which also highlighted the United States Coast Guard's awareness of the need for improved fire safety [3].

2. "GASEOUS EXTINGUISHING SYSTEMS LEAK – IT'S OFFICIAL"

Cylinders leak and accidentally discharge. CO_2 can cause fatalities if it does. It is known that pressure gauges fail and valves too; WIKA Instrument discovered 25% of pressure gauges failed in an audit of 250 plants. [4] In Germany, they constantly monitor at the cylinders' weakest point – the cylinder neck. This is a haphazard approach to the complex engineering world a dynamic pressurised fire system exists in.

Gaseous extinguishants have a specific design purpose as an effective medium for the extinction of flammable liquid fires and fires in the presence of electrical and ordinary Class A common combustible, Class B flammable liquid and Class C electrical hazard fires. The appropriate fire authority, the health and safety authorities and insurers all have valuable input to the consideration of which extinguishant to use, in addition to the prevailing national standards and statutory regulations of each country. These are very highly pressurised systems. A cup of water stands at 1bar pressure – approximately 14.5 psi. Gaseous systems vary in pressure from FM-200® or NOVEC™ 1230 at 25-30 bar, CO_2 at 50 bar or 720 psi, to Inergen at 300 bar or 4,500 psi. In engineering terms, they are "dynamic" systems, not passive ones, and thus ought to be constantly monitored and integrated into a business management system because it is likely they will leak.

It is not widely spoken about and rarely, if ever, do news articles discuss the leaking of gaseous extinguishing systems at sea. However, industry leaders know these systems risk leak of discharge during the installation, commissioning and maintenance of pressurised gaseous extinguishing systems because the regulations underpinning these systems require agent leak identification at an interval of every 6 months. The industry would not put in place regulations without having first had to react to issues they discovered. Standards are typically reactive not proactive, thus giving further evidence to the need for this issue to be published and for a push for calling for its constant monitoring. Any other pressurised system is permanently monitored but not the active fire system which "stays in its own shadows". Arguments suggest that fire industry manufacturers are keen to keep the information about leaking systems confidential, negating customer awareness so as not to reveal the underlying engineering risk.

2.1 A TYPICAL SHIP'S EXTINGUISHING SYSTEM

In terms of ship's extinguishing systems there exist two broad categories: sprinkler systems and gas systems (commonly CO_2 but also Novec™1230 and FM-200® among others). While the former can suffer leakage, the latter can cause catastrophic effect given the high pressures. An average ship's CO_2 system comprises between 200 and 600 cylinders each containing 45KG of CO_2 under high pressure. One of the highest probabilities of discharge occurs during their maintenance. These clean agents are designed to operate in limited spaces where there is a need for speed of suppression given the asset risk and where the space is occupied by people. They must extinguish a fire without damaging the asset they protect. They must extinguish the fire and enable operational continuity of the vessel whilst power is maintained. They must be easily maintained in-situ. They must comply with NFPA 2001 standards demanding fast discharge in 10 seconds and fire extinguishing within 30 seconds. Protecting humans and equipment alike whilst doing so in an environmental manner regarding ozone depletion and governed by the Montreal Protocol. They must be non-flammable and non-toxic. Contained in a design parameter which is "state of the art" and deliver confidence to the operator that it delivers "best fire safety practice".

2.2 FIRE SAFETY REGULATIONS

This paper will outline the core regulation about clean agent extinguishing systems enshrined in BS EN ISO 14520 -1:2015(E). This regulation relates not just to land based systems but also to vessels. Below is a short summary of some of the key regulations to demonstrate that they recognise that marine fire cylinders leak as an operational fact.

- ISO 14520-1:2015(E)
  - 6.2.4.2 Means shall be provided to indicate that each container is correctly charged.
  - 9.2.1.1 At least annually, or more frequently as required by the authority, all systems shall be thoroughly inspected and tested for proper operation by competent personnel.
  - 9.2.1.3 The storage container contents shall be checked at least every six months as follows: a) Liquefied gases: for halocarbon agents, if a container shows a loss of agent in quantity of more than 5 % or a loss of pressure (adjusted for temperature) of more than 10 %, it shall be refilled or replaced. b) Non-liquefied gases: for inert gas agents, pressure is an indication of agent quantity. If a container shows a loss of agent quantity or a loss of pressure (adjusted for temperature) of more than 5 %, it shall be refilled or replaced.
  - Annex F b) Every 6 months: 5) for liquefied gases, check weigh or use a liquid level indicator to verify correct content of containers; replace or refill any showing a loss of more than 5 %
- NFPA 12 code 4.8.3.4.2 If, at any time, a container shows a loss in net content of more than 10 percent, it shall be refilled or replaced.
- International Maritime Organisation (IMO) Safety of Life at Sea (SOLAS) International Fire Safety Systems (FSS) Code Chapter 5, 2.1.1.3 Means shall be provided for the crew to safely check the quantity of the fire-extinguishing medium in the containers.
- IMO MSC Circular 1120
  - 7.3.2.6 CO_2 cylinders are normally filled up to two-thirds of the height of the cylinder, however, the level will change according to the ambient temperature. Means should be provided to verify the liquid level in all the cylinders, either by weighing the cylinders or by using a suitable liquid level detector. Means for checking the quantity of medium in containers should be so arranged that it is not necessary to move the containers completely from their fixing position. This is achieved, for instance, by providing hanging bars above each bottle row for a weighing device or by using suitable surface indicators.
  - 7.3.7.3 CO_2 even makes allowance for crew to undertake this work if suitably qualified IMO MSC/Circular 1318 the MCA adopts the following policy for this Circular: iv) MSC/Circular 1318 paragraph 4) Monthly inspections may be performed by competent crew members, e.g. crew who have undertaken an STCW Advanced Fire Fighting course.

2.3 PRICE IS KING

Traditionally, the maritime industry treats fire protection systems as a necessary expenditure, rather than a means to safeguard valuable crew and cargo and maintain business continuity of the vessel. The competitiveness of the free market places great pressure on cost cutting: to deliver systems which often only minimally comply with regulations, and deliver asset protection at the most economical budget. With the value of assets, vessels and importance of business continuity growing, insurers are asked to underwrite almost incalculably high risk.

Given both the crew lives and cargo at stake, it seems unfathomable that regulations do not mandate fire systems should be permanently monitored rather than certified typically just once a year, particularly since it is a regulatory obligation to ensure that crew can check these themselves. The definition of a free market is an idealised form of a market economy in which buyers and sellers are allowed to transact freely based on a mutual agreement on price without state intervention in the form of taxes, subsidies or regulation. Unable to operate in an idealised economy but one that is regulated, the only reason why price dominates the environment is either because Government and Regulators are standing aside from the creation of an environment in which safe engineering goes rewarded rather than prejudiced, or because the industry itself is unaware of new technology that will help them meet both the spirit and letter of the regulation. This Ungoverned Space can be addressed today, with minimal education and cost, to create a "safety first" culture where owners and operates are rewarded for pursuing above and beyond regulation compliance.

2.4 ACCIDENTS: REPORTED & UNKNOWN

In aerospace, if a fault occurs on an aircraft that information is quickly and openly shared with Airline Operators, Civil Aviation Authorities and Engineering Organisations. In contrast in shipping, unless a fatality occurs, it is left un-reported. This does not seem accurate or correct considering the aforementioned assumption that systems leak and protect space integrity changes, How many ships sailing with partially-filled, over-filled or empty cylinders, how many accidental discharges or slow seepages are occurring and whether protect space integrity is tested and maintained after every significant change to the space, are left un-shared and unknown. Here are some high-profile accidents related to fire to emphasise the need for improved fire systems monitoring:

- September 2004 – Hong Kong - A Routine Inspection of the Fixed CO2 Fire Extinguishing System that led to the Death of Four Officers [5]
- November 2008 - Akula II K-152 Nerpa- At least 20 people have died in an accident on a Russian nuclear submarine when a fire extinguishing system (Halon) was activated by mistake [6]
- May 2010 - Marsol Pride, uncontrolled release of fire-extinguishing gas into engine room, Tui oil and gas field [7]
- August 2011 - Accidental discharge of carbon dioxide on board SD Nimble resulting in serious injury to a shore-based service engineer at Her Majesty's Naval Base Faslane [8]
- July 2014 - Port Hedland, Western Australia -A fire started in the engine room of the bulk carrier Marigold, while it was loading a cargo of iron. Firefighting by the ship's crew included activating the Halon gas fixed fire suppression system for the engine room. However, a full release of Halon gas did not occur, nor was the engine room effectively sealed [9]
- February 2015 – Twentynine Palms, California - 22 US Marines Injured when a halon-containing fire extinguisher went off [10]

2.5 ANECDOTAL EXPERIENCE

Industry experience and anecdotes, whilst not independently reviewed, lend themselves to the argument that the industry is not appropriately educated about the issues of fire safety in terms of the extinguishing systems and protected space integrity.

- Systems portrayed and installed by contractors as NOVEC™ 1230 but filled with sand or water.
- High pressure gas systems without the means to actuate them.
- Cylinder pressure gauges sticking in position under humidity or mechanical fatigue.
- Safety pins being retained in position in the cylinder valves after installation.
- Marine CO2 systems with 20% of the CO2 cylinders installed on commercial shipping being empty or partially-filled.
- Over-filled and under-filled cylinders.
- Pipework and cylinders freshly painted but with severe internal corrosion leading to particulates of rust which block the discharge nozzle mechanism.
- Room integrity testing with questionable results and with the room integrity remaining un-monitored after testing.
- Liquefied Gas Extinguishants being confused by installers and customers with Inert gas systems.
- There exists a lack of understanding of the organic compounds of some liquid extinguishants and their corrosive effect on the cylinder in the event of condensate ingress.
- Shipping companies not implementing the FSS code of the IMO SOLAS regulations.
- Bathroom weighing scales being chained to the CO2 cylinders in an effort to comply with IMO SOLAS FSS Code regulations. But no un-qualified Officers or Crew are qualified to shut-down, dis-mantle, weigh and re-install a CO2 cylinder

2.6 BEST PRACTICE: ULTRASONIC TECHNOLOGY OFFERS OPPORTUNITY FOR IMPROVEMENT

The most critical point of a pressurised cylinder is the meeting of the cylinder neck with its valve. For over 100 years the industry has monitored and permanently weighs (even suspends) cylinder pressure and weight of contents at that critical point. The assumption is that the cylinder pressure gauge is of high, almost MIL-SPEC quality, but they are often not and commercial pressures under a bidding process can incline some to select low-end budget, minimally compliant gauge mechanisms. Technologies will soon exist to monitor both liquefied gaseous content and non-liquefied gaseous pressure safely - from the *external sides* of the cylinder *rather than within it* – in both fixed and portable forms. Currently, there are fast, accurate and reliable methods using ultrasonic technology. For example, for checking contents by liquid level indication. Once able to monitor the contents, the mass/weight of the liquefied extinguishant can be calculated. If one can monitor the pressure of the gas on top of the liquefied extinguishant then one can also monitor the pressure of an Inert gas which is in an entirely vaporous form. Today ultrasonic technology can be utilised not only for handheld inspection tools used by marine surveyors, chief engineers and crew, but also for both constant monitoring, delivering remote diagnostics and data to a shore-based operator.

These technologies offer great opportunity to the industry. These methods enable the vessel owner, manager and operator to go above and beyond minimal regulation compliance to achieve best standards of safety at sea in terms of fire safety. They enable customers to have confidence that a fire company has installed a gaseous extinguishing system with a semi-autonomous monitoring capability to safeguard it. The customer-supplier relationship becomes an embedded one, with additional value, creating revenue reward. It is in the public interest. What CEO of a shipping company today wants to be shown to have installed a fire system that is left standing and unsupervised over thousands of miles of open seas and often in extreme weather states for 364 days of the year until the return of the marine servicing company for its annual certification check?

3. COMPARTMENT ROOM INTEGRITY MONITORING – THERE IS A PROBLEM THERE TOO

This is essential under ISO 14520 where gaseous extinguishing systems have to be designed in relation to the discharging agent hold-time (if the room cannot hold the agent because of leaks the agent will disperse and not extinguish the fire) and discharging agent peak pressure (if the pressure is too high for partition walls or suspended ceilings they will be blown apart or damaged and possibly destroying the room integrity). At the design stage of a fire extinguishing system protected spaces are tested for room integrity by positively pressurising a room and detecting escaping pressure, to verify that the room itself into which the gaseous extinguishant discharges on actuation, can both hold the agent after its discharge and hold its pressure on actuation. The fire system is then installed and commissioned. An enclosure within a ship at sea turns and bends as any structure does. It ages and leak sites develop. Regulations may require the protected space to be tested every year, but it should be tested after every major change to the room.

3.1 REGULATIONS FOR MAINTAINING PROTECTED SPACE ROOM INTEGRITY

NFPA 2001 and ISO 14520 are the standards to adhere to with relevance to the total flooding of clean agent extinguishing systems. Both NFPA 2001 and ISO 14520 rely on prediction equations such as the FSSA equations to predict the Hold Time and the maximum positive and negative Peak Pressure during discharge. Both standards recommend using Door Fan Testing to analyse the enclosure and retrieve the necessary values. It is imperative to maintain a certain enclosure integrity as these clean agents' function on the flooding principle for it to work correctly.

IMO SOLAS Chapter II-2 Regulation 10, G23.3 states, "The Regulations require that 85% of the required concentrations for machinery spaces and cargo pump rooms is achieved in such spaces within two minutes." If the room contains leakages, slowing down the process to achieve the required flooding concentration, how is the CO2 extinguishing system able to achieve the desired concentration in the time limit proposed? If room integrity is not secure, wrong arrangements may be made to increase the rate of CO2 discharge when in fact the problem is related to the room integrity itself.

ISO 14520

- 9.2.4.1 At least every 12 months it shall be determined whether boundary penetration or other changes to the protected enclosure have occurred that could affect leakage and extinguishant performance. If this cannot be visually determined, it shall be positively established by repeating the test for enclosure integrity in accordance with Annex E.
- 9.2.4.2 Where the integrity test reveals increased leakage that would result in an inability to retain the extinguishant for the required period, remedial action shall be carried out.
- 9.2.4.3 Where it is established that changes to the volume of the enclosure or to the type of hazard within the enclosure, or both, have occurred, the system shall be redesigned to provide the original degree of protection. It is recommended that the type of hazard within the enclosure, and the volume it occupies, be regularly checked to ensure that the required concentration of extinguishant can be achieved and maintained.

Pressure Systems Safety Regulations 2000: Except for pressure vessels at sea, then all those installed in the UK will come under the Pressure Systems Safety Regulations 2000 and as such, other that the ISO standards, the user/owner of such installations have certain responsibilities which are summarised in a user/owner flowchart below:

These regulations recognise the risks to a vessel or building structure ageing and generating leak sites. The case is clear that the industry should support constant monitoring, above and beyond full regulation compliance, as the risks are so clearly described in the core standards.

3.2 INDUSTRY PROOF

A vessel generates leak sites due to load states, sea states, wind states, and dynamic movement. The severity is amplified within a vessel structure constantly changing by varying sea, wind, load states, cargo types and dynamic stresses. There is a great deal of bending and deformation that naturally occurs in ships during travel. The dynamic movement of a vessel is illustrated through analysis of stills from a video [11] showing the stress and effect on a vessel during passage from Suez Canal to Singapore in severe weather conditions) as shown by the analysis in the inset 1, 2 and 3 in Figure 2 below [12].

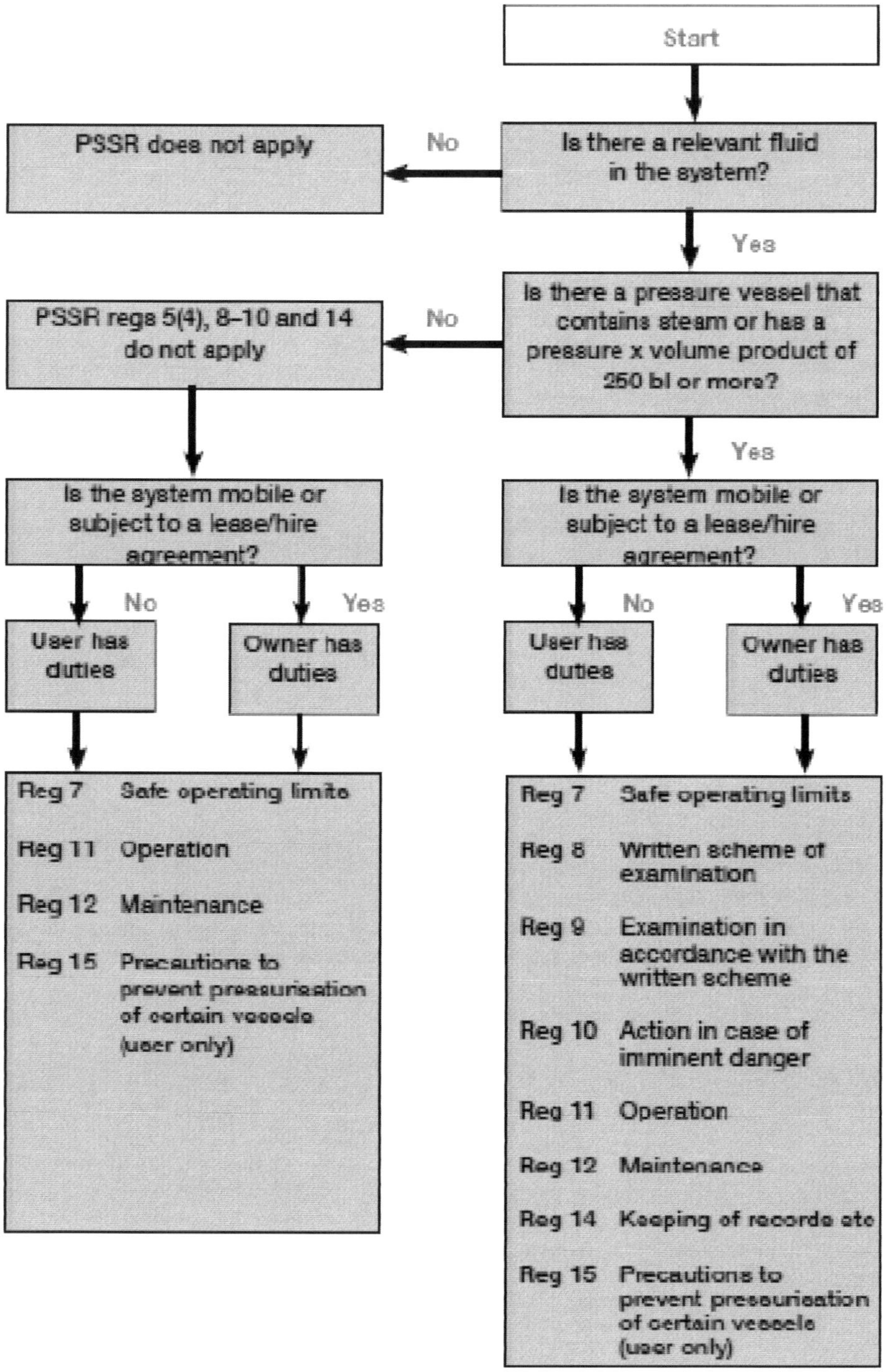

Figure 1: User/Owner Responsibilities

Inset 1: This still (1m:20s) shows wire frames highlighting rectangular sections along the vessel which will be used for comparison with Inset 2 (below)

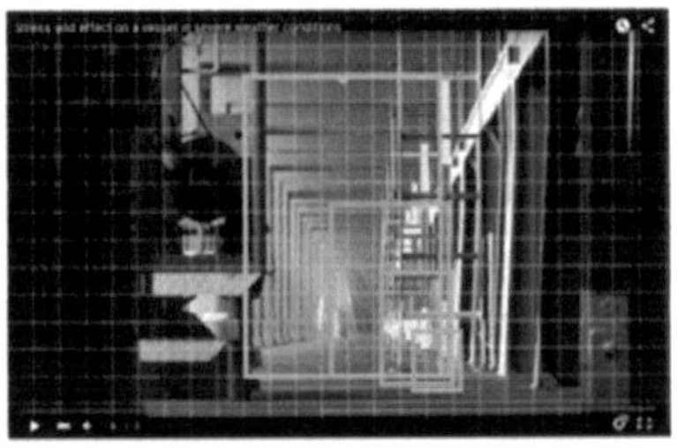

Inset 2: Still from (1m:30s) with rectangular boxes, combined with grids to highlight the movement

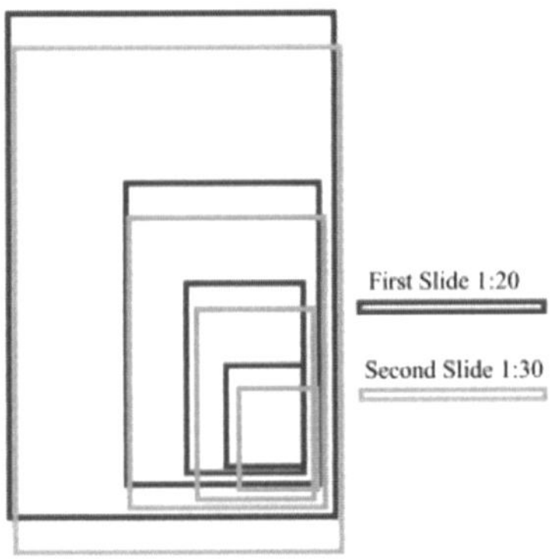

Inset 3: This image overlays the rectangular sections from Inset 1 and 2 to show the difference between 1m:20s and 1m:30s.

Figure 2: Vessel bending in severe weather

Considering the importance of maintaining room integrity for the gas to suppress a fire within 2 minutes, protected spaces onboard vessels should be tested far more regularly and frequently with the new technology now available. New technology is dual-use: to protect crew when entering spaces protected by CO2.

4. ULTRASONICS AS LEADING TECHNOLOGY SOLUTIONS

One of the sciences being harnessed by innovators in fire safety is that of ultrasound: i.e. acoustic (sound) energy. Sound is vibrations that propagate as a mechanical wave or pressure and transmit through solid, liquid or gaseous mediums.

Dolphins and whales can communicate at sea over long ranges, as sound travels more efficiently through liquids than air. In the air bats navigate by airborne ultrasound. In physiology sound is the reception of such waves and in psychology their perception by the brain. This is why a human being listening with one ear angled from the other (and therefore at a different distance from the sound source to the other) "hears" sound as "one sound". Ultrasound waves are a high frequency above the human audible range of approximately 20 to 20,000Hz. [13]. Ultrasonic methods can range from 40,000 Hz used in watertight integrity testing [14], 1 MHz used in liquid level detection [15], to diagnostic ultrasound devices employing frequencies of 2 – 15 MHz [16]. Although the shipping world merely uses it as a tool to gauge thickness, it has seen far more varied use across military, medical and industrial fields. With science at the company's core, Coltraco Ultrasonics is an example of a company aiming to replicate nature and use these fundamental physical principles to design and manufacture products and systems that can be used by the Royal Navy, Engineers, Crew and customers alike. E.g. to identify that difference in a cylinder containing liquefied agent and in room integrity testing. New SmartShip® technology utilizes ultrasonics to enable operators to implement a "safety first" culture, with the addition of constant monitoring.

4.1 ULTRASONICS AND ATTENUATION

Attenuation is the loss of intensity of sound as it passes through a medium and is a combination of absorption and scattering [17]. For ultrasonic waves, absorption is the dominant cause of attenuation. Absorption can occur because of friction between air molecules which generates heat and through relaxational processes where sound energy is transferred to the air molecules and the molecules dissipate the energy through rotation and vibration. The attenuation over a distance is quantified by the attenuation coefficient, α. See in the equation below:

$$A = \alpha \times l \times f$$

where A is the attenuation, α is the attenuation coefficient, l is the length it travels into the medium and f is the frequency of the incident ultrasound beam.

Attenuation coefficients vary widely between materials, for example, air has an attenuation coefficient (at a frequency of 1MHz) of α = 164MHz · m (for 20°C) [18]. while water is α = 0.22MHz·m [19]. The attenuation is dependent on many other factors including the frequency of the wave and the medium in which the wave is propagating. As can be seen from the equation above, the attenuation is linearly dependent on the frequency of the incident ultrasound beam. This explains why ultrasound, with a high frequency of 1MHz is attenuated so greatly.

It is also known that ultrasound intensity decays exponentially due to attenuation:

$$I = I_0 e^{-\alpha x}$$

where I is the intensity, I_0 is the initial intensity, α is the attenuation coefficient and x is the distance travelled by the wave.

As higher frequency results in higher attenuation, the equation shows that ultrasound intensity will decrease exponentially. This will occur much faster compared to audible sound of lower frequency due to the greater attenuation in ultrasound. To address the "Ungoverned Space", the attenuation of ultrasound waves is discussed with relevance to two main applications: liquid level detection, pressure monitoring and protected space integrity / leak detection.

4.2 HOW IT WORKS TO IDENTIFY LIQUID LEVEL AND MONITOR PRESSURE

As explained above, higher frequency ultrasound has a greater attenuation. To measure the liquid level, the attenuation should be greater, to differentiate the different phases within the cylinder being measured. This technology utilises the attenuation of the ultrasound as it travels through different materials. It is the extent of this attenuation (most importantly its energy dissipation) that differentiates between different materials.

Instrumentation which relies on the attenuation of ultrasound is the Portalevel™ MAX handheld liquid level indicator by Coltraco Ultrasonics and Permalevel™ Multiplex constant monitoring system. These convert electrical energy into ultrasonic waves through the reverse piezoelectric effect. Piezoelectricity is the appearance of an electrical potential across a crystal when it is subjected to mechanical stress, and is a reversible effect. A crystal which exhibits this effect by becoming charged when it is compressed, twisted or distorted is a piezoelectric crystal. Similarly, passing electricity through a piezoelectric crystal causes the crystal to vibrate back and forth; this is the reverse piezoelectric effect [20].

The liquid level of a cylinder can be found because the ultrasonic signal received from the vapour section of the cylinder is different to the liquid section. (The exact location of the liquid level is then pinpointed by finding the point at which the readings change). The difference is the response of liquids and gases due to their physical properties. In a liquid, the molecules are all free to move but still are mostly in contact with each other. In a gas, none of the molecules are touching each other – there are large gaps between then – i.e. gases are much less dense than liquids. The large difference in density suggests that the liquid and gas sections disperse ultrasound very differently (the ultrasonic waves travel very differently through the two media). These different media possess different attenuation coefficients which then affects the overall attenuation of the ultrasonic signal resulting in distinct values that can be used to interpret the liquid levels.

At sea whether liquefied or non-liquefied systems are used they are pressurised and the gas pressure affects the ultrasound travelling in it. So, whether liquid level or pressure monitoring the capability to monitor marine fire systems at sea now exist.

4.3 HOW IT WORKS TO MONITOR PROTECTED SPACES INTEGRITY AND IDENTIFY LEAKS

It has already been established that ultrasound has been effective in detecting leaks relating to watertight integrity in ships. Research which Coltraco carried out showed that "a linear relationship was found between signal leakage out of a certain sized hole (in a watertight box) and the quantity of water able to enter that hole' [21].

"The theory relating the area of a leak site to volume flow rate through said leak is well understood. The volume flow rate, Q, through a hole is given by:

$$Q = A\sqrt{2gh}$$

Where A is the area of the hole, g is the acceleration due to gravity on Earth and h is the height of the head of water above the leak. Therefore, we are to expect a linear relationship between flow rate and hole area."

The research further suggests that there is a linear relationship between the flow rate of water through a leak and ultrasound pressure measured through the leak. An increase in signal received is proportional to the quantity of water which could enter through the leak. This has important implications as it allows the user of the Portascanner™ to operate the equipment in confidence that a spike in the signal does indeed imply that there is a leak.

Taking this equation and substituting h with parameters applicable to gases, we get the following:

$$Q = A\sqrt{\frac{2\ P_i}{\rho_i}}$$

Where Q and A are defined as above and P_i is the pressure within the enclosure and ρ_i is the density of air leaking out of the enclosure.

Just as how the flow rate of water through a leak has a good fit linear relationship to the ultrasound pressure measured through the leak, we would expect the same behaviour for gas leaking through the aperture as shown in the closely related equations. This would allow effective leak detection in an enclosure needed to contain the clean agent to extinguish fire. Any spike in the ultrasound signal would suggest a leak beginning to form

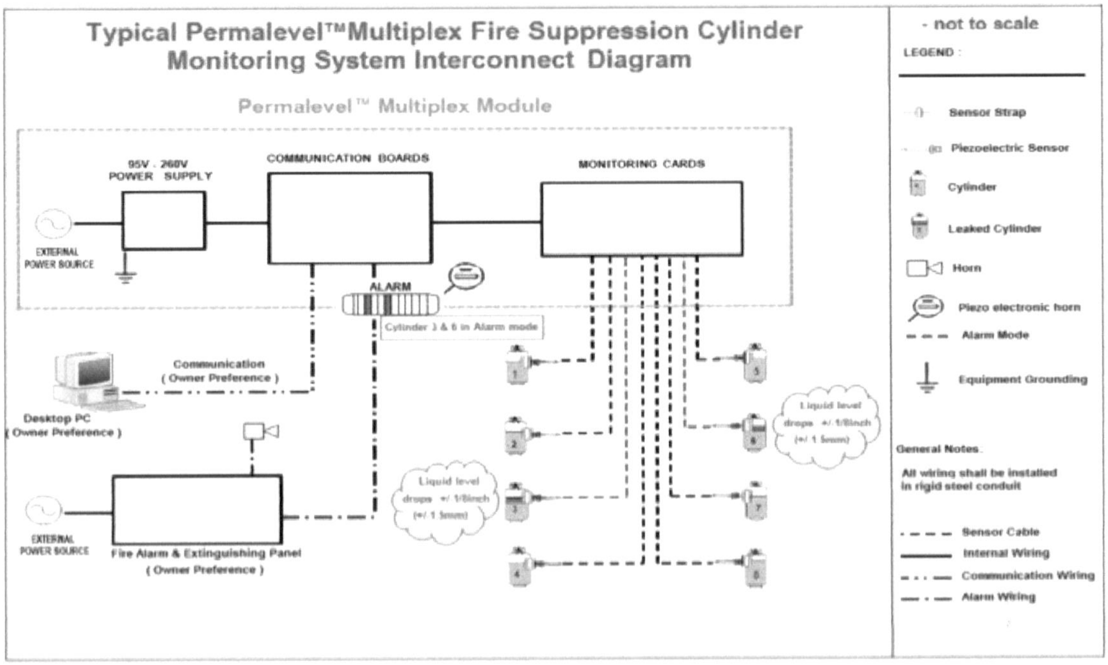

Figure 3: Permalevel® Multiplex Network Capability

in the enclosure. Ultrasonic Technology is highly advantageous because it accurately and rapidly detects the exact leak locations and size. Although models vary, generally the equipment is comprised of two main components: a generator and a receiver. The ultrasound generator emits a modulated signal of a specific frequency of ultrasound. By utilizing a lower frequency ultrasound, attenuation in air is reduced to allow effective wave propagation through any leaks in seals. The receiver then picks up the signal leaking through the seal and converts it into a result indicating room integrity.

5. SAFESHIP® TECHNOLOGY TO CLOSE THE GAP OF THE UNGOVERNED SPACE

Uwe-Peter Schieder, IUMI Loss Prevention Committee vice says, "[improving fire safety] is an issue of improving the technology" [22]. The "white heat" of technology shines as a beacon of hope and justifies customers actively engaging in constant monitoring of fire systems and protected spaces as an integrated and essential element to their business activity. The UK shipping industry is accepted as a world leader in its understanding of good engineering practice. The UK is the 5th largest contributor to the IMO by GT and the 3rd largest contributor to the IMO overall. Technologies exist right now that can mathematically and far more efficiently test fire systems contents and room/protected space integrity. Technology exists to quickly and accurately identify pipework corrosion. Gases under pressure can be monitored and liquefied gases can be constantly monitored. Contents and pressure can be simultaneously monitored offering remote diagnostics and remote monitoring capabilities and can extend to the delivery of autonomous systems in relation to them.

5.1 MATHEMATICS MUST BE AT THE CORE

Customers depend on the industry to deliver fire engineering to minimise risk. Insurance companies like P&I Clubs underwrite that risk. But the mathematics of fire engineering failure are high: whether in the application and understanding of the formulas they use to calculate design concentrations of gases or flow rates, or in the deployment of fundamental engineering principles to protect dynamic pressurised systems and the structures they are working so hard to protect against the risk of fire. It is important that the fire protection of a vessel be considered as a whole. These problems can be solved by applying fundamental scientific and engineering principles. The whole drive of civilisation is to continually strive to understand and apply the fundamental Laws of Science and application of Physical Principles. Mathematics is required to ensure safe and accurate standards are put in place and implemented. British Standards are some of the most rigorous in the world. The British Standards Institute suggest the UK generates over 50% of all European Standards. BS EN ISO 14520 -1:2015(E) reasonably assumes that the execution of its provisions is entrusted to people qualified and experienced. Furthermore, the regulatory bodies and influencers should reflect on past standards to check they are relevant and correct. Where the maths is wrong or where there are gaps – as proven by the ungoverned spaces explored in this paper – the regulators must accept the need for amendments and updates.

Influencing the owners, operators and crew, the regulators need to show a fuller understanding of new innovative systems and methods available.

5.2 TECHNOLOGICAL ANSWERS

Coltraco Ultrasonics is a leading example of a manufacturer operating in 108 countries, with equipment onboard over 10,000 vessels worldwide, who is innovating to improve safety at sea. Constant monitoring of fire extinguishing systems by Permalevel® Multiplex. This is a fixed system to monitor the fire system cylinders 24/7 with data-logging and autonomous monitoring. Signals from these fixed monitoring sites can be monitored centrally on the bridge and in the ship's technical office concurrently. Please see a rough schematic of the Permalevel® Multiplex network for a simplified outline of the potential connectivity in Figure 3.

Handheld inspection of the cylinders to comply with the IMO SOLAS FSS regulation above, and many others from ISO, NFPA, and so on, is available via UL, ABS & RINA approved liquid level indicator Portalevel™ MAX. Converting the known liquid level into the weight/mass of agent is achieved through Portasteele™ Calculator – another of the world's first ever such technology. Coltraco's technology will monitor both liquid content and gas pressure safely from the external sides of the cylinder rather than within it.

5.3 CONSTANT MONITORING IS REQUIRED

Marine gaseous fire systems deliver the very self-contained resilience we need at sea. But resilience means the permanence of capability and functionality. That permanence can only be delivered by constant monitoring of the systems that enable it. It is essential that marine gaseous extinguishing systems be carefully maintained to ensure instant readiness when required. Routine maintenance is liable to be overlooked or given insufficient attention by the owner of the system. It is neglected to the peril of the lives of seafarers and at the risk of crippling financial loss to the vessel or critical offshore infrastructure. It is usually the case of systems like these that they are out of sight and out of mind, as they are in awkward, rarely-visited spaces. Add to this the priceless nature of data and whether the special hazard is a container ship, offshore platform or submarine, the critical need for constant monitoring is evident.

The importance of maintenance of such systems cannot be emphasised enough. Installation and maintenance should only be done by qualified personnel. Inspection preferably by a third party, should include an evaluation that the extinguishing system continues to provide adequate protection for the risk (protected zones, as well compartments built for room integrity, can change over time as they age or are modified). But as importantly the marine gaseous system should be maintained not just in relation to itself but regarding the "Protected Space" into which its extinguishing agents discharge on actuation. If that space cannot "hold" the agent on discharge, or its walls and bulkheads or indeed its distributing pipework, cannot sustain the pressures of the gaseous agent on discharge, then the agent will disperse and the fire grow unimpeded. So constantly monitoring the Compartment Room Integrity too is now possible.

6. CONCLUSION

Having systems that operate transparently will work not just to convince a vessel owner that his asset is in good hands, but also to reassure the crew that their safety is taken seriously by both their employer and the marine servicing company. The science of a marine gaseous extinguishing system is a complex one. The mathematics that underpin its science are demanding ones. But in simple terms gaseous systems are pressurised and in that they are dynamic not passive ones. They are there to protect critical marine structures in a safe and expeditious manner in the only way that a gaseous system can, complemented by the integrity of the protected space. One would not imagine an alarm system exists without monitoring it 24/7, but why is the industry still leaving unattended the very automatic gaseous clean agent extinguishing systems that protect us at sea when even the core regulations of its installation and maintenance specifically allude to its potential to accidentally discharge or leak its contents? Fortunately lead elements of the critical infrastructure community are asking the same question. So are the world's insurance companies. But should our own industry not answer the question by implementing constant monitoring 24/7/365 before it is asked of them by the very people it is helping protect?

7. DISCLAIMER

The opinions expressed are those of the author.

8. ACKNOWLEDGEMENTS

The author would like to thank those at Coltraco Ultrasonics for in-house research and to Durham University Physics Department who undertook research for Coltraco Ultrasonics. Thanks also goes to Prof Paula Chadwick, Dr Douglas Halliday of the Physics Department at Durham University and to Prof Steve Dixon and Mr Jack Dixon of the Centre of Industrial Ultrasonics at the Department of Physics and Engineering at Warwick University. Thank you also to industry professionals who have taken the time to review some of this paper prior to publishing.

9. REFERENCES

1. FM Global in Food Factory Design: Can fire risk be designed out?, *Food*

Manufacturer.co.uk, URL last accessed 19.12.2016 <http://www.foodmanufacture.co.uk/Manufacturing/Food-factories-can-fire-risk-be-designed-out>.

2. 'November 2016 Newsletter', *IMRRA*, URL last accessed 15.12.2016 <http://www.marinerating.com/documents/IMRRA-Newsletter-201610.htm>.

3. Gallagher, J., *IHS Fairplay*, article published 09.12.2016, URL last accessed 15.12.2016 <http://fairplay.ihs.com/safety-regulation/article/4279126/fire-risk-grows-as-ships-get-bigger>.

4. Deane, J., WIKA Instrument in "Don't Neglect Pressure Gauges", *Chemical Processing*, 29.10.2013, URL last accessed 19.12.2016 <http://www.chemicalprocessing.com/articles/2013/8-causes-of-pressure-gauge-failure/>.

5. Leung, H.K., 'A Routine Inspection of the Fixed CO3 Fire Extinguishing System that led to the Death of Four Officers!', *Marine Department, Hong Kong Special Administrative Region*,. PDF last accessed 19.12.2016 <http://web.directemar.cl/apec/papers/HongKong/FixedCO2.pdf> & <http://www.mardep.gov.hk/en/publication/pdf/mai040927.pdf>.

6. 'Twenty die on Russian submarine', *BBC News*, 09.11. 2008, URL last accessed 19.12.2016 < http://news.bbc.co.uk/1/hi/world/europe/7718156.stm>.

7. Report 10-203: Marsol Pride, uncontrolled release of fire-extinguishing gas into engine room, Tui oil and gas field, 'Investigation 10-203, Report', *Transport Accident Investigation Commission*, 27 May 2010.

8. 'Accidential discharge of carbon dioxide during fixed CO2 fire extinguishing system test on tug SD Nimble with 1 person injured', *Marine Accident Investigation Branch*, URL last accessed 19.12.2016 <https://www.gov.uk/maib-reports/accidental-discharge-of-carbon-dioxide-when-testing-the-fixed-co2-fire-extinguishing-system-on-tug-sd-nimble-at-hm-naval-base-faslane-scotland-with-1-person-injured>.

9. 'Investigation Report 312-MO-2014-008: Engine room fire on board *Marigold* at Port Hedland, WA on 13 July 2014', Australian Government Transport Safety Bureau, URL last accessed 19.12.2016 <https://www.atsb.gov.au/publications/investigation_reports/2014/mair/312-mo-2014-008/>.

10. Lyons, K., '22 Marine Injured in Training Accident at California Base Due To Fire Extinguisher Discharge', *Bustle*, URL last accessed 19.12.2016 < https://www.bustle.com/articles/64147-22-marines-injured-in-training-accident-at-california-base-due-to-fire-extinguisher-discharge>, 13.02.2015.

11. Tuxen, C., 'Stress and effect on a vessel in severe weather conditions', *YouTube*, 8th April 2014, URL last accessed January 2016 https://www.youtube.com/watch?v=PmlTk_3NN_g&feature=youtu.be>.

12. Dixon, J., Internal Correspondence, *Coltraco Ultrasonics*, 2016

13. Ihara, I., 'Ultrasonic Sensing: Fundamentals and its Applications to Nondestructive Evaluation', in Mukhopadhyay, S., and Huang, R., ed., *Sensors: Advancements in Modeling, Design Issues, Fabrication and Practical Applications*, 1st ed. Berlin: Springer, 2008, pp.287 - 305.

14. 'Portascanner® Brochure', *Coltraco Ultrasonics*, 2016.

15. 'Portalevel® Brochure', *Coltraco Ultrasonics*, 2016.

16. Laugier, P. and Haiat, G., 'Introduction to the Physics of Ultrasound' in: Laugier, P., and Haiat, G., ed., *Bone Quantitative Ultrasound*, 1st ed. Springer Netherlands, 2011, pp.29 - 45.

17. Rossing, T. D., *Springer Handbook of Acoustics*, 2007.

18. Jakevicius, L., and Demcenko, A., 'Ultrasound air coefficient dependence on air temperature in closed chambers', *Ultragarus Journal*, 2008.

19. Goldenberg, D., et al, 'A Review of Tissue Substitutes for Ultrasound Imaging', *Journal Ultrasound in measurement and Biology*, 2010.

20. Katzir, S., 'The Beginnings of Piezoelectricity: A Study in Mundane Physics – Volume 246', *Boston Studies in the Philosophy and History of Science*, Springer Science & Business Media, 2007.

21. SSmith, 'Research: SSmith report', *Coltraco Ultrasonics*, 3 March 2014, NOTE: Confidential.

22. Gallagher, J., 'Fire risk grows as ships get bigger', *IHS Fairplay*, 09.12.2016, URL last accessed 15.12.2016 <http://fairplay.ihs.com/safety-regulation/article/4279126/fire-risk-grows-as-ships-get-bigger>.

9.1 ADDITIONAL DOCUMENTS USED

1. Research conducted by Coltraco Ultrasonics at 34. Durham University 2014-current.

2. Research conducted by Coltraco Ultrasonics at 36. Warwick University 2015.

3. Lampert, E., 'Tankers sailing with above average risk at a 6 month high', *Tanker Shipping & Trade*, 14.12. 2016, URL last accessed 19.12.2016 <http://www.tankershipping.com/news/view,tankers-sailing-with-above-average-risk-at-a-sixmonth-high_45827.htm>.

4. Frozee, J., 'CO2 Flooding System – Fixed FireFighting on Ships', *Marine Engineering*, 25.10.2015, URL last accessed 19.12.2016 <http://marineengineeringonline.com/co2-flooding-system-on-ships/>.

5. 'Hydroelectric Research and Technical Services Group', *Alternate Methods of Measuring Liquid Levels in CO2 Bottles*. CO2 System Operation and Maintenance, US Department of the Interior Bureau of Reclamation, 2005, pp.35 - 37.

6. 'Maintenance', *Fire Extinguisher Guide*, URL last accessed 19.12.2016 <http://www.fireextinguisherguide.co.uk/maintenance/>.

7. Eckert, C.A. 'Supercritical fluids as solvents for chemical and materials processing', *Nature*, 1996, pp. 383, 313–318.

10. AUTHORS BIOGRAPHY

Dr Carl Stephen Patrick Hunter BA(Dunelm) Hon DSc FRINA FIMarEST MRAeS is CEO and Managing Director of Coltraco Ultrasonics, a British designer and manufacturer of portable and fixed monitoring systems for the naval, shipping, offshore, energy and fire sectors. He is a former GreenJacket Officer in the British Army and a Graduate and Honorary Doctor of Science from the University of Durham, a Fellow of the Royal Institution of Naval Architects, a Fellow of the Institute of Marine Engineers and Member of the Royal Aeronautical Society, Royal Institute for International Affairs, Royal Society of Asian Affairs and Royal United Services Institute. Dr. Hunter is a Council Member of the British Naval Equipment Association, a member of the Export Council of the Fire Industry Association(FIA) and the FIA's Working Group Gases which is the Lead UK Fire delegate organisation to British Standards and BSI the lead fire delegate to the International Standards Organisation in Gaseous Extinguishing Systems and an active member of the Society of Maritime Industries the lead British marine organisation and voice of the marine industry in the UK. His interests are in Physics, International Business, the Constitution, South Asia, International & Strategic Affairs, Defence, the British Commonwealth and bringing UK Scientists into business. Dr Carl lives in London and Somerset in England, travels extensively in Asia, Europe and the Middle East, has been married for 27 years to Dorothy and a proud Father of four Children.

RESPONSE TIME ANALYSIS CHALLENGES ON A SMART SHIP - AN INITIAL CASE STUDY

D. Armanes, American Bureau of Shipping, Greece

SUMMARY

Integrated Automation Systems (IAS), generally considered as the stepping stone for Total Ship Automation and Autonomous Vessels, provide today the platform for further cost reduction and advanced functionality on-board modern vessels. Although, there are many best practices to be followed during testing, commissioning and maintenance time, during analysis and design time there is an increasing response time complexity with any new integrated process. Consequently, sound methods and tools are increasingly important for their precise and reliable verification. This paper explores the fundamental notions in response time analysis by using a well know IAS, sufficiently generalized, as an example. The process identifies initial engineering data for the analysis and proposes simplistic behavioural and architectural models, annotated with these data for faster response time verification.

1. INTRODUCTION

Ship Automation and Control can be categorized over several disciplines of the Control theory. Although not all automation and monitoring tasks have hard Real-Time requirements, methods to assess the time predictability of Integrated Automation Systems (IAS) implementations, as defined in ABS Rules for Building and Classing Steel Vessels, Part 4, Chapter 9, Section 4, are required so as Classification Societies can review the inherent design risk in a timely manner. This is mainly due to the fact that IAS are being considered as the stepping stone for Total Ship Automation and Autonomous Vessels, fundamentally changing the integration depth of several ship operations. A deployed IAS, covers many aspects of such operations and may include the propulsion plant operation, power management operation on the auxiliary engines, auxiliary machinery operations, cargo on-and-off-loading operation, navigation and administration of maintenance and purchasing of spares. In this paper we present an example response time theoretical analysis of a well-known IAS for drill-ships, focusing on just a small subset of its functionality commonly known as Alarm and Monitoring. This example is further used to illustrate several points that can be clarified if design and operational data are being submitted towards a Smart Ship design verification and inevitably its risk assessment.

2. LITERATURE REVIEW

There is a well-established research community involved in the response time analysis of programmable electronic systems (PES). In the United States, past IEEE proceedings accommodated a state of the art position paper for the trends and challenges of system level design [1]. In a pivotal paper, Prof. Eduard A. Lee from University of California, Berkeley discussed the problem with multi-threaded systems [2]. His work inspired an abundance of concurrent research efforts regarding the required programming model for distributed, real-time systems, including formal system modelling and verification like the ones described in [3] and [4], eventually resulting to a holistic problem statement in [5]. A combined perspective of representatives from the University of Virginia, the University of Pennsylvania, the University of Texas and Carnegie Mellon University regarding the advantages offered to the society by such systems, was presented in [6]. For the general public to better understand such advantages, three application areas were considered: assisted living, emergency response systems for natural or man-made disasters, and protecting critical infrastructures at the national level. Later, in a fundamental overview of networked, closed-loop control systems, R. A. Gupta and Prof. Mo-Yuen Chow from Advanced Diagnosis, Automation, and Control Laboratory, North Carolina State University, provided for the latest research trends [7]. In an effort to clarify the opportunities and challenges of embedded controllers, Defence Advanced Research Projects published a position presentation [8], drawing conclusions by comparing the breadth of impact, embedded software has on their scope of work (from avionics to micro-robots), against the automotive, the consumer electronics and plant automation industries. Eventually, the Department of Defence, sponsored Carnegie Mellon University for a technical forecast on superior data collection, data fusion, analysis and use of information to meet mission objectives. This goal depends on increasingly complex systems characterized by thousands of platforms, sensors and decision nodes connected through heterogeneous wired and wireless networks, effectively named Ultra-Large-Scale Systems [9]. The lessons learned from this study was to avoid the obvious approach of a system-of-systems and focus on socio-technical ecosystems instead. A lesson that can be adopted during a Smart Vessel design, validation and third party verification as it is implied by the recent memorandum of understanding between Hyundai Heavy Industries (HHI), Microsoft, Intel and SK Shipping [10]. In Europe, Artist, the Network of Excellence on Embedded System Design, published the Roadmap on Real-Time Techniques in Control System implementation several years ago, to drive research efforts towards the interactions between the control, computing, and communication subsystems, the influences these interactions have on control performance, as well as different ways to handle the interactions [11]. Currently, the industrial association of Artist, called Artemis-IA, numbers 177 industrial partners without including however some distinct marine automation vendors, that could provide invaluable input in defining the unique marine integrated automation challenges of the Smart Vessel against other industrial applications to the academic community.

Evidently, the response time analysis of PES, has been a very promising research topic for decades due to its multidisciplinary aspects. Consequently, it is not possible to provide the vast amount of citations, specifically from Asia, in the limited space of this paper. We do use selected publications from Asia as references throughout this study however. With our apologies to numerous researchers, we take another step in this research area, from a classification society perspective. Below we revisit the problem of alarm response time analysis, considering the state of the industry designs. In our view, this is the simplest example of response time verification without losing generality of more strict integrated, sequence control logic response time requirements, inherently present in Smart Vessels design requirements. Needless to say, for correlation of safety, maintenance and operational data in the time and frequency domains, these response time requirements become extremely tight (subject to the plant dynamics), unless some form of distributed clock synchronization protocol was considered or network delay compensation provided, during the network design phase.

3. PROBLEM STATEMENT

Let R, be the Alarm and Monitoring System's worst-case latency between the instant that the plant changes state and the instant the alarm is published on the Human Machine Interface (HMI) screen, without considering any configurable delay from the crew. The random, alarm state change event, will be detected by the Programmable Logic Controller (PLC) maximally at the periodic scan time T_{PLC}. This is due to the fact that the PLC program will be taking into account the data just received from the plant, while, almost concurrently, the plant will be changing state to an alarm condition. This alarm state will be acknowledged by the PLC at subsequent processing durations. Let D be the processing delay (i.e. the time between the reception of the input data in the remote I/O registers and the time where the output alarm condition is received by the HMI). It is the key problem to assess T_{PLC} and D against class requirements for R, considering a number of scheduling constrains and the timing characteristics of networks' components involved in the path. To do so, architectural and behaviour models at certain levels of abstractions are required. The following section proposes such models in the most simplistic form by means of traditional block and timing diagrams in an attempt to appeal both to marine engineers and other stakeholders alike not very familiar with the involved concepts.

4. BUILDING AN IAS TIMING MODEL

As a case study we will use a well know IAS product integrating thruster automation and Alarm/Monitoring capabilities for a group of Variable Frequency Drives (VFD) on a drillship. The subject system is a PLC based solution, using a single, multitasking processor with a real-time, multithreaded operating system for the plant automation needs. Two periodic automation tasks are running pre-emptively with a Rate-Monotonic Scheduling (RMS) algorithm in the Programmable Electronic Unit (PEU), with independent access to the Input/Output Direct Memory (IODM). In addition, three independent communication controllers for remote I/Os and HMI connectivity implemented in a Field Programmable Gate Array (FPGA) are used for the three types of networks present on the drillship: an RS-483 based remote I/O network, a PROFIBUS DP network utilizing a Master-Slave configuration and an Ethernet based redundant type. For the Ethernet based network, three fiber optic, second level switches are used, one for each physically independent process segment. The HMI application runs on a Microsoft Windows Operating System utilizing a Client-Server communication approach with the Alarm Database. For the purpose of this study we will consider only the I/O data transfer latency from the remote I/O cards and not the sensors' delay or the I/O polling interval of such sensors which may include any signal averaging algorithms. This is due to the fact that the plant dynamics in this case, are at the order of microseconds while the PLC scan time is at the order of milliseconds, so only the remote I/O communication protocol polling time may introduce some considerable response time delay as explained in the next section. Since there is only one master and one slave in this network, any network propagation delay is considered negligible but maximum cable lengths are used for the calculations in any case. Consequently, the following Equation 1 will be used, as presented in [12].

Equation 1: Processing Delay

$$D = E(I_n) + E(X_n) + E(O_n)$$

Where,

D: Processing Delay

$E(I_n)$: Worst case reception time of n Input. Subject to PROFIBUS DP bit rate, process data amount and controllers technology

$E(X_n)$: Worst case execution time of n Alarm Processing Task. Subject to microprocessor architecture, scheduling and multitasking choices

$E(O_n)$: The worst case transmission time of n Output between the PLC and HMI over the redundant Ethernet

As discussed earlier, we will only consider the alarm and monitoring WCRT class requirements in this paper. Any control feedback signals, although generally modelled, will not be consider any further. In the remainder of this paper the term *plant* will be used for the mechanical thrusters, VFD and steering gear assembly. The

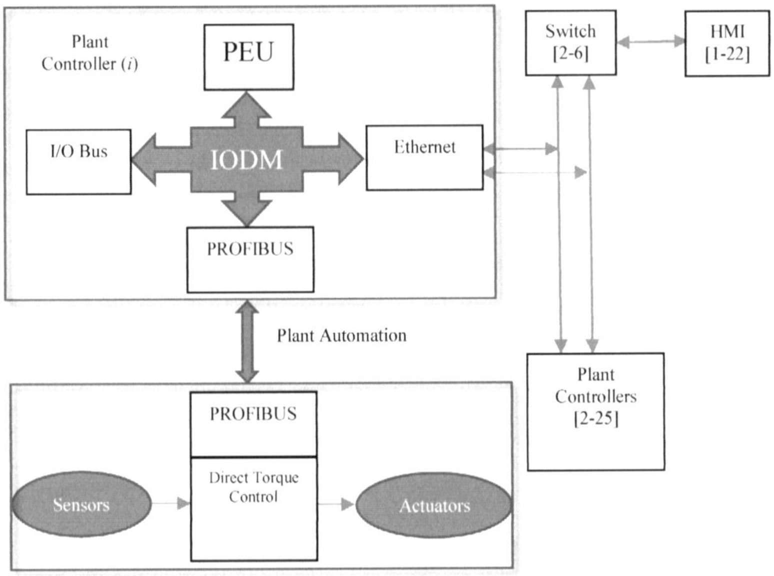

Figure 1: IAS Architectural Model

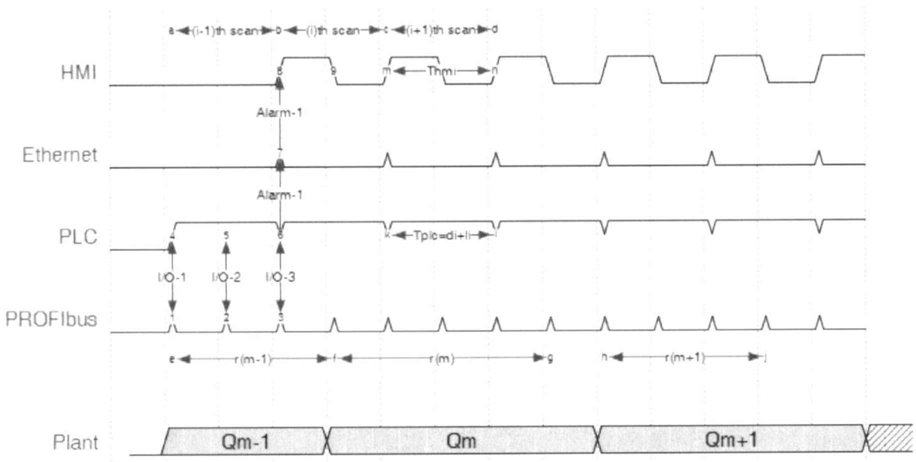

Figure 2: Initial behavioural model

architectural model of the Integrated Automation System is presented in Figure 1, below. The initial behavioural model of the IAS is given in the following digital timing diagram, depicted in Figure 2.

5. CALCULATION OF $E(I_n)$

For the PROFIBUS DP network the parameters in Table 1 are used. Due to limited access to the actual configuration files, any diagnostic messages, telegram retries and additional acyclic telegrams are not part of this study. Therefore, only the ideal single Master, single Slave system is addressed. For simplicity we focus on the below PROFIBUS DP Telegrams (SD) in this paper and the Send and Request Data service only:

- Telegrams without data field (SD1)
- Telegram with variable length of 4 to 249 bytes and therefore a payload in the range 1 to 244 bytes with Service Access Points (SD2)
- Telegram with fixed data length of 8 bytes data (SD3)

Table 1: PROFIBUS DP known parameters

Communication Method	Master – Slave
Network Speed	Min:9.6Kbits/sec, Typical: 1500 Kbit/sec Max:12,000Kbits/sec
Data Transfer Size	Up to 244 Bytes
Transmission Media Technology	RS-485
Number of Nodes	1
Scan Rate	20Hz
Maximum Distance	9.6Kbits/sec:1000m, 12,000Kbits/sec:100m

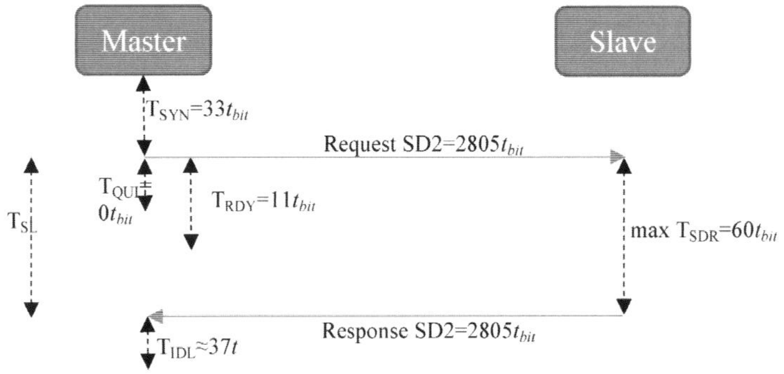

Figure 3: PROFIBUS DP cycle time

T_{SDR}: Station Response Time
T_{SYN}: Synchronization Time
T_{SL}: Slot Time
T_{QUI}: Quiet Time
T_{RDY}: Ready Time
T_{IDL}: Idle time until the master is ready for the next message cycle. For masters based on ASPC2 ASICs, 37t_{bit} is used according to [13].

Each byte in the telegram is transferred as 11 bits according to the UART coding. We will be considered with the theoretical worst case response time and as a result, a cyclic payload of 244 Bytes is assumed on an SD2 telegram, (i.e. 2684 bits). In reality, the payload will be at the order of 1-20 Bytes. One SD2 Telegram with Service Access Points uses 11 additional UART Bytes for structure, addresses and error detection i.e. 121 bits. We therefore have 2805 bits per SD2 Telegram. The transition time of a bit, t_{bit}, varies depending on the selected bit rate. Usually this will be decided during commissioning of the system and it will not be available during design verification. Consequently, the theoretical worst case is derived by the minimum bit rate at 9.6Kbits/sec. Therefore, t_{bit} = 0.104 msec. The cycle time of a PROFIBUS DP message using default values for all timing parameters, is presented below in Figure 3.

According to [13], on any PROFIBUS DP network, certain time ratios must be respected to ensure that a telegram sequence can function correctly. After a synchronization pause lasting T_{SYN} to mark the beginning of a sequence, the Request Telegram (SD2) is sent by the master to the slave. The slave should start its response no sooner than min T_{SDR} (SDR = Station Delay of Responder), but no later than max T_{SDR}. The min T_{SDR} time must include at least the T_{QUI} time of the in-line amplifier and repeater and the T_{RDY} time of the master. In modern PROFIBUS DP systems, min T_{SDR} = 11t_{bit} but for this study the maximum T_{SDR} for bit-rates less than 187.5Kbits/sec is used. The master waits for a response from a slave for a maximum period of T_{SL} (SL = slot). This time is determined by max T_{SDR}. Since there is only one Master and one Slave on the bus under study, no token exchange is require. Moreover direct input from Human Machine Interfaces (HMI) is not present in the design. Therefore, the *Bus Cycle* time is the same as the *Message Cycle* time and it is calculated from the below Equation 2.

Equation 2: PROFIBUS DP Cycle Time

$$T_{Message\ Cycle} = T_{SYN} + 2 \times SD2\ Telegram + maxT_{SDR} + T_{IDL}$$

Since the PLC I/O scan rate is 20Hz, as per Table 1, the theoretical worst case response time of the PROFIBUS DP in this case is approximately ten times more than the one implied by the application. Consequently, the actual bitrate and the SD2 payload should be provided during design phase, for accurate conclusions of $E(I_n)$. For the Remote I/O Bus, there are limited known parameters and due to the proprietary nature of the protocol used, it will not be addressed any further.

6. CALCULATION OF $E(X_n)$ UNDER THE TRADITIONAL PIPELINE OPERATION

In the absence of exact PROFIBUS DP parameters we assume that the $T_{Message\ cycle}$ is the one specified by the PLC I/O scan time, i.e. $T_{Message\ cycle}$ = 50msec. As this is half the specified PLC scan time, T_{PLC}, it is not possible for data received by the current I/O cycle to be taken into account in the current PLC calculations. Consequently, in this case the previous data are used as per Figure 4 below which increases the fidelity of the Behavioural Model in Figure 2 by taking into account the Architectural Model in Figure 1. A single, non-preemptive task is assumed at this stage for the traditional pipeline operation.

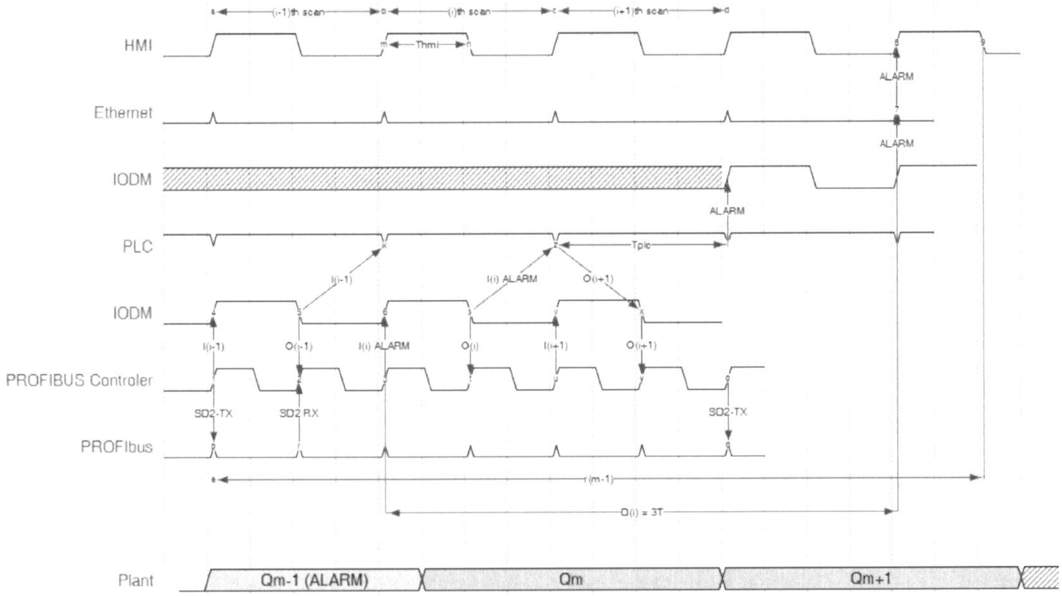

Figure 4: Behavioural Model with IODM cycles

At the i_{th} scan, the PLC processing task starts before input data, $I(i)$ ALARM have been written in the Input-Output Direct Memory (IODM). The output data transmission, $O(i)$, starts also before the completion of the i_{th} PLC processing algorithm. As a result, the i_{th} PLC processing algorithm is executed with the data stored in IODM from the $I(i-1)$ bus cycle. The output data of the i_{th} PLC processing algorithm are transmitted over to the plant at the $O(i+1)$, bus cycle. Let's assume that the plant is going into an alarm state, $Q(m-1)$, at the exact point in time when the $(i_{th}-1)$ PLC processing algorithm starts. This is the worst case scenario according to the problem statement. The alarm will reach the IODM after a fraction of time T_{PLC} and it will processed at the next PLC scan, $(i+1)$. The alarm condition will be written in IODM and let, pessimistically, be accessed by the Ethernet controller at the worst case scenario after a time equal to the PLC scan period to account for all previous application Ethernet frames not related to alarms. Assuming no Ethernet delay due to bus traffic, the alarm will require $R_{database} + R_{HMI}$ before it is published to the crew. The worst case response time in this case for the state change, $Q(m-1)$, is calculated by Equation 3.

Equation 3: Worst case response time

$$r(m-1) = R = T_{PLC} + E(I_n) + E(X_n) + E(O_n) + R_{database} + R_{HMI}$$

Where,

$R_{database}$: Database worst case response time
R_{HMI}: HMI worst case response time
$E(I_n) = E(X_n) = E(O_n) = T_{PLC}$

7. CALCULATION OF $E(O_n)$ WITHOUT ETHERNET LOAD

In the above section, $E(O_n)$ was based on the assumption that the Ethernet controller will require a certain amount of time to access the new data in the IODM, upper bounded to the PLC scan period, without considering any Ethernet network delays. For a more accurate calculation, both the actual upper bound of the Ethernet controller need to be estimated and the actual network delay. The below section address the latter issue following an end-to-end network delay methodology for simplicity. The former should be considered as a given design specification. The main difference between PROFIBUS DP and Ethernet for the scope of this paper is the one provided in [14]: "Because PROFIBUS uses a cyclic (periodic) polling mechanism between masters and slaves, it is also deterministic. That is, the behaviour of a PROFIBUS system can be reliably predicted over time. In fact, PROFIBUS was designed to guarantee a deterministic response. To contrast, CAN and Ethernet are event-driven bus systems and consequently form non-deterministic systems." Although, this statement is subject to much debate based on the deployed protocol stack and services, we will consider it as true in this case since the actual Ethernet protocol stack of the drillship was not disclosed before the completion of this study. The choice for an Ethernet based process network is better justified in [15]. In Switched Ethernet, as it is the case for the investigated system, each process controller in the network has its own, full-duplex connection to a central switch. This allows the bus arbitration to occur within the switch, as they can buffer incoming data until it can be transmitted further. However according to [16], the buffering can result in serious delays, especially in highly utilized networks and in some cases, hub-based networks outperform switched networks at low loads, due the lack of switching delays. In IAS, the process network worst case load is an important factor in the response time analysis of the selected system since it is shared between all integrated processes. Consequently, due to lack of actual bus utilization data, is going to be assumed that all integrated processes are involved in a concurrent alarm transmission scenario with no additional network load present. From the submitted design documents, there is a case where 25 heterogeneous process controllers, share the same second level switch towards 22 heterogeneous HMI's. The switching technology of choice is the Store-and-Forward, where the entire incoming packet is buffered and checked for corrupted bits before an attempted to retransmit it, is initiated. In [17], a delay analysis of Networked Control Systems based on 100Mbits Switched Ethernet is provided, based on the assumption that there is no switch buffer overflow and the cable lengths between the switch and the nodes are of the same length. We adopt the former assumption for a full-duplex 10BASE-T/100BASE-T/1000BASE-X SFP Ethernet, compatible with the IEEE 802.3 standard but not the latter as the propagation delay D_p, although proportional to the cable length as depicted in Equation 4, is insignificant for the alarm and monitoring response time analysis.

Equation 4: Propagation Delay

$$D_p = d/s$$

Where,
d: The distance between the Process Controller and the Ethernet Switch,
s: The wave propagation speed, approximately 2×10^8 m/sec for both the standard Cat 5 UTP cable and the fiber optic cable.

Consequently, for the maximum 10BASE-T cable length of 100m, $D_p = 0.556\ \mu sec$. For simplicity, it will not be considered any further.

For the network described above, the Maximum Transmission Unit (MTU) of 1500 Bytes is defined by the standard. The overall Ethernet packet length as per IEEE 802.3 specification is therefore 1518 Bytes, considering the overhead but without including the optional 802.1Q tag field. For a more tight worst case MTU, specific alarm frame payloads per process controller should be used.

A two layer switching circuitry is used with mixed bit rates, therefore we have the following store-and-forward transmission delays calculated by Equation 5 and presented in Table 2.

Equation 5: Transmission Delay

$$D_T = N/R$$

Where,
D_T: Transmission Delay
N: Number of bits per frame
R: Rate of transmission

Ethernet Switches must allow a minimum idle period between transmissions of Ethernet packets known as the Inter-Packet Gap (IPG). A brief recovery time between packets allows devices to prepare for reception of the next packet. The standard minimum IPG is $96t_{bit}$, shown individually for each applicable bit rate in Table 2.

Table 2: Transmission Delays with IPG

	Transmission Delays		IPG
$D_{T10Mbps}$	1.21msec	$D_{T10MbpsIPG}$	$96 \times 1bit / 10Mbps$
$D_{T100Mbps}$	0.121msec	$D_{T100MbpsIPG}$	$96 \times 1bit / 100Mbps$
D_{T1Gbps}	0.0121msec	$D_{T1GbpsIPG}$	$96 \times 1bit / 1Gbps$

The architectural model of the network is depicted in the Figure 5 below. The arrows signify parallel transmission as opposed to the lines that signify a serial transmission. The theoretical worst case scenario for the model below, is calculated by the longest transmission path when the Process Controllers in the A_1 segment transmit

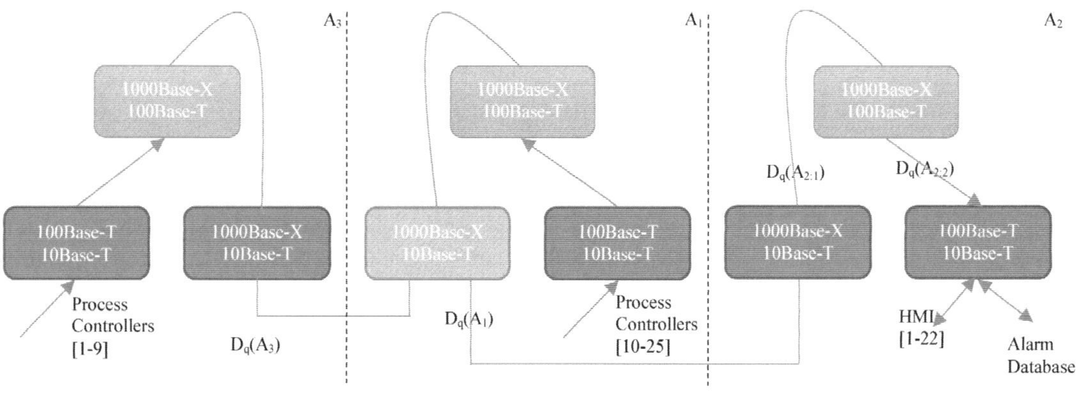

Figure 5: Process Network Architectural Model

simultaneously their alarm frames after time $D_{T10Mbps}$, from the time the Process Controllers in segment A_3 start to transmit simultaneously their respective alarm frames. In this case, the buffer of the orange switch in the A_1 segment, will have to store and forward all 25 frames sequentially to the A_2 segment. Therefore, the queue delay for A_1, $D_q(A_1)$ is calculated by the below equation.

Equation 6: A_1 segment queue delay

$$D_q(A_1) = 25 \times D_{T10Mbps} + 24 \times D_{T10MbpsIPG} = 30.25\ msec$$

For the A_3 segment, the queue delay is calculated as follows.

Equation 7: A_3 segment queue delay

$$D_q(A_3) = 9 \times D_{T1Gbps} + 8 \times D_{T1GbpsIPG} + 9 \times D_{T10Mbps} + 8 \times D_{T10MbpsIPG} = 11.07\ msec$$

For the A_2 segments, the respective queue delays are calculated as per the equations below.

Equation 8: $A_{2:1}$ queue delay

$$D_q(A_{2:1}) = 25 \times D_{T1Gbps} + 24 \times D_{T1GbpsIPG} = 0.304 msec$$

Equation 9: $A_{2:2}$ queue delay

$$D_q(A_{2:2}) = 25 \times D_{T100Mbps} + 24 \times D_{T100MbpsIPG} = 3.048 msec$$

Equation 10: $A_{2:3}$ queue delay

$$D_q(A_{2:3}) = 25 \times D_{T10Mbps} + 24 \times D_{T10MbpsIPG} = 30.48 msec$$

Equation 11: A_2 queue delay

$$D_q(A_2) = D_q(A_{2:1}) + D_q(A_{2:2}) + D_q(A_{2:3}) = 33.83 msec$$

Finally the $E(O_n)$ can be calculated by the longest transmission path delay and the Ethernet controller worst case access delay, as per below.

Equation 12: $E(O_n)$ delay

$$E(O_n) = T_{PLC} + D_{T10Mbps} + D_{T100Mbps} + D_{T1Gbps} + D_q(A_3) + D_q(A_1) + D_q(A_2) = 176.5\ msec$$

Consequently, in this theoretical worst case scenario, $E(O_n)$ is bounded at 8.8% of the maximum latency permitted by ABS for the Alarm and Monitoring System of the IAS implementation. If ABS maximum latency requirements are to become more demanding, as per suggestions from the industry, this percentage could be doubled.

8. ESTIMATING $R_{database}$ AND R_{HMI}

Class requirements for the response time of alarms impose real-time constrains in the transactions of the Alarm Database as well. Consequently, database transactions require temporal validity as a late commit may result in missing the response time deadline for the alarm. Quality of Service (QoS) guarantees are, therefore a valuable component in the response time analysis and overall performance of the system. QoS guarantees are more realistic than worst case execution times at this level due to the dynamic variation of the database workload and the access patterns of transactions. For the IAS under study there are two distinct transaction types. The first is the alarm update transaction, containing the temporal alarm data objects and the second is the HMI refresh query. The former transaction has a higher priority than the later. Also the operation of the former transaction is always a *write* where the operation of the later transaction is always a *read*. There is a certain execution time associated with each operation and the execution time of a transaction is the sum of the execution time of all its operations. However, it is very rare to be provided explicitly in real applications. As a result, the notion of *Execution Time Estimation Error*, as defined in [18] is used, having normal distribution with a mean of 20% and a standard deviation of 10%.

For the purpose of this theoretical study, let's assume that the time required for one *read* or *write* operation is the maximum of 2.4 msec considering an additional average *Execution Time Estimation Error* [18]. Furthermore, based on the average number of monitored processes per process controller, we assume that each Alarm Ethernet frame contains 150 Alarm status updates of 10 Bytes each. In a best-effort protocol all transactions are admitted and therefore the final *write* operation will occur with a database delay depicted below by Equation 13.

Equation 13: Database write delay

$$R_{write} = 25\ F \times 150 \frac{W}{F} \times OT = 9 sec$$

Where,
F: Number of Alarm Frames
$\frac{W}{F}$: *Write* operations per frame
OT: *Write* Operation Time

If the alarm status refresh query request is defined as a minimum Ethernet packet of 64 bytes and the 22 heterogeneous HMIs have been organized to access the Alarm server synchronously following a Client-Server model, the last refresh query request will reach the database after the time specified below.

Equation 14: Client Request Delay

$$D_q(A_{2:client}) = 22 \times D_{T10Mbps} + 21 \times D_{T10MbpsIPG} = 27\ msec$$

It will remain at the incoming buffer until the request query is processed by the database *read* operation in a First-In-First-Out (FIFO) order, assuming a uniprocessor environment. The time required from the database server to respond with the data required by the last HMI in the network is given by Equation 15.

Equation 15: Database Read Delay

$$R_{read} = 22\,F \times 150\frac{W}{F} \times OT = 7.92sec$$

The network delay for the last Alarm frame towards the last HMI is given in Equation 16.

Equation 16: Server Response Delay

$$D_q(A_{2:server}) = D_{T10Mbps} + D_{T10MbpsIPG} = 1.22\,msec$$

Therefore, $R_{database}$ and R_{HMI} are calculated in Equation 17 and Equation 18 respectively.

Equation 17: Database Delay

$$R_{database} = R_{write} + R_{read} = 16.92sec$$

Equation 18: HMI Delay

$$R_{HMI} = D_q(A_{2:client}) + D_q(A_{2:server}) = 28.22\,msec$$

Finally,R can be derived by substituting the above results in Equation 3. In the discussed worst-case theoretical scenario above, it was found to be 870.8% of the maximum latency permitted by ABS for the Alarm and Monitoring System of the IAS implementation.

9. CONCLUSIONS AND FUTURE WORK

Aiming to increase the awareness of both the benefits and the associated risks of Integrated Automation Systems leading to Total Ship Automation and eventually to Autonomous Ships, this paper introduces a few simplistic aspects of the response time analysis concept. A theoretical study of a well-known IAS system was provided based nonetheless on a few assumptions that require further investigation. For example, the multitasking environment of the PLC was not taken into account. Rate Monotonic Scheduling (RMS) is a well-defined algorithm that can yield better $E(X_n)$ estimations if included in the analysis. More tight estimation methods for $E(X_n)$ were proposed for PLCs in [19] and [20]. More recent advances, are using commercial tools that take into account both the architecture of the underling micro-processor and the unique features of the compiler involved [21]. For calculating the $E(O_n)$ on the process network, the response time of the embedded Ethernet controller is important. In addition, a very simplistic end-to-end network delay methodology was used. Stochastic Network Calculus is a formal alternative utilized in avionics switched Ethernet [22], better suited for classification submittals. In a real-time database, tardy transactions (transactions that have missed their deadlines) are aborted upon their deadline miss. A major performance metric is the Miss Ratio (MR) and several other transient performance metrics such as overshoot and settling time that are adopted from control theory for a real-time system performance specification, since average MR is not sufficient for dynamic system verification due to time-varying workloads [23].

In such integrated automation realities, ABS envisions Smart Ships having both a higher degree of decision making support for the crew in real-time and a reactive intelligence, when response times are too short for human intervention, in order to better promote the security of life and property and preserve the natural environment. Sound methods and tools are increasingly important for precise and reliable verification but also for the inherent risk assessment required by their operators, their charters, the port state controls and the general public.

10. ACKNOLEDGMENTS

The author would like to thank Steven O'Day, Principal Engineer at ABS for his valuable help and insight. He would like also to extend his gratitude to Christopher Serratella and Michael Wasicek, both directors at ABS, for their continuing trust and support.

11. REFERENCES

1. A. Sangiovanni-Vincentelli, 'Quo Vadis, SLD? Reasoning About the Trends and Challenges of System Level Design', *Proceedings of the IEEE, vol. 95, no. 3, pp. 467-506*, 2007.

2. E. A. Lee, 'The Problem with Threads', *University of California, Berkeley*, 2006.

3. E. A. Lee and Y. Zhao, 'Reinventing Computing for Real Time', *University of California, Berkeley*, 2006.

4. C. Stergiou, 'Schedulability Analysis and Verification of Real-Time', *Electrical Engineering and Computer Sciences University of California, Berkeley*, 2013.

5. E. A. Lee, 'Absolutely Positively on Time: What Would It Take?', *Embedded Computing*, 2005.

6. J. A. Stankovic, I. Lee, A. Mok and R. Rajkumar, 'Opportunities and Obligations for Physical Computing Systems', *IEEE Computer Society , pp. 23-31*, 2005.

7. R. A. Gupta and M.Y. Chow, 'Networked control system: Overview and research trends', *IEEE Transactions on Industrial Electronics, vol. 57, no. 10, p. 2527–2535*, 2010.

8. J. Sztipanovits and S. Sastry, 'Embedded Software: Opportunities and Challenges', *Department of Defense*.

9. L. Northro, et al., 'Ultra-Large-Scale Systems', *Carnegie Mellon, Pittsburgh*, 2006.

10. P. Technology, 'Biggest Shipbuilder in Microsoft-Intel Plan', *[Online]. Available: https://www.porttechnology.org/news/biggest_shipbuilder_in_microsoft_intel_master_plan?utm_source=Port+Technology+Newsletter&utm_campaign=6d7a731942PortTechnology_Daily06_07_2016&utm_medium=email&utm_term=0_89349292d5-6d7a731942-235035601* ,[Accessed 18 08 2016].

11. EU/IST FP6 Artist2 NoE, 'Roadmap on Real-Time Techniques in Control System Implementation', *[Online]. Available: http://www.artist-embedded.org/artist/-Roadmaps-.html* , [Accessed 18 08 2016] .

12. S. Jeong et al., 'Response Time Driven Scheduling for Programmable Logic Controllers with Network Based I/O systems', *Real-Time Systems, vol. 25, pp. 67-91*, 2003.

13. M. Felser, 'PROFIBUS Manual', *[Online], Available:http://profibus.felser.ch/en/index.html?uart_codierung.htm*, 2013.

14. Acromag, 'BusWorks 900PB Series - Technical Reference', *ACROMAG INCORPORATED*, 2002.

15. B. Galloway and G. P. Hancke, 'Introduction to Industrial Control Networks', *IEEE*, 2012.

16. R. Vi´egas, R. A. M. Valentim, D. G. Texira and and L. A. Guedes, 'Analysis of protocols to ethernet automation networks', *SICE-ICASE International joint Conference, p. 4981 – 4985*, 2006.

17. M. Li, 'Delay Analysis of Networked Control Systems Based on 100MBit Switched Ethernet', *The Scientific World Journal*, 2014.

18. Y. Wei et al., '*QoS Management in Distributed Real-Time Databases*', *[Online], Available: https://www.cs.virginia.edu/~stankovic/psfiles/rtss03.pdf*, [Accessed 18 08 2016].

19. Koo, K. and Kwon, W. H., 'Predicting Execution time of relay ladder logic for Programmable Logic Controller', *Proceedings of the 96' ETFA, , pp. 670-675*, 1996.

20. Koo, K. and Kwon, W. H., 'Worst-case timing prediction of relay ladder logic by constraint analysis', *Proceedings of the 3rd RTCSA, pp. 180-186*, 1996.

21. Hanxleden, R. V., et al., 'WCET TOOL CHALLENGE 2011: REPORT', *In Proceedings of the 11th International Workshop on Worst-Case Execution Time (WCET) Analysis*, 2011.

22. Ridouard, F. et al., 'Stochastic network calculus for end-to-end delays distribution evaluation on an avionics switched Ethernet', *5th IEEE International Conference on Industrial Informatics*, 2007

23. Lu C. et al., 'Feedback control real-time scheduling: Framework, modeling, and algorithms.', *Journal of Real-Time Systems, Special Issue on Control-theoretical Approaches to Real-Time Computing, pp. 85-126*, 2002.

12. AUTHORS BIOGRAPHY

Demetres Armanes is currently a senior research engineer at the American Bureau of Shipping (ABS). He holds a Ph.D. in Intelligent Control and a B.Eng. in Electrical and Electronics Engineering from The University of Liverpool. He worked as a Chartered Engineer for five years, designing Marine Programmable Electronic Systems, prior joining ABS Nautical Systems in 2010 at the role of Regional Account Manager, responsible for a portfolio of 42 clients in the EMEA region.

13. DISCLAIMER

The views and opinions expressed in this article are those of the author and do not necessarily reflect the position or views of American Bureau of Shipping.

HANDLING BIG DATA IN SHIP PERFORMANCE & NAVIGATION MONITORING

L P Perera, SINTEF Ocean, Trondheim, Norway.

SUMMARY

Vessels are equipped with on-board internet of things (IoT) to observe various ship performance and navigation conditions. These IoT connected to ship navigation and automation systems collects and exchanges data among vessels and shore based data centres. Vessel navigation strategies are often based on such information to satisfy various energy efficiency and emission control standards in shipping. However, such IoT may create various shipping industrial challenges under large-scale data sets, so called "Big Data" challenges. This study illustrates to overcome Big Data challenges under the proposed data handling framework with various data analytics for on-board vessels and shore based data centres. The basis for these data analytics consists of data driven models developed under an engine-propeller combinator diagram of a selected vessel. The respective data analytics of sensor faults detection, error compression & data recovery, integrity verification & regression, and visualization & decision support are developed along the same.

1. INTRODUCTION

1.1 INTERNET OF THINGS

Ship navigation will be based on vessel performance and navigation data collected by various on-board internet of things (IoT), sensors and data acquisition systems, in the future [1]. Such IoT connected to ship navigation and automation systems collects and exchanges data among vessels and shore based data centres. This network connectivity in the ocean should consist of appropriate maritime infrastructure for proper internetworking in shipping as an information industry. Such infrastructure creates opportunities for direct integration of ship performance and navigation information into computer-based systems, including digital models, estimation algorithms, resulting improved efficiency, accuracy and economic benefit to the industry.

The outcome of such opportunities will eventually be evaluated under integrated bride systems (IBSs), where appropriate navigation strategies to overcome emission control based energy efficiency measures in shipping especially under emission control areas (ECAs) should be executed [2]. Intelligent decision support capabilities under IBSs eventually influence on such ship navigation strategies. Hence, the maritime information and communication technology (ICT) infrastructure, IoT and "Big Data" will play an increasingly important role in shipping in the coming decade, where various ship-owners have already established their own initiatives for digitalizing the industry. A comprehensive overview of ship performance and navigation conditions can be observed in the future IBSs under such facilities and that information should use to achieve the respective digitalization objectives (i.e. safety and efficiency).

1.2 ON-BOARD DATA HANDLING

The data handling process can play an important role in the same situations, where the respective industrial digitalization activities should facilitate with accurate ship performance and navigation information. However, on-board systems may encounter various industrial challenges in handling Big Data with ship performance and navigation information: model uncertainty, data quality & quantity, information visualization challenges and high computational power, etc. Hence, appropriate industrial solutions to handle such data handling challenges should also be considered. One should note that conventional approaches with empirical mathematical models may fail to facilitate with such data handling challenges in shipping. Therefore, a new data handling framework is proposed in this study to overcome the same.

The basis of the same framework consists of developing data driven models to capture realistic ship performance and navigation conditions with various data analytics to improve the quality and quantity of the respective data sets. It is noted that appropriate methodologies to develop such data driven models have not been presented in the recent literature. Furthermore, some data driven models may fail due the data quality and quantity issues. Such models are derived by observing the structure of the respective data sets in various industries and that may lead to erroneous models in some situations. Therefore, this study proposes to use domain knowledge (i.e. ship operational and navigation conditions) to improve the accuracy of such data driven models.

It has shown that the same data driven models under a high dimensional parameter space can facilitate adequate solutions towards Big Data challenges (i.e. data quality and quantity issues) in the industrial digitalization process. Furthermore, the same models may present superior performance under ship energy efficiency and reliability applications [3]. The proposed data handling framework with data driven models and analytics to handle such challenges in the shipping industry is presented in Figure 1. One should note that the basis for these data analytics is data driven models and that are developed under an engine propeller combinator diagram of a selected vessel [4]. The respective details on such data driven models are presented in the following sections to overcome the respective industrial digitalization challenges.

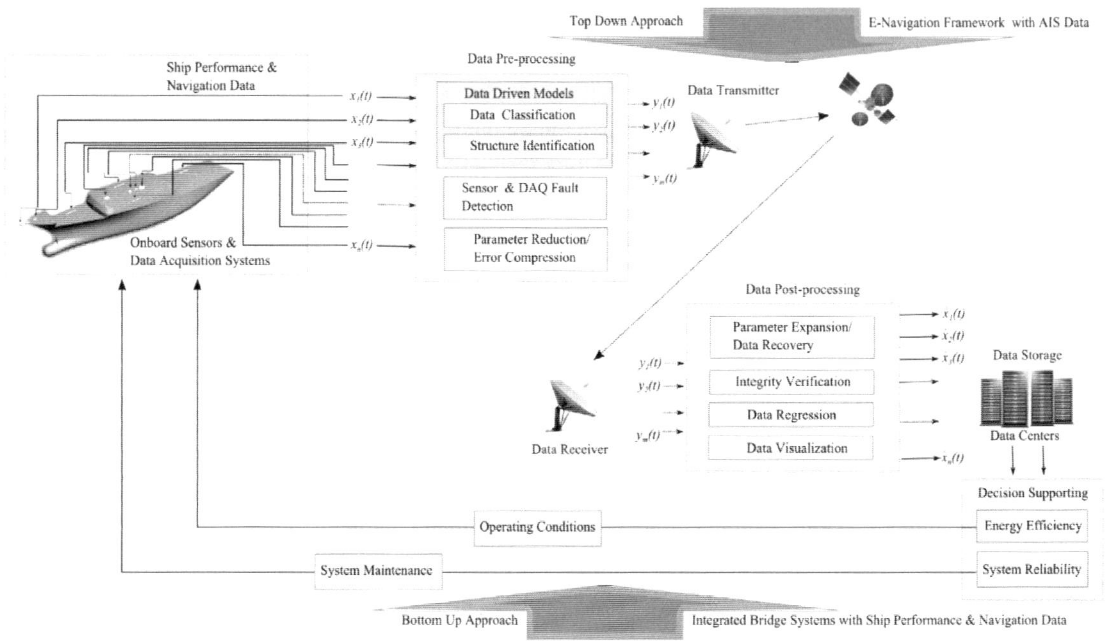

Figure 1: Data Handling Framework.

2. DATA ANALYTICS

2.1 DATA HANDLING FRAMEWORK

Ship performance and navigation data collected IoT transfer though the framework and that consists of various data handling layers (see Figure 1). That is supported by a top down approach of e-navigation framework with AIS data [5] and a bottom up approach of integrated bridge systems with ship performance and navigation data. The data handling layers, i.e. data analytics, are categorized under two sections of pre and post processing. Data pre-processing is categorized as an on-board application consisting sensor & DAQ fault detection and parameter reduction/error compression layers supported by the data driven models. Data post-processing is categorized as a shore based application consisting parameter reduction/data recovery, data integrity verification & regression and data visualization & decision supporting layers supported by the same data driven models.

2.2 DATA PROCESSING

The basis for both data pre & post-processing is data drive models derived by identifying the proper structure for ship performance and navigation data sets. The data structure can be supported by the domain knowledge (i.e. ship operational and navigation conditions) as mentioned before. Detecting sensor and DAQ fault situations and removing such erroneous regions from the data sets are done as the second layer. That identifies sensor and DAQ fault situations through the same data structure. Various out-layers (i.e. high dimensional boundaries) with respect to the data structure can be identified in this approach as sensor and DAQ faults [6]. The proposed data driven models can be used to develop such high dimensional boundaries for data out-layers and that can eventually be used as fault detection thresholds.

The last layer in data pre-processing is parameter reduction/error compression and the first layer in data post-processing is parameter expansion/data recovery. Those steps improve the data quality and reduce the data quantity. The same data structure (i.e. data driven models) are considered for these layers [7]. One should note that the parameter reduction layer can also compress erroneous conditions and the parameter expansion layer can also recover such erroneous conditions in the data sets. The pre-processed parameters communicate through the transmitters in much smaller improved data sets. The transmission process can use additional technologies, where the size of the data sets can be further reduced.

The same are obtained by shore based data centres through the receivers and transfer though the post-process. The integrity verification layer is incorporated with other data sources (i.e. AIS, weather data, etc.) to verify the quality of measured ship performance and navigation parameters. Therefore, additional erroneous conditions (i.e. communication errors, etc.) can identify and isolate from ship performance and navigation data by this layer. It is noted that during the parameter reduction and expansion processes, some variations can be introduced into the estimated data sets. Therefore, the regression layer is introduced in this data handling framework to reduce such parameter variations. That is incorporated with various smoothing algorithms to reduce the respective parameter variations.

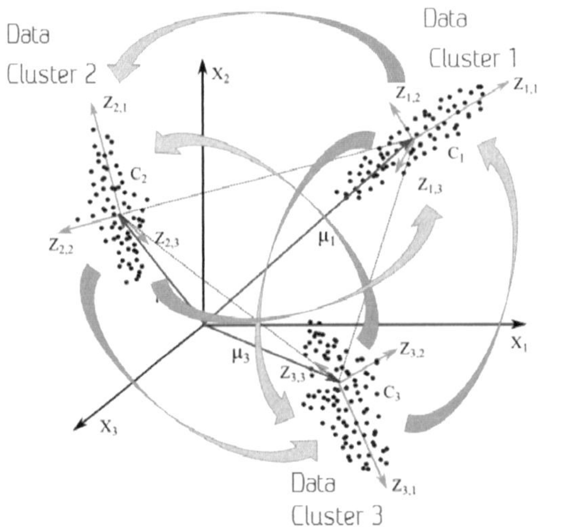

Figure 2: Data Driven Models

2.3 DECISION SUPPORTING

The improved ship performance and navigation data sets should be visualized, appropriately by the visualization layer of the data handling framework. It is expected that the data structure (i.e. data driven model) can be visualized under this layer. The structure to such data driven models can vary along the time and that also can relate (i.e. quantify) to vessel operating and navigation conditions. Vessel energy efficiency and system reliability conditions can be obtained from such information to support the respective decisions support features. That is done by the decision supporting layer of the same framework. Hence, the proposed data driven models, the respective data structure, can facilitate all layers in the data handling framework.

2.4 DATA DRIVEN MODELS

Conventional mathematical models with various ship performance and navigation parameters may fail to handle Big Data in shipping. Hence, this section elaborates the development steps of high dimensional data driven models to overcome such industrial challenges [8]. That can often be categorized as digital models due to their discreteness and that supports digitalization applications of shipping. A simplified view of the proposed data driven model in a three-dimensional space is presented in Figure 2. In general, such data driven models consist of various data clusters within the data set and each data cluster may have a structure. That structure may create a vector filed to represent the respective vessel performance and navigation conditions (i.e. Data cluster 1, 2 & 3). One should note that the same vector field introduces the respective structure of the data set. The initial data driven models can be derived offshore data centres and the same can be implemented on-board vessels to further improve such models. Therefore, the improved models can be used in this framework to support the respective data handling layers. It is believed that such data driven models in real-time may have additional flexibility in handling Big Data sets. The respective data clusters should group around the operating and navigation conditions of vessels (i.e. engine-propeller operating points, trim-draft combination points). This data drive model is developed in a three-dimension vector space with the right-hand coordinate system of $X_1X_2X_3$. One should note that X_1, X_2 and X_3 represent the parameters of the respective data set. It is assumed that this data set consists of the respective data clusters C_1 , C_2 and C_3 and that are distributed with the respective mean vectors μ_1, μ_2 and μ_3.

Vessel operational and navigation conditions can jump from one data cluster to another under such parameter space categorized as digital models. This approach can also classify as the linearization (i.e. piecewise linearization) of ship behaviour under various operating and navigation points in a high dimensional space. Hence, the data structure can relate to ship energy efficiency and system reliability conditions, where the structural changes under various time-horizons can be observed. The first step in developing such data driven models is to identify the respective data clusters in a high dimensional space. The second step is to identify the respective structure of each data cluster. The respective structural vectors of the i-th data cluster is represented as $Z_{i,1}$, $Z_{i,2}$ and $Z_{i,3}$. One should note that such vectors, the respective mean and structural vectors, (i.e. the respective data structure) can be used for both energy efficiency and system reliability applications under the data visualization layer.

However, there are various challenges one may encounter during the development phase of such data driven models. These clusters may exist in a high dimensional space and difficult to visualize from the respective data sets. In general, the number of data clusters in a data set is unknown (i.e. difficult to visualize). That can create the main difficulty in creating such data driven models for ship performance and navigation conditions. Even though, there are many machine learning algorithms are presented to cluster data sets, those can often fail to identify the proper number of data clusters with the data set, since some data are scattered. Hence, this study proposes to incorporate domain information, ship operational and navigation conditions, to calculate the number of data clusters. One should note that the structure of each data cluster should be visualized to extract the respective information on ship performance and navigation conditions that are capsulated into each data cluster.

2.5 INFORMATION VISUALIZATION

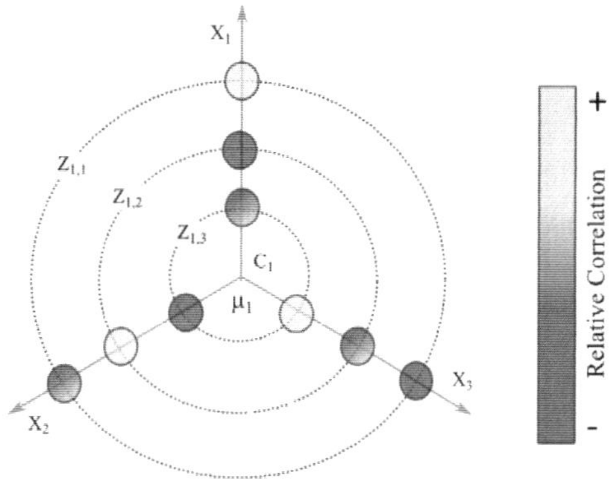

Figure 3: The structure for data cluster 1

The last layers are data visualization and decision supporting in this framework to support ship energy efficiency and system reliability applications. Energy efficiency applications consist of identifying optimal vessel operating conditions to reduce the overall fuel consumption and emissions of vessels. System reliability applications consist of identifying the health conditions of on-board systems to reduce the operating costs of vessels. To implement both energy efficiency and system reliability applications, ship performance and navigation data should be visualized, appropriately. Such visualization methods can eventually be influenced on decision supporting features in shipping. A simplified version of such visualization method is presented in Figure 3. The data structure of a selected data cluster (i.e. data cluster 1) is presented in this figure, where the relative correlation values among ship performance and navigation parameters in a high dimensional space are visualized. One should note that covariance values among ship performance and navigation parameters are used to derive the structure of each data cluster. That information is extracted and visualized as the relative correlation relationships among the same parameters.

The respective vector structure is represented as $Z_{1,1}$, $Z_{1,2}$ and $Z_{1,3}$ in this figure. The largest and smallest covariance values of the data cluster are represented by $Z_{1,1}$ and $Z_{1,3}$, respectively. Hence, the most and least important information of the data cluster is accommodated in $Z_{1,1}$ and $Z_{1,3}$, respectively. It is noted that sensor noise and other erroneous conditions can often accommodate into the least important vectors. $Z_{1,1}$ shows that: when parameter X_1 increases, then parameter X_3 decreases, and parameter X_2 may not vary that much with respect to other parameter variations. Hence, the relative correlation information can transfer into ship energy efficiency or system reliability applications.

3. SHIP DATA

3.1 VESSEL PARTICULARS

A ship performance and navigation data set of a selected vessel is considered in this study to develop the proposed data handling framework. The data set is collected from a bulk carrier with following particulars: ship length: 225 (m), beam: 32.29 (m), gross tonnage: 38.889 (tons), deadweight at max draft: 72.562 (tons). The vessel is powered by 2 stroke main engine with maximum continuous rating (MCR): 7564 (kW) at the shaft rotational speed of 105 (rpm) and a fixed pitch propeller with 4 blades (i.e. diameter 6.20 (m)). Furthermore, the vessel has two auxiliary engines with MCR: 850 (kW) at the respective shaft rotational speed of 800 (rpm). The vessel is equipped with IoT that collects the following parameters: propeller shaft rpm (rpm), engine power (kW), propeller shaft torque (kNm), engine fuel consumption (kg/s), fuel density (kg/m3), fuel temperature (C), fuel mass counter (kg), ship speed over ground (SOG) (Knots), speed through water (STW) (Knots), course (deg.), vessel position (longitude in deg., latitude in deg.), relative wind speed (m/s), relative wind direction (deg.), and aft and fore drafts (m). The respective ship performance and navigation data set consists of a selected set of such parameters considered for the data handling framework.

3.2 ENGINE PROPELLER COMBINATOR

The ship performance and navigation data set as statistical distributions (i.e. histograms) for each parameter are considered to identify vessel operating and navigation conditions and the results are presented in [9]. The respective parameters of engine power, shaft speed and fuel consumption are considered and the results show that the main engine was operating around three engine modes (i.e. three mean shaft speed values), therefore other engine related parameters (i.e. engine power and fuel consumption) have similar data distributions. It is concluded that the parameters of engine speed, power and fuel consumption should be considered as three data clusters for the data driven models. The reason for such selection is that ship performance and navigation data are clustered around the mean engine operating points (i.e. engine modes). However, possible clustering relationships among other parameters have not observed at this stage. Hence, the engine-propeller combinator diagram, where engine power and shaft speed values are combined, is used as the basis for the proposed data driven models. Furthermore, other ship performance and navigation parameters are also clustered around engine operating modes, initially.

3.3 DATA DRIVEN MODEL

The ship performance and navigation parameters that are clustered around three engine modes are presented in

Figure 4. The figure is used to identify the frequent engine operating regions, i.e. the respective engine-propeller combinator diagram, as the basis for the data handling framework, where the data driven models are derived. Three data clusters are identified in this figure and that are denoted by cluster A_1, A_2 and A_3 (see Figure 4). One should note that these data clusters relate to three main operating modes of the main engine. Therefore, the vessel performance and navigation parameters are clustered under these engine operating modes and those clusters create a meaningful structure for the data driven models. Furthermore, these data clusters create the respective discreteness within the data sets and such discreteness supports towards digital models in shipping. A statistical distribution of two parameters of engine power and shaft speeds is presented in the same figure. The top-left plot of the figure represents the contour plot of the same statistical distribution, where the most frequent operating regions of the engine propeller combinator diagram are identified. The top-right plot of the figure represents the same contour plot with the respective engine fuel consumption of the vessel. One should note that the higher engine power regions coincides with the higher fuel consumption regions in this plot.

The parameters in the respective data clusters relate to the following parameter range : main engine (ME) power from 3000 (kW) to 8000 (kW) and shaft speed from 80 (RPM) to 120 (RPM). In general, that is identified as the most frequent engine propeller operating region. The data points around the zero speed-power values representing slow ship manoeuvring conditions are ignored from the data analysis, therefore considerable sensor and DAQ noise conditions (i.e. erroneous data intervals) are removed. Then, the data classification step along the combinator diagram is executed and the results are presented in Figure 5.

3.4 DATA CLUSTERING

The respective data clusters on the engine propeller combinator diagram are identified by Gaussian mixture models (GMMs) with an expectation maximization (EM) algorithm [10-11]. The GMMs consist of three multivariate Gaussian distributions and that relate to the respective engine operating modes presented as contour plots in Figure 5 with the respective data points (i.e. in the engine propeller combinator diagram). One should note that these GMMs are used to classify the respective data points and the EM algorithm is implemented to calculate the respective parameters of the same. These GMMs are denoted as multivariate Gaussian distributions with the respective mean and covariance values. Hence, the first step of data driven model has completed by the data clustering step. However, the structure of each data clustered should be investigated to complete the proposed data driven models.

3.5 STRUCTURAL IDENTIFICATION

The respective data structure on the engine propeller combinator diagram are identified by Principal Component Analysis (PCA) [12]. PCA identifies the largest covariance directions among the respective data set that are orthogonal to each other by using a statistical procedure. Hence, the respective principal components (PCs) are linearly uncorrelated variables derived through possibly correlated variables of the data set. The data structure (i.e. the largest covariance directions that are orthogonal to each other) is identify as the respective PCs of the cluster as presented in Figure 6. The data set used in this step consists of 10 ship performance and navigation parameters: STW (Knots), SOG (Knots), ME power (kW), shaft speed (rpm), ME fuel consumption (cons.) (Tons/day), auxiliary (aux.) fuel consumption (cons.) (Tons/day), average (avg.) draft (m), trim (m), and relative (rel.) wind speed (m/s) and direction (deg).

One should note that the high dimensional vector space presented in Figure 3 is further illustrated in this figure. Figure 6 represents a 10 dimensional vector space, where the respective PCs are in a polar coordinate system demonstrated. The top and bottom PCs are dentoed by Z_1 and Z_2, respectively. Each PC is by a dotted circle presented, where the top PC, i.e. Z_1, has the highest radius. Each axis that intersects the dotted circles represents a parameter from the ship performance and navigation data set. The respective vector components of each PC are by coloured circles presented, where the circle radius represents the significance of each component with respect to other components within the same PC. This figure also represents an overview of the relative correlations among the respective parameters of ship performance and navigation data.

One should note that high positive and negative correlations represented by yellow and blue colour large circles, respectively (see the colour bar in Figure 6). The 1st PC represents: when avg. draft (high) increases, STW (medium) decreases, shaft speed (medium) decreases, SOG (high) decreases, and trim (medium) decreases. The 1st PC shows that ship resistance has increased due to the draft increments, where vessel STW and SOG values are also decreased. The same conditions have decreased shaft speeds due to high engine loading conditions. Furthermore, the draft increments are compensated by the trim adjustments of the vessel. In general, when vessel avg. draft increases, then trim increases, STW decreases, and SOG decreases. Hence, the 1st PC with their vector components is in this section summarised and other PCs can also be derived, similarly.

4. DATA ANALYTICS

4.1 FAULT DETECTION

The next layer in the data handling framework consists of detecting sensor and DAQ faults and isolating such

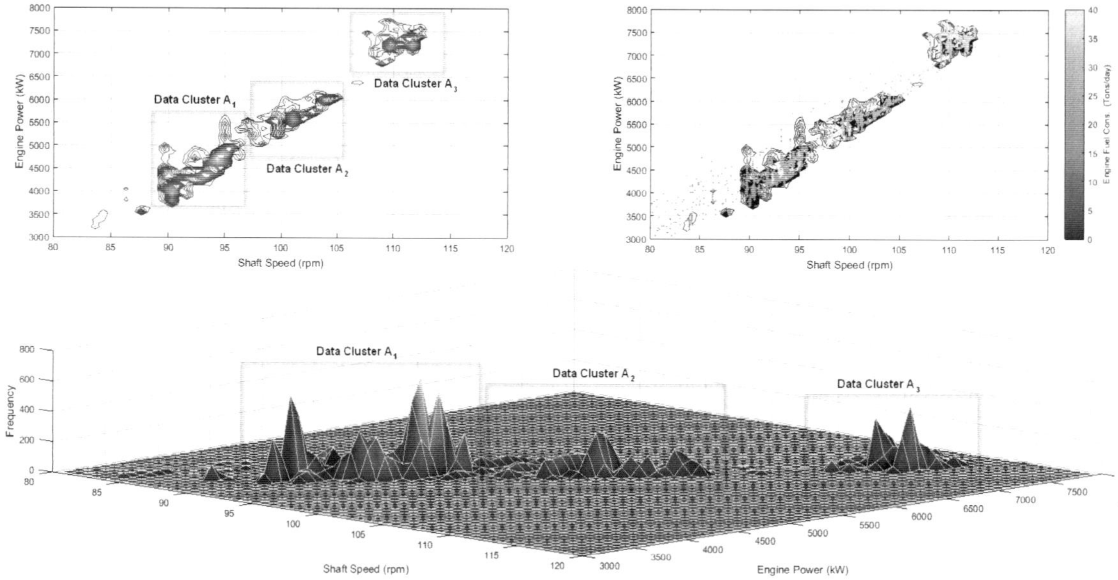

Figure 4: Data Clustering in Engine Propeller Combinator Diagram

erroneous regions from ship performance and navigation data. The initial sensor and DAQ faults can be observed by checking each parameter behaviour with respect to its mean and variance values. Each parameter in ship performance and navigation data sets may have its variation region, if the parameter is going beyond the same range is a fault. Complex sensor and DAQ faults are identified by checking the respective covariance values among the ship performance and navigation parameters. However, advanced relationships among ship performance and navigation parameters can identify such faults. Hence, the proposed data driven models, i.e. the data structure, can be used in complex sensor and DAQ fault situations [6]. The respective outliers of each data cluster in a high dimensional space are observed to capture such erroneous data regions as discussed before. Any data point that positions beyond such outliers is in the category of sensor and DAQ faults often observed under the bottom PCs. This approach consists of projecting the data set into the bottom PCs and observing such outliers.

Data cluster A_3 classified in the engine and propeller combinator diagram, is considered for this step. Figure 7 presents the results of sensor and DAQ fault detection with the respective data outliers, where the selected set of ship performance and navigation parameters is presented in the top 5 plots and the detected fault situations (i.e. fault alarm) are presented in the bottom plot. One should note that presentation of these plots is with respect to the number (No.) of data points (i.e. sample number). The detection of two fault situations in this data set is framed by two windows. In the first fault, several parameters (STW, shaft speed, and auxiliary engine fuel consumption) with some unusual behaviour (i.e. a sudden drop in the parameter value) are detected. In the second fault, (i.e. an interval), the auxiliary engine fuel consumption with considerably higher values are detected.

4.2 PARAMETER REDUCTION-EXPANSION

The parameter reduction and expansion steps are the last layer in data pre-processing and the first layer in data post-processing, respectively. Data cluster A_3 classified in the engine and propeller combinator diagram, is considered for these steps, where the same data structure is used. The ship performance and navigation data set may consist of additional parameters that have not considered for the data clustering process. Such parameters should along the same engine operating modes be classified, appropriately. The reduced data cluster consists of a new basis that is a linear combination the parameters of the initial data cluster. One should note that erroneous data regions in the initial data cluster can be compressed by this step. The parameter reduction step should be done with respect to the PCs and that can preserve the important information (i.e. the variance values) of the data cluster. The PC structure developed in the previous section is applicable under these layers of the data-handling framework.

As the next step, the reduced data cluster is transmitted by the respective communication networks to shore based data centres with the data expansion step. One should note that erroneous data regions can be recovered by the same step. The same PC structure used under the previous step is in the expansion step used. Comparing the input-output data sets reveals the success in the parameter reduction-expansion process. Figure 8 presents the results of such compression-expansion process with respect to the number (No.) of data points (i.e. sample number) with cluster A_3, before (BCD) and after (ACD) compression (i.e. actual data cluster) and expansion (i.e. estimated data cluster) layers, respectively. One can observe that some parameter variations between the actual and estimated data clusters. The input to the same process is 9 ship performance and navigation parameters and that has compressed into 7 new parameters by considering the top 7 PCs of the data cluster in this situation. Hence, the data cluster is projected into the top 7 PCs by the compression layers. The compressed data cluster consists of 99.5 % of actual ship performance and navigation information under the top 7 PCs. Hence, the new data cluster preserves 99.5% of actual ship performance and navigation information with the compression ratio of 22% (i.e. 7/9 parameters).

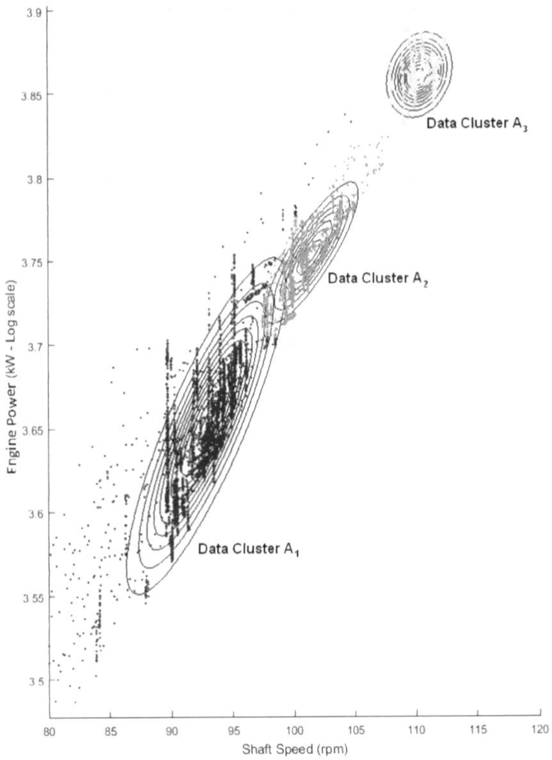

Figure 5: Engine Propeller Combinator Diagram with GMMs

4.3 INTEGRITY VERIFICATION-REGRESSION

The next layer of the data-handling framework consists of data integrity verification & regression. That can recover the parameter variations reported in the previous results. Furthermore, the integrity of ship performance and navigation data sets can improve by comparing it with other data sources (i.e. AIS, weather data, etc.). Hence, this layer can remove additional erroneous regions on the data sets. Erroneous data regions communicated by other networks are possible to identify by shore-based data centres in such situations. Furthermore, weather information along ship routes is available from the respective service providers, and that information can improve the integrity of ship performance and navigation data by this layer.

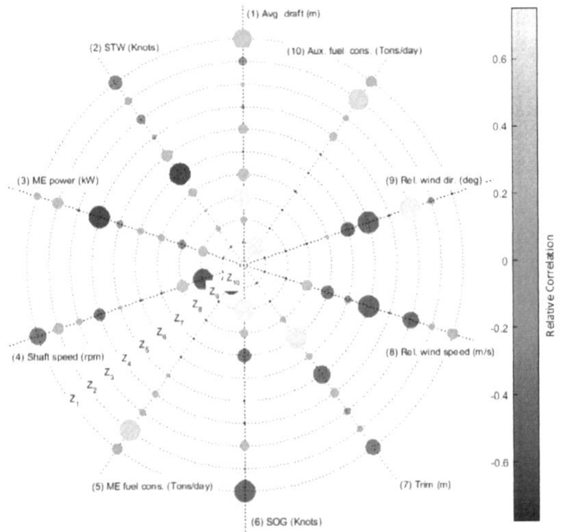

Figure 6: Principal Components of Data Cluster A_3

The next layer of data post-processing is data regression. The estimated data clusters of ship performance and navigation information may have parameter variations (see Figure 8), in which should eliminate by various smoothing algorithms. Therefore, the accuracy of estimated ship performance and navigation data can be further improved to reduce the parameter fluctuations by this layer.

4.4 VISUALIZATION-DECISION SUPPORT

The last layer of this data handling framework is data visualization and decision supporting that can be into energy efficiency and system reliability further divided. At this layer, appropriate visualization methods should reveal the improved quality in ship performance and navigation data by the proposed data handling framework. Less data scattering situations in these clusters due to the proposed data cleaning layers are expected. These data clusters consist of a higher dimensional space, where the respective PCs can also improve data visualization and decision supporting features as presented in Figure 6. These data clusters may project into the respective combined PCs to improve the visibility of vessel performance and navigation conditions as the future of this study.

5. CONCLUSIONS

An appropriate data handling framework is presented in this study and that consists of pre and post processing steps [13]. The proposed pre and post-processing steps can handle Big Data challenges on-board as well as shore based data centres under IoT as presented in this study. The pre-process can improve the quality and reduce the quantity of ship performance and navigation information. The post-process can further improve the quality and visualize the information, appropriately. Both processes consist of data handling layers that are vital to the proposed data handling framework. Hence, the pre-process can reduce the computational burden on the on-board crew, where improved and reduced data sets can be delivered to shore based data centres. The post-process can reduce the complexities on handling large-scale data sets, where high skilled crew to analyse such data sets must not always be presented in the data centres.

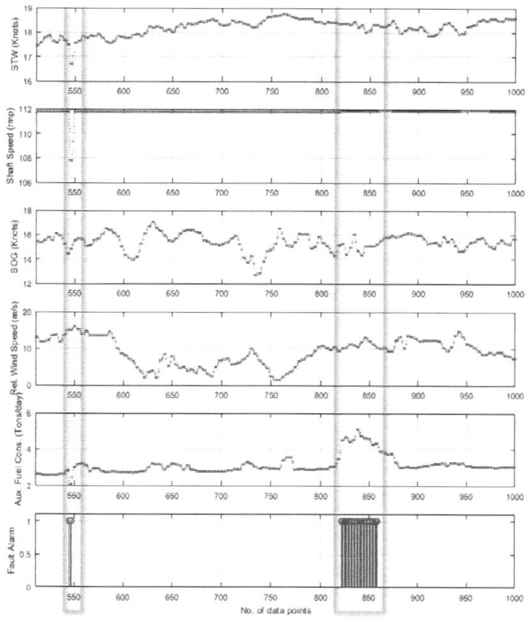

Figure 7: Sensor and DAQ fault detection of Data Cluster A_3

Data driven models in shipping facilitate the basis for such data handling framework. Those models are derived by identifying the respective data clusters and the structure of each data cluster of ship performance and navigation parameters. Vessel operating and navigation conditions are used to derive a meaningful structure into these digital models. Such ship performance and navigation models consist of discrete data clusters in a high dimensional space and categorize as digital models for shipping. Hence, this approach opens a novel path towards digitalizing sea going vessels under high dimensional space. Furthermore, the distribution of data clusters, the structure of each data cluster and their variations along the time line can be used to visualize time-varying vessel performance and navigation conditions. Such data driven models under the proposed data handling framework may have special features of self-learning (i.e. data clusters and the structure of each data cluster), self-cleaning (i.e. sensor and DAQ fault removal and compression, data recovery, data regression & integrity verification), self-compression & expansion (i.e. parameter reduction and expansion) as presented in this study. Furthermore, that have a multi-purpose structure for both energy efficiency and system reliability applications in shipping.

6. ACKNOWLEDGEMENTS

This work has been conducted under the project of "SFI Smart Maritime (237917/O30) - Norwegian Centre for improved energy-efficiency and reduced emissions from the maritime sector" that is partly funded by the Research Council of Norway.

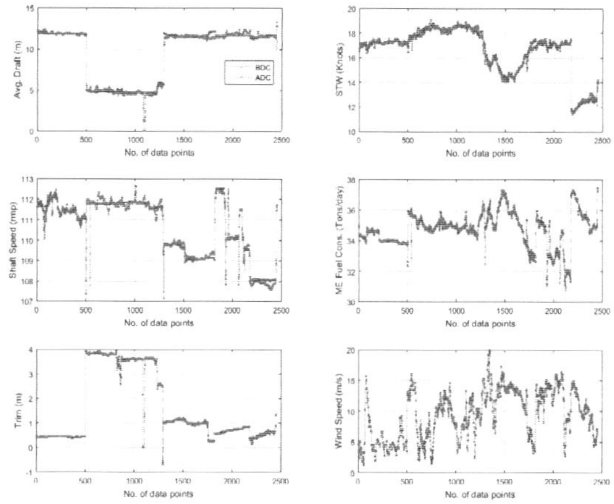

Figure 8: Data compression and Expansion of Data Cluster A_3

7. REFERENCES

1. L.P. Perera and C. Guedes Soares, "Weather Routing and Safe Ship Handling in the Future of Shipping," Journal of Ocean Engineering, 2017 (DOI: 10.1016/j.oceaneng.2016.09.007).

2. L.P. Perera and B. Mo, "Emission Control based Energy Efficiency Measures in Ship Operations," Journal of Applied Ocean Research, vol. 60, 2016, pp. 29-46.

3. O.J. Rodseth, L.P. Perera, and B. Mo, "Big data in shipping - Challenges and opportunities," the 15th International Conference on Computer Applications and Information Technology in the Maritime Industries (COMPIT 2016), Lecce, Italy, 2016.

4. L. P. Perera, and B. Mo, Data Analytics for Capturing Marine Engine Operating Regions for Ship Performance Monitoring, the 35th International Conference on Ocean, Offshore

and Arctic Engineering (OMAE 2016), Busan, Korea, (OMAE2016-54168), 2016.

5. IMO, "Development of an e-Navigation Strategy," IALA and e-navigation, sub-committee on Safety of Navigation, Report of the Correspondence Group on e-Navigation, NAV/53/13/3, 2007.

6. L.P. Perera, "Statistical Filter based Sensor and DAQ Fault Detection for Onboard Ship Performance and Navigation Monitoring Systems," the 8th IFAC Conference on Control Applications in Marine Systems (CAMS 2016), Trondheim, Norway, 2016.

7. L.P. Perera, and B. Mo, "Data Compression of Ship Performance and Navigation Information under Deep Learning," the 35th International Conference on Ocean, Offshore and Arctic Engineering (OMAE 2016), Busan, Korea, June, 2016 (OMAE2016-54093).

8. L.P. Perera and B. Mo, "Data Analytics for Capturing Marine Engine Operating Regions for Ship Performance Monitoring," the 35th International Conference on Ocean, Offshore and Arctic Engineering (OMAE 2016), Busan, Korea, 2016 (OMAE2016-54168).

9. L.P. Perera and B. Mo. Marine Engine Operating Regions under Principal Component Analysis to evaluate Ship Performance and Navigation Behavior, the 8th IFAC Conference on Control Applications in Marine Systems (CAMS 2016), Trondheim, Norway, 2016.

10. A. Ng, "Mixtures of Gaussians and the EM algorithm," Lecture notes on Machine Learning, 2015.

11. T.K. Moon, "The expectation-maximization algorithm," Signal Processing Magazine, IEEE , vol. 13, no. 6, 1996, pp. 47-60.

12. J. E. Jackson, "Principal components and factor analysis: part i-principal components." Journal of Quality Technology, vol. 12, no. 4, 1980, pp. 201–213.

13. J. Zhang, F. Y. Wang, K. Wang, W. H. Lin, X. Xu and C. Chen, "Data-Driven Intelligent Transportation Systems: A Survey," IEEE Transactions on Intelligent Transportation Systems, vol. 12, no. 4, 2011, pp. 1624-1639.

8. AUTHORS BIOGRAPHY

Lokukaluge P. Perera received the B.Sc. and M.Sc. degrees in Mechanical Engineering and Systems & Controls from the Oklahoma State University, USA, in 1999 and 2001, and the Ph.D. degree in Naval Architecture and Marine Engineering from the Technical University of Lisbon, Portugal, in 2012, respectively. He has won Doctoral and Postdoctoral Fellowships from the Portuguese Foundation for Science and Technology in 2008 and 2012. Currently, he is a Research Scientist at the SINTEF Ocean, Norway. His research experience, prior to the SINTEF Ocean, includes the Centre for Marine Technology and Engineering (from 2008 to 2012), Portugal and the Advanced Technology Research Center (from 1998 to 2001), USA. His industrial experience includes Wartsila Finland, Finland (from 2012 to 2014). His academic experience includes the Naval & Maritime Academy, Sri Lanka as the Director of Academic Studies (from 2005 to 2008) and the Ocean University of Sri Lanka, Sri Lanka as the Acting Head/Lecturer (from 2003 to 2005). Furthermore, Dr. Perera was a visiting lecture (from 2001 to 2005) for: University of Ruhuna, University of Moratuwa, Open University of Sri Lanka, and Colombo International Nautical & Engineering College, Sri lanka.

ACOUSTIC EMISSION TECHNIQUES MONITORING CRITICAL VESSEL EQUIPMENT: THE ALTERNATOR BEARING MONITORING SYSTEM

Cristiano Garau Lead Application Engineer, Parker Kittiwake
Neil Randall, Technical Specialist, Parker Kittiwake

SUMMARY

The current economic climate in the shipping industry has forced ship owners and operators to rethink their organisation and budgets in a continuous effort to rationalise costs and prevent down time. Protecting vital equipment and maximising a vessel's operational efficiency is now more than ever, a key priority. It is within this context that effective condition monitoring tools and techniques have become increasingly important to help the industry to manage, avoid or mitigate these costly issues. Over the last decade, new condition monitoring technology has been developed and refined. Their use is advancing at rapid pace in the shipping industry.

The practice of condition monitoring has evolved over the years from the days of engineers relying solely on their experience, senses and intuition, to the aid provided by the sophisticated online sensing technology available today. This technological advancement has enabled ship owners and operators to plan and manage maintenance requirements with the least possible impact on the vessel schedule and helping in minimising the associated costs.

This paper will explain how to employ an effective condition monitoring technique based on acoustic emissions. Owners and operators will the tools they need to avoid catastrophic damage on key vessel equipment such as the main generator by proactively prompting maintenance tasks and ultimately prevent unexpected downtime. To illustrate a real world application, this paper will also include a case study with Gram Car Carriers AS, explaining how the company has benefited from employing acoustic emissions techniques in preventing alternator bearings failure.

1. INTRODUCTION

Acoustic Emission (A.E.) is a Condition Monitoring (C.M.) technique based on the detection of the high frequency sounds produced by impacts and friction generated within a defective bearing or gear as it is condition deteriorates. The elastic waves (not airborne in the audible range) generated in this condition travel readily from the bearings and gears through the structure of the machine. Here they are detected by the A.E. sensors mounted on the surface of the monitored asset.
The aim of this technique is to detect faults at a much earlier stage than other techniques based on more macroscopic changes, such as temperature changes and rms vibration level, so that a prompt remedial action can be taken in order to minimize the possibility of a sudden failure or the occurrence of secondary damage.

Vibration Monitoring (based on audio frequency detection) captures a wide spectra of frequencies typically high signal levels at low frequencies; this very useful technique although being the most used C.M. technique suffers from a low signal to noise ratios which implies the use of complex signal manipulation techniques such as Fast Fourier Transform (F.F.T). The A.E technique employ sensors that operate at higher frequency detection than standard accelerometers, this choice has the distinct advantage on most rotating machinery of having a high signal to noise ratio.

Figure 1[1] shows the response of an accelerometer (normally operating in a 2Hz-20 kHz spectrum) Vs an A.E. sensor (normally operating in a 10-100 kHz spectrum) to minor impacts generated by a small solder bead on a plate to which a running motor is attached. This graphs shows how the A.E. detected signal has a higher signal to noise ratio compared to Vibration data analysis for the detection of minor impacts in a noisy background. The same concept also applies in the cases of a frictional source.

It is important to observe that in taking A.E. measurements the choice of detection frequency is important but not critical. The output from the A.E. sensor can be processed in a number of different ways; in this paper data will be shown either as signal envelope or standard parameters.

The main advantages when enveloping an A.E. signal are based on a lower sampling rate required, lower cost of data capture and overall easier data to handle. However this signal manipulation technique requires some compromises in terms of lost raw frequencies, although the useful information is in the way the signal is modulated not in the raw frequency detection.

Because of the nature of the measurement, A.E. sensors do not detect the low frequency vibrations associated

[1] Acoustic Emission as a Basis for the Condition Monitoring of Industrial Machinery
Trevor J. Holroyd (PhD BSc FlnstNDT)

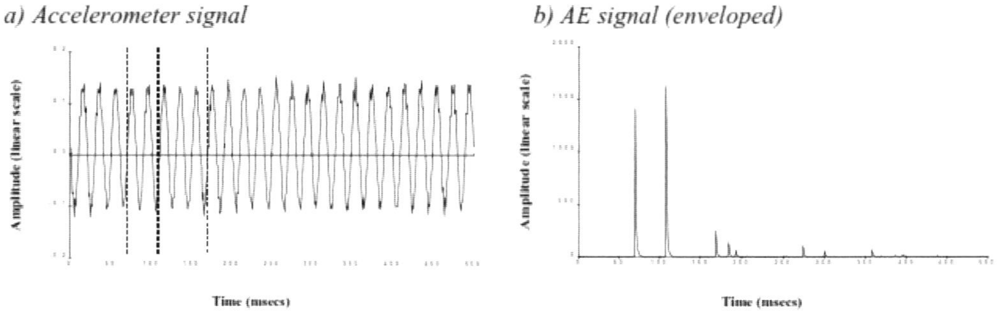

Fig 1 - Accelerometer and AE signals from impacts on a plate attached to a running motor.

with the normal running of the ship. These sensors are also immune to the high levels of audible noise normally present in the engine room.

The main A.E. technique employed in this paper is simple to use and gives a clear indication when the monitored component is developing a fault. The A.E. signature detected by the sensors is analysed in terms of two parameters: Distress® and dB Level. These parameters are used in combination to trend the degradation of the component.

Distress® is a measure of the transient activity (impacts and frictions) generated by the monitored bearing or gear. The parameter is extremely sensitive to the early stages of wear and it can be used also to detect lubrication problems before any permanent damage occurs. Distress® is a dimensionless parameter with a scale of 0-100, capable to track the progression of any damage that may occur.

dB Level is the mean level of the A.E. signature, it is used when damage becomes more severe and can track it to final failure. dB Level is normally expressed in a 0-100 scale and it is influenced by the rotational speed of the monitored component: as a rule of thumb, doubling the rotational speed increases the dB level of 12 dB, halving the rotational speed decreases the dB level of 12 dB.

For rotational speeds greater than 35 rpm using the Distress® and dB Level allows the operator to have a direct and easy way to assess the asset bearing condition. In general a value of Distress® less than 5 is indicative of a bearing in very good condition. A Distress® value between 5 and 10 is also indicative of a bearing in good health. A value of Distress® 10 or above should be considered suspect indicating possible lubrication problems and/or ongoing damage.

dB level, because its dependence on rotational speed, is used comparatively as trending parameter. A sharp increase of dB level of 25 units is normally indicative of an incumbent failure. Fig 2 shows how Distress® and dB may vary through the life time of a component.

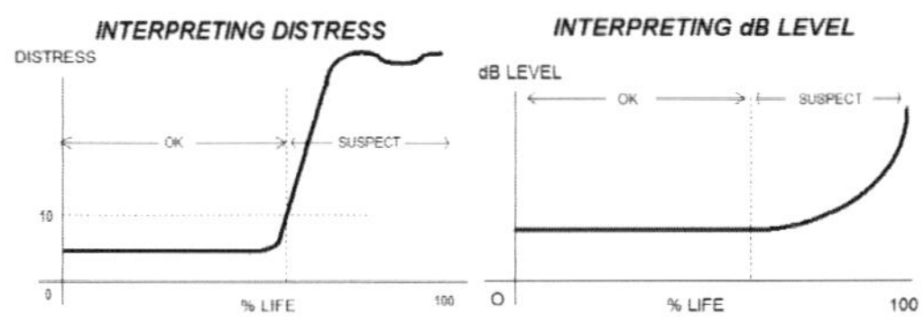

Fig 2: Idealised plots of how Distress® and dB Level vary throughout the life of a typical machine during the operational life of a component

2. CASE STUDY – GRAM CAR CARRIERS AS

Gram Car Carriers AS (GCC) is a substantial tonnage supplier to the world's major operators of sea-borne vehicle transportation. The Company owns a quite diversified fleet of PCC (Pure Car Carrier) RORO (Roll On-Roll Off) vessels that deliver vehicles globally.

GCC has installed Parker Kittiwake's Alternator Bearing Monitoring System (ABMS) on 4 sister vessels: Viking Constanza, Hoegh Caribia , City of Oslo and Viking Odessa. The vessels, are PCC RORO with a gross tonnage of 20209, able to carry up to 2000 cars at the time. The vessels propulsion is provided by two 6 cylinder medium speed MAK diesel engines coupled to the input shafts of a RENK gearbox linked to the propulsion shaft. The gearbox also has an auxiliary output shaft linked to the vessel's main generator. Two auxiliary generators can be used when the vessels are in port during the cargo loading/offloading operations. Three of these vessels, all made in Japan between 2009 and 2010 by a top quality shipyard, suffered alternator bearing failures within the first year of life. These failures were catastrophic and a real logistical nightmare to resolve. This was translated in significant costs to the Owners (well into 6 figures) including hire and cost of additional tugs (due no power available to operate the electric bow thrusters). Once repaired, all ships then ran smoothly until 2015 when another bearing failure was experienced with catastrophic consequences. Further checks on the three remaining sister vessels revealed similar failures starting to develop, leading to those assets having to come out of service again. The cause and failure mode of these events were not fully understood and very difficult to predict. Different causes could have led to these failures such as the use of incorrect materials, misalignment issues, alternator design issues, vibrations, vessel structural problems etc.

DNV-GL was also asked to advise on such failures however it was very difficult for Class to pinpoint the exact causes of the issues and suggest an adequate corrective action.

At that stage, for obvious reasons that are not just driven by economic factors, Gram Car Carriers decided to consider the use of online condition monitoring techniques. The choice was narrowed down between acoustic emission and vibration monitoring. After careful consideration GCC, in light of the previous offline experiences, characteristics and design of the vessels opted to trial the Alternator Bearing monitoring System based on acoustic emission sensors.

3. MONITORING ALTERNATOR BEARINGS WITH ACOUSTIC EMISSIONS SENSORS

The chosen trial vessel was the Viking Constanza. The vessel had a newly fitted set of alternator bearings so it would provide a good baseline in the case of a sudden change in the monitored parameters. The Alternator Bearing Monitoring System is designed around a data acquisition system that displays and logs the data streamed by the A.E. sensors in real time 24/7. The system supports the Modbus over TCP/IP protocol so can be linked to the vessel's LAN (Local Area Network) and it is able to trend the Distress® and dB parameters in graphs, providing also visual alarms set on previously determined levels.

Two A.E. Sigma sensors were installed on the drive end (DE) and non-drive end (NDE bearings) as shown in Figure 3. The A.E. Sigma Sensors provide Distress® and dB parameters but are also able to capture the full A.E. signature of the monitored component. This information can be analysed using F.F.T. in order to provide a more in depth analysis of the root causes of a developing failure.

The first three month of data from the Viking Constanza proved immediately to be very useful in showing the vessel's Owners how the system was able to detect the stress on the bearings given by normal operational events such as bad weather or manoeuvring in shallow water in Bremerhaven. The system was also able to detect the effective lubrication of the bearings. All this new wealth of information convinced GCC to implement the Alternator Bearing Monitoring System in the remaining three vessels. The ability to easily confirm that the monitored asset is behaving as expected is very important. It allows the operator to avoid unnecessary

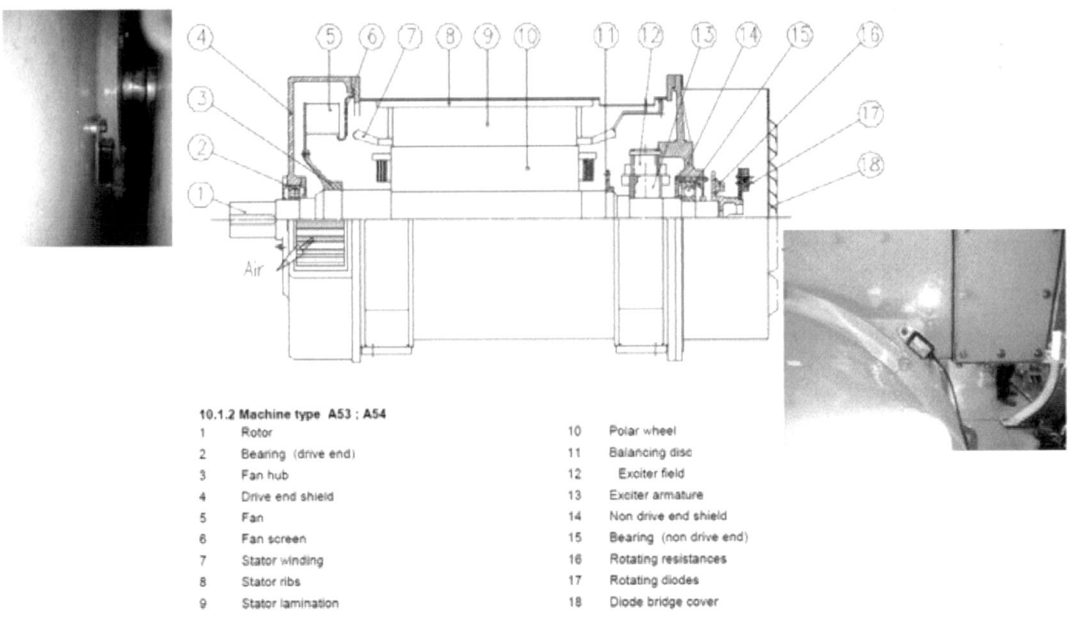

Fig 3: Installation points on the main generator DE and NDE bearings of the A.E. Sigma Sensors

maintenance and adopt a condition based maintenance approach rather than one based on time schedule. This clearly reduces costs but also avoids the possibility of introducing faults to a system that is actually running perfectly well.

Cross referencing data from the four vessels allowed a direct comparison between the assets showing a significant deviation from the expected values on the City of Oslo.

The Viking Constanza, Hoegh Caribia and Viking Odessa all recorded for either DE and NDE bearings low levels of Distress® (<10) and reasonable levels of dB.

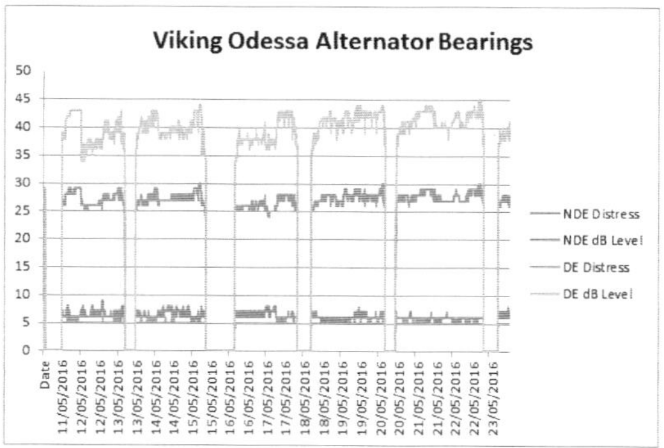

Fig 4: Viking Odessa data

The data from the City of Oslo, since the day of commissioning on the 14/05/2016, was rather different as shown in Fig 5. Distress® for the NDE bearing was on average well above 20, for DE bearing was above 10. dB levels for both bearings were also higher if compared to the other vessels reinforcing the suspect that one of the two may be seriously damaged. On the 17/05/2016 both bearings were re-greased but such task did not show any noticeable changes in the data recorded.

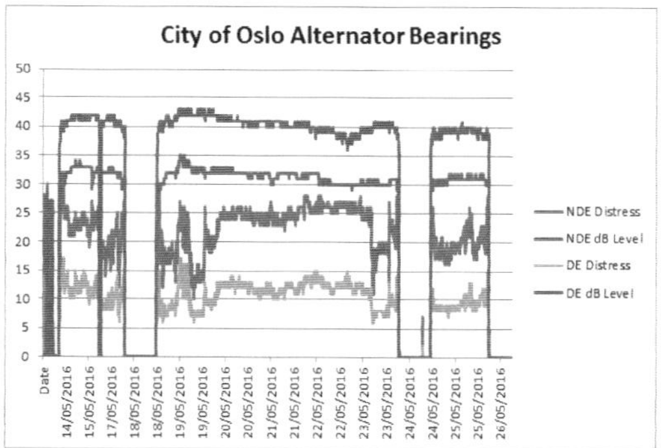

Fig 5: City of Oslo data

After a month in operation with slowly deteriorating conditions GCC decided to change the NDE bearing on the 17/06/2016: NDE bearing changed shows a moderate damage to bearing surfaces and no significant reduction in Distress® readings. GCC then plans to change DE Bearing. Figure 6 shows the timeline of fault detection and how the issue was fixed tracking the changes in A.E. signature

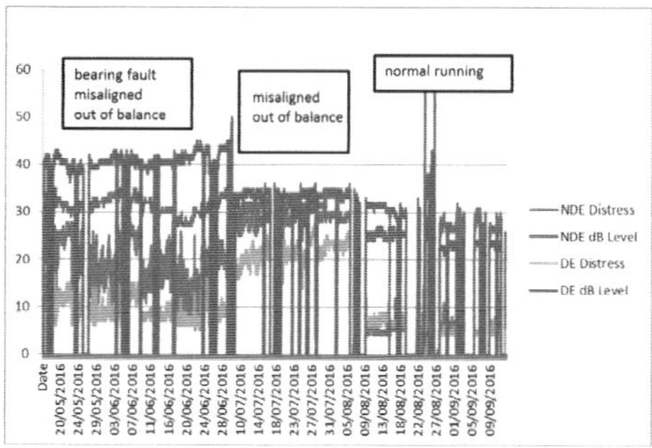

Fig 6: City of Oslo data showing detecting the fault and correcting the issues

On the 01/07/2016 the DE bearing shows the signs of rapid deterioration increasing Distress® of 10 units and an increase of 6dB over only 2 hours. The recoded bearing temperature also increases from 61 to 83° C but significantly later if compared to A.E. At this point, the Owner made the decision to put the alternator offline and run the vessel on the auxiliary generators until the DE bearing could be replaced. On the 8/07/2016 the DE bearing is changed. The inspection confirms the inner race to be severely damaged by metal fatigue as shown in figure 7. The dB Level which is used to track severe damage has reduced by 16 dB. This is a significant reduction, however the Distress® readings on both bearings are now higher than before that this maintenance task was carried out. From the data it was evident that the mechanism causing the damage was still present even if both bearings were replaced.

Fig 7: DE bearing showing inner race metal fatigue

On the 09/07/2016 it was decided to investigate further using the diagnostic feature available in the Alternator Bearing Monitoring System: the A.E. Sigma Sensors can capture the full A.E. signature from the bearings. This data can be analysed via F.F.T. to give more information about the root causes of the fault. Fig 8 shows the waveform from the NDE bearing taken on the 28/07/2016.

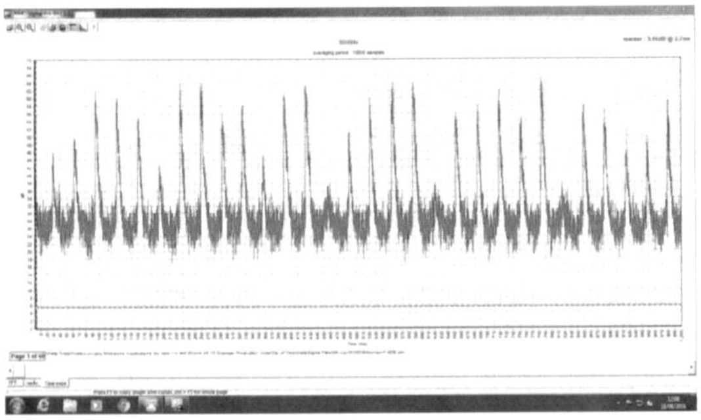

Fig 8: NDE bearing A.E. signature taken on the 28/07/2016

From this data it is possible to clearly see the repetitive impacts occurring every 30 times per second. This matches the shaft rotation frequency and indicates that there is an alignment/balance problem. The F.F.T. spectra of this signal is shown in Fig 9 confirms clearly the presence of a 30 Hz harmonic.

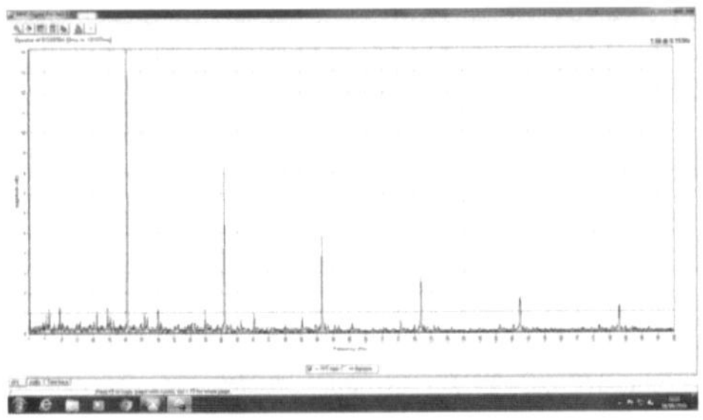

Fig 9: NDE bearing A.E. signature FFT taken on the 28/07/2016 showing 30 Hz peak

On the 22/07/2016 a vibration data analysis of the asset confirms that the relative positioning of the gearbox and alternator have changed since the installation hence corrective measure will be required.

On the 08/08/2016 a specialist of Laser realignment and balancing is called in. Figure 6 shows how, after re-alignment and re-balance, Distress® readings immediately dropped to below 10 and it is also recorded a slight reduction in dB Level.

The data of the 15/08/2016 confirms that the data from the City of Oslo are now in line with those from the other sister vessels.

4. CONCLUSIONS

The proven benefits of a condition based, rather than a planned time based maintenance strategy are: reduced costs and increased reliability. Asset owners like Gram Car Carriers AS as well as operators can now focus on the most important functions of the system and only intervene when required. They can also avoid maintenance tasks that are at best unnecessary and at worst damaging.

The installation of online condition monitoring tools, like the Alternator Bearing Monitoring System, used in combination with experience and asset knowledge, is the key to productive, condition based maintenance. By effectively analysing and using the readily available data provided by systems like the ABMS, the asset Owner can make educated decisions, saving time and reducing downtime. It quickly becomes clear that that the savings delivered far outweigh the initial capital investment required to set up these useful tools.

5. ACKNOWLEDGEMENTS

A very special thank you to Mr. Neil Randall for the excellent work done in the Gram Car Trial and for data analysis provided for this paper.

The authors would like to thank Gram Car Carriers AS for granting permission to use A.E. data extracted from the vessels mentioned in this paper, special thanks to Mr. Bryan Slaughter.

SHIP MACHINERY CONDITION MONITORING USING PERFORMANCE DATA THROUGH SUPERVISED LEARNING

C Gkerekos, I Lazakis and **G Theotokatos,** University of Strathclyde, Glasgow, UK

SUMMARY

This paper aims to present a methodology for intelligent monitoring of marine machinery using performance data. Monitoring of machinery condition is a crucial aspect of maintenance optimisation that is required for the vessel operation to remain sustainable and profitable. The proposed methodology will train models pertinent to specific machinery components using pre-classified performance data and then classify new data points using the models developed. For this, measurements are suitably analysed and processed to retain most of the information (variance) of the original dataset while minimising number of required dimensions. Finally, new data are compared against the models developed to evaluate their condition. The above will provide a flexible but robust framework for the early detection of emerging machinery faults. This will lead to minimisation of ship downtime and increase of the ship's operability and income through operational enhancement. Case studies that show initial results obtained through main engine data are included.

NOMENCLATURE

INCASS	Inspection Capabilities for Enhanced Ship Safety (EU FP7 Project)
C.F.W.	Cooling Fresh Water
C.W.	Cooling Water
L.O.	Lube Oil
MCR	Maximum Continuous Rating
M/E	Main Engine
NAOME	Naval Architecture, Ocean and Marine Engineering
NN	Neural Network
OEM	Original Equipment Manufacturer
PCA	Principal Component Analysis
SCADA	Supervisory Control and Data Acquisition
SRM	Structural Risk Minimisation
SVM	Support Vector Machine

1. INTRODUCTION

Ships are a significant asset of the global goods transportation system as over four-fifths of merchandise are carried by sea [1]. Current financial situation of the shipping industry combined with an average global merchant fleet vessel age of almost twenty years [1] make clear that a high level of operations optimisation is required for the vessel to remain sustainable and profitable.

Maintenance of a ship's machinery components can substantially affect the ship's sustainability and profitability. Meanwhile, current maintenance state-of-practice in shipping offers ample room for improvement. As such, the introduction of novel methods of monitoring the condition of machinery equipment, suggesting suitable maintenance actions, and scheduling those actions in an optimised fashion is significant.

Three main maintenance types exist: reactive, preventive, and predictive. Reactive maintenance concerns maintenance that is only performed once a component fails completely. Preventive maintenance refers to maintenance that happens at a fixed frequency, usually following Original Equipment Manufacturers (OEMs) recommendations. Preventive maintenance offers many benefits compared to reactive. However, preventive maintenance also exhibits multiple shortcomings such as high maintenance costs and significant (planned) downtime. An optimised maintenance scheme would offer extended machine lifespan, coupled with reduced maintenance costs and downtime. Such schemes are usually classified under predictive maintenance. Nevertheless, prerequisite for the development of any predictive maintenance system is a condition monitoring framework that can accurately estimate the condition of monitored systems, subsystems and components. This framework takes as input several measurements and analyses them appropriately to return an estimation of their condition as output. Two customary sets of measurements for condition monitoring are performance and vibration measurements. Performance measurements are a valid basis for the estimation of reciprocating machinery while offering the additional benefit of automatic acquisition in most applications. In the case of rotating machinery, vibration measurements can offer good insights but usually need to be manually acquired using specialised equipment.

While predictive maintenance is widely used in other fields such as nuclear power production and aerospace, there are not many applications in the marine field. Currently, most maintenance actions carried on board vessels can be classified as preventive maintenance. Hence, this paper aims to present the development of a framework concerning the processing of performance data and training of appropriate models for the condition monitoring of marine machinery.

Section 1 introduces the paper's scope and motivation of research. Section 2 refers to the research background. Section 3 elaborates on the proposed methodology concerning dimensionality reduction, data processing, and model training. Section 4 details the setup of multiple case studies used to validate the proposed methodology focusing on different main engine (M/E) components. Section 5 presents and discusses the results obtained through these case studies. Finally, in section 6, overall conclusions are provided along with further research steps.

2. RESEARCH BACKGROUND

In general, three types of maintenance are applicable for machinery applications: reactive, preventive, and predictive or condition-based.

Reactive (also known as run-to-failure, breakdown or corrective) maintenance concerns maintenance that is only performed following the complete failure of a component. At that point, no repairing is possible and the component is replaced by a new one [2]. In some cases, repairing is possible, albeit with a significantly increased cost as a large spare-parts inventory is required [3, 4]. This method of maintenance offers provides the longest time between shutdowns but failures are catastrophic and can possibly affect multiple components and/or machines [4]. Hence, reactive maintenance is mainly applied to relatively not expensive and non-critical machines or where redundancies have been implemented so that production is not interrupted.

Preventive maintenance refers to maintenance that happens at a fixed frequency, usually following Original Equipment Manufacturers (OEMs) recommendations. Compared to reactive maintenance, preventive maintenance offers significant increase in machine lifespan. This is because the probability of catastrophic failures is diminished. Additionally, preventive maintenance is more cost-effective as the number of components or machine that need complete replacement is reduced. Moreover, as a considerable tranche of maintenance is performed as a precaution and before the perception of any defects, unplanned downtime is reduced. Preventive maintenance generally aims to provide such maintenance intervals that only 1-2% of machinery experience failures between maintenance intervals [4]. Thus, the clear majority of machines would be able to continue working without maintenance for multiple maintenance intervals. This introduces increased "infant mortality" in machines, due to faults that would otherwise have been avoided [4]. Infant mortality concerns both failures caused by faulty replacements and by general tampering during maintenance activities. Besides, excessive maintenance causes significant, albeit planned, down- time. On top of that, unexpected failures still occur as maintenance happens at a fixed frequency, without taking into consideration the actual machine condition.

Predictive maintenance provides a more intelligent method of maintenance planning. There, present and past condition of each component is taken into consideration to offer bespoke maintenance scheduling for each component and each machine. Predictive maintenance requires a higher expenditure at installation but over an extended period, becomes more economical than preventive or reactive maintenance. Especially in industries where machines are expected to run for long periods without any shutdowns, it has been shown that predictive maintenance can reduce relevant costs by up to 65% [5]. Furthermore, in terms of downtime, planned downtime is minimised to the bare necessary minimum and unplanned is almost diminished. This optimised maintenance scheduling permits the maximisation of machine lifespan. Still, while predictive maintenance proves to be more economical during a machine's lifespan, results take years to show.

2.1 MAINTENANCE IN THE MARITIME SECTOR

In sectors such as defence, aviation, manufacturing, automobile, and nuclear power production, maintenance focus has recently shifted from reactive to preventive/predictive. Ship maintenance amounts to 10-15% of the shipping company direct operating costs [7]. However, in the maritime sector, ship maintenance has been considered an area of needless expenditure and advanced monitoring methods have not yet been widely applied [8]. Nevertheless, some attempts towards predictive maintenance in shipping have been made in the past few years. For example, a methodology where vibration data are combined with performance data (cylinder pressures) for the condition monitoring of a main engine has been suggested [9]. Accordingly, a thermodynamic model of a main engine has been developed to perform condition monitoring using cylinder pressure traces [10]. Besides, a self-learning algorithm for fault diagnosis in the combustion system of a marine diesel engine has been developed [11]. Furthermore, a self-learning model for the condition monitoring of ship machinery based on vibration measurements was developed in [20].

2.2 PERFORMANCE MONITORING OF MACHINERY

Performance monitoring of machinery is a problem that requires the development of a suitable model. This model can either use a first-principles analysis (i.e. white-box model), or use a more 'brute-force' approach by developing a model using self-learning algorithms coupled with an acquired dataset (i.e. black-box model). Often, a combination of both techniques is applied, leading to grey-box models. [12] developed a framework for the analysis of data acquired through wind turbine SCADA (Supervisory Control and Data Acquisition) bus measurements to perform condition monitoring based on correlations between measurements. [13] and [14] both presented an overview of Support Vector Machine (SVM) techniques for fault diagnosis and monitoring in engineering applications.

3. SUGGESTED METHODOLOGY

The methodology elaborated in this paper concerns a) the description of a suitable pre-processing technique for the acquired dataset and b) the development of a self-learning model that can estimate whether a given data point corresponds to a reference (nominal) condition considered during training. As such, a self-learning model can be trained without the need of obtaining data corresponding to faulty conditions. A visual representation of the suggest methodology is presented in Fig. 1.

A number of performance measurements related to Main Engine operation are used as input. These are depicted in Table 1. Additionally, engine MCR and vessel speed were taken into consideration. This permits the development of a model that performs accurately in varying operating conditions.

Table 1: Performance measurements utilised as input for model training.

Component	Measurement description
Cylinder 1-8	Exhaust Gas Outlet Temperature
	Scavenging Air Temperature
	Jacket C.F.W. Outlet Temperature
Thrust Bearing	L.O. Outlet Temperature
Fore Camshaft Bearing	Temperature
Scavenging Air Manifold	Pressure
Air Cooler	C.W. Inlet Pressure
Fuel Oil	Inlet Pressure
	Inlet Temperature
Cylinder Jacket	C.F.W. Inlet Pressure
Turbocharger	L.O. Inlet Pressure
Piston	Cooling Oil Inlet Pressure
Lube Oil	Inlet Pressure
	Inlet Temperature

3.1 DATA PRE-PROCESSING

Acquired datasets need to be pre-processed before being used for model training so that any erroneous or missing measurements are rectified. Data pre-processing is currently a hot topic in data mining and predictive analytics, with cutting edge research focusing on optimisation and automatisation of pre-processing.

In cases of smaller, manually acquired datasets, an alternative pre-processing is performed using visual red-flags. Such red flags are elaborated in [23] and summarised in Table 2.

Methods for imputing missing or erroneous data points are described in [23, 24]. A straight-forward approach, especially valid when large datasets are available, is to completely discard any instance that contains missing features. Alternatively, any missing feature can be replaced by the mean or mode (i.e. most commonly found) value of that feature, taking into consideration the whole dataset. Besides, a regression model can be trained using the remaining data points as training and then use known instance features as input so that missing features are estimated as model output.

Table 2: Examples of data pre-processing red flags [23].

Problems	Metadata	Examples/Heuristics
Illegal values	Cardinality	e.g., cardinality (gender) > 2 indicates problem
	Max, min	Max, min should not be outside permissible range
	Variance, deviation	Variance, deviation of statistical values should not be higher than threshold
Misspellings	Feature values	Sorting on values often brings misspelled values next to correct values

3.2 DIMENSIONALITY REDUCTION

Different performance measurement variables are usually correlated. This undesirably augments model's complexity. At the same time, increases the number of data points required for training as the number of data points should exceed the number of features [16, 17]. In order to facilitate model training, dimensionality reduction techniques such as Principal Component Analysis (PCA) are applied [18, 19].

PCA provides the orthogonal transformation of possibly correlated variables into a set of linearly uncorrelated ones (principal components). To perform PCA on a given dataset, the following steps are required [21]:

- The mean of each data dimension is calculated and subtracted from the original dataset to obtain the adjusted dataset.
- The dataset's covariance matrix is calculated.
- Eigenvalues and unit eigenvectors of the covariance matrix are calculated.
- Eigenvectors are sorted by eigenvalue, highest to lowest. A number n of features is selected based on the explained variance – complexity trade-off and a feature vector is obtained by combining the first n eigenvectors.
- The post-PCA dataset matrix is obtained by multiplying the transpose of the feature vector by the transpose of the adjusted dataset matrix.

By only retaining a few features, the dimensionality of the modelling is reduced providing a basis for better results given a limited number of samples n. Nevertheless, at the same time there exists the inherent trade-off where some variance from the dataset is lost. As such, an optimal number of principal components should be selected so that most of the dataset's variance remains explained while the total number of features is reduced.

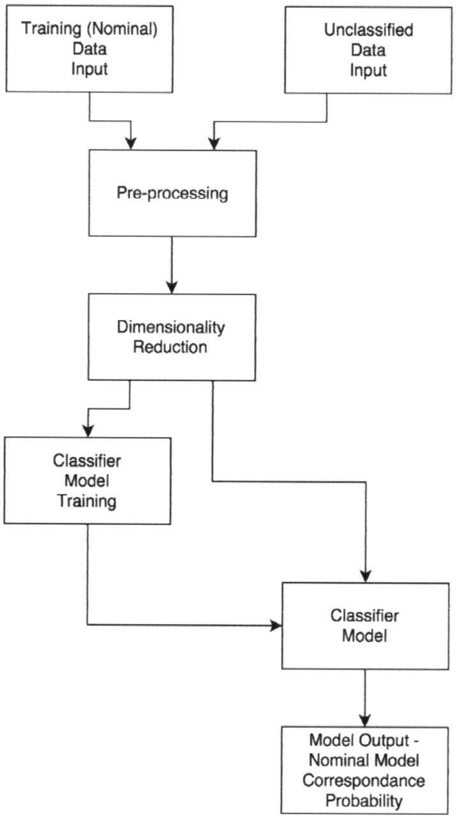

Figure 1: Methodology visual representation

3.3 MODEL TRAINING

Once certain features are obtained through PCA, these features are used as input for model training. Model in this case is a classifier that returns as output the probability that a given data point corresponds to the reference (nominal) dataset used for training. Using a training dataset that is representative of typical vessel operation and that is spanning different operational profiles, this classifier can identify abnormal patterns.

Most common self-learning classifiers are based on Neural Networks (NN), Support Vector Machines (SVM) and Decision Trees.

Compared to other pertinent algorithms, SVM is the most suitable when treating small datasets [13]. SVMs are based on Structural Risk Minimisation (SRM) that leads to balancing model complexity against overfitting [22]. Another advantage is that any optimisation minimum achieved through SVM will be a global minimum, something that is not necessarily true when treating NN minima.

Traditionally, SVMs aim to classify a dataset used as input into two of more classes. This is done by developing a separating hyperplane that can classify input data into different classes. This can either be done linearly or through a non-linear kernel function. In the case of non-linear kernel functions, input data are mapped to a high-dimensional feature-space where linear classification is possible.

In the case of the described methodology, one-class training is used. There, the dataset used for training is considered part of the single class and a small number of measurements (typically in the range of 5%) are classified as outliers to define the boundaries of the class. This algorithm builds a model where for each set of points used as input, a number in the 0-1 range is output. This number expresses the probability of the set of points corresponding to the defined class.

The exact mathematical formulation of SVM is not elaborated in this paper but the reader can refer to either Vapnik's seminal work [22] or any book in Data Mining, e.g. [21].

4. METHODOLOGY APPLICATION

In this section, four case studies are presented. Each case study performs a sensitivity analysis by considering an abnormal measurement that corresponds to a specific machine component and evaluating the model results. The dataset used for these applications was acquired on board a 4500 TEU containership as part of EU FP7 INCASS (Inspection Capabilities for Enhanced Ship Safety) Project measuring campaign [15]. Measurements were obtained hourly during a two-day period. During that time, vessel was super slow steaming with a relatively constant speed of approximately 11.5 knots, corresponding to 12% of engine MCR (Maximum Continuous Rating) point. As the dataset used was manually acquired with no missing values and no sensor malfunctions, no pre-processing was required.

4.1 CASE STUDY 1 (CAMSHAFT BEARING TEMPERATURE)

This case study evaluates the performance of the developed model by increasing the camshaft bearing temperature that is used as one of the algorithm inputs while maintaining all other variables in their original range. As such, while the reference (nominal) dataset has an average camshaft bearing temperature of 46 °C, gradually increasing temperatures of up to 92 °C are used as input.

4.2 CASE STUDY 2 (M/E CYL #1 EXHAUST GAS TEMPERATURE)

Following the case study set up described above, this second case study varies the temperature of the exhaust gas of a selected cylinder. The reference (nominal) dataset has an average temperature of 260 °C for the exhaust gases of this cylinder. Temperatures are gradually increased up to 520 °C and model results are evaluated.

4.3 CASE STUDY 3 (M/E LUBE OIL INLET TEMPERATURE)

In this third case study, lube oil inlet temperature is varied as part of the performed sensitivity analysis. The reference (nominal) dataset has an average temperature of 45 °C. Temperatures are gradually increased up to 58 °C and model results are again evaluated.

4.4 CASE STUDY 4 (M/E FUEL OIL INLET TEMPERATURE)

In this final case study, fuel oil inlet temperature is varied as part of the performed sensitivity analysis. The reference (nominal) dataset has an average temperature of 137 °C. Temperatures are gradually increased up to 178 °C and model results are again evaluated.

5. RESULTS

This section presents and discusses the results obtained through the case studies described above, in Section 4. Results of all case studies are visualized, showing the algorithm output for different average input measurements. Additionally, upper acceptable limit and alarm threshold values are depicted for each component, as provided by OEM.

Overall results obtained follow OEM recommendations. In some cases, the algorithm seems to be too aggressive, offering a probability drop inside the acceptable range. That can be attributed to the inherent algorithm function: estimating the probability of a measurement corresponding to a reference condition. As the training dataset in the case of these case studies only considers a very narrow, super-slow steaming operational profile, any measurements beyond that (albeit acceptable overall) do not match this profile. Additionally, whereas OEM values reflect overall limits that are not revised depending on machinery operating profile, algorithm output takes that into consideration.

5.1 CASE STUDY 1 (CAMSHAFT BEARING TEMPERATURE)

In this case study, model performance in varying camshaft bearing temperatures was evaluated. As camshaft bearings offer no redundancies, accurate monitoring of condition is crucial for same operation of vessel. The model accurately returns a probability drop as temperature increases (Fig. 2). A value of around 35% is returned at the OEM upper acceptable level, with 15% returned at the OEM alarm threshold. This demonstrates algorithm's ability to attribute lower correspondence probabilities as unclassified input data starts to diverge from the data used for training. In general, these probabilities can be converted into binary classification using a suitable sigmoid function [21]. In that case, probabilities over 50% will be given value 1, denoting an input dataset with a high similarity to the nominal one, and probabilities below 50% will be given value 0, denoting substantial disparities between the two datasets.

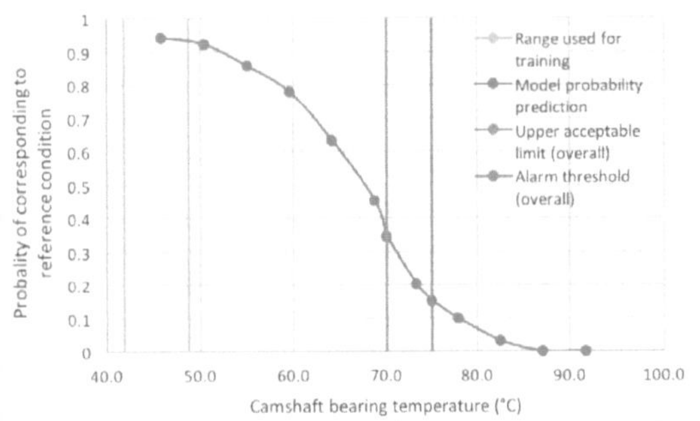

Figure 2: Camshaft bearing temperature sensitivity analysis

5.2 CASE STUDY 2 (M/E CYL #1 EXHAUST GAS TEMPERATURE)

In this case study, model performance in varying exhaust gas temperatures of a specific cylinder is evaluated. Input for all other cylinders is retained nominal. In this case, low values of 5-10% are returned for OEM acceptable limit and alarm threshold (Fig. 3). However, there exists a significant drop inside acceptable range. This can be attributed to the limited training dataset. Additionally, as noted above provided limits concern overall values while for this case study, the ship is super-slow steaming. Furthermore, in this case cylinders #2-#8 retain nominal exhaust gas temperature values. As such, a disparity, not present in training dataset, appears.

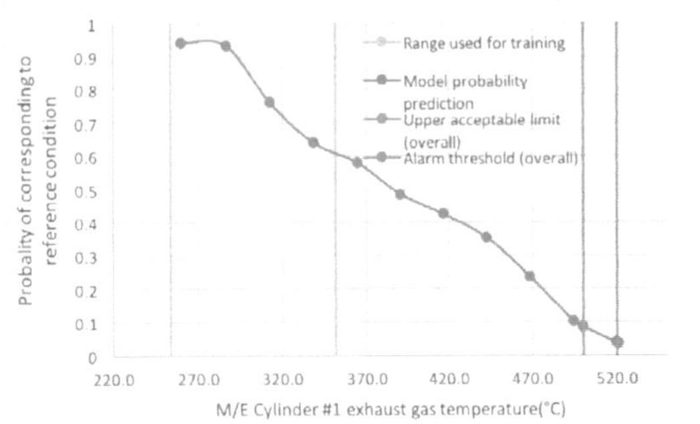

Figure 3: M/E Cylinder #1 exhaust gas temperature sensitivity analysis

5.3 CASE STUDY 3 (M/E LUBE OIL INLET TEMPERATURE)

This case study evaluates model performance in varying lube oil inlet temperatures. This component is peculiar in the sense that there exist both an upper and a lower acceptable limit. The nominal dataset included values in the whole acceptable range. As such, the model performs well, with a steady drop beyond acceptable range limits and with a probability of 30% at alarm threshold (Fig. 4).

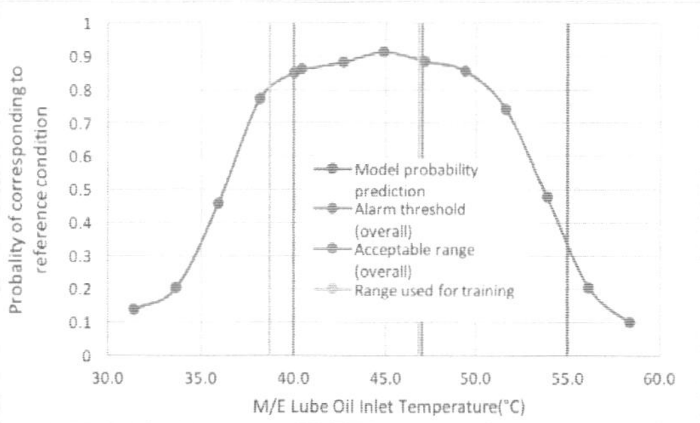

Figure 4: M/E lube oil inlet temperature sensitivity analysis

5.4 CASE STUDY 4 (M/E FUEL OIL INLET TEMPERATURE)

This case study evaluates model performance in varying fuel oil inlet temperatures. This does not reflect a possible M/E fault per se but demonstrates framework's performance when measurements of multiple subsystems are combined. This is valuable in identifying the root cause of malfunctions at system level. As with case study 3, there exist both a lower and a higher temperature threshold. Acquired values used for training span the bulk of acceptable range and values beyond limits are accurately identified as non-nominal (Fig. 5).

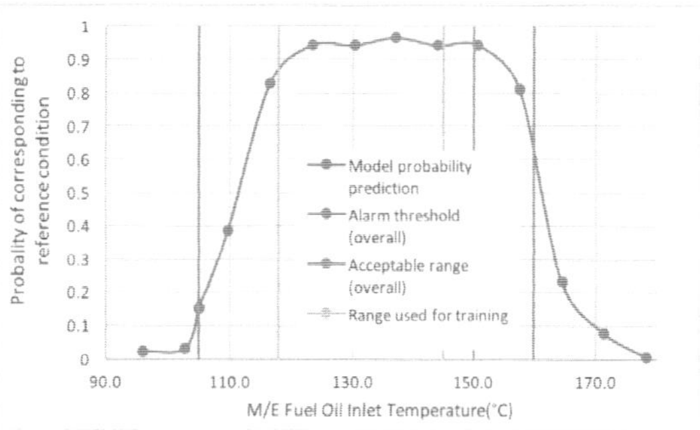

Figure 5: M/E fuel oil inlet temperature sensitivity analysis

6. CONCLUSIONS

This paper aims to present an initial framework for the processing of performance data to monitor the condition of ship machinery. First, an overview of the current state of research in the field of maritime maintenance and condition monitoring was provided. Then, proposed methodology was elaborated and showcased through several case studies. These case studies showcased the model performance while simulating faults in different subsystems.

In conclusion, future research steps include model training using a bigger, more comprehensive dataset. Besides, the development and implementation of a Decision Support System providing guidance with regards to the selection of optimal maintenance actions is proposed. Additionally, the use of vibration alongside performance measurements will be considered so that a more robust model is obtained.

7. ACKNOWLEDGEMENTS

The work presented in this paper is partially funded by INCASS project. INCASS Project has received research funding from the European Union's Seventh Framework Programme under grant agreement no 605200. This publication reflects only the authors' views and European Union is not liable for any use that may be made of the information contained within.

8. REFERENCES

1. United Nations, 'Review of Maritime Transport', *United Nations Conference on Trade and Development*, 2015.

2. Mohanty, A. R., 'Machinery Condition Monitoring', *CRC Press*, 2014.

3. Scheffer, C. and Girdhar, P., 'Practical Machinery Vibration Analysis and Predictive Maintenance', *Newnes*, 2004.

4. Randall, R. B., 'Vibration-based Condition Monitoring: Industrial, Aerospace and Automotive Applications', *John Wiley & Sons*, 2011.

5. Neale, M. et al., 'A Guide to the Condition Monitoring of Machinery, *Tech. Rep., Committee for Terotechnology, London, G.B.*, 1979.

6. Lazakis, I. et al., 'Increasing ship operational reliability through the implementation of a holistic maintenance management strategy', *Ships and Offshore Structures*, 2010.

7. Stopford, M. 'Maritime Economics', *Routledge*, 2008.

8. Lazakis, I. and Olcer, A., 'Selection of the best maintenance approach in the maritime industry under fuzzy multiple attributive group decision-making environment', *Proceedings of the Institution of Mechanical Engineers, Part M*, 2015.

9. Chandroth, G., 'Condition monitoring: the case for integrating data from independent sources', *Journal of Marine Engineering & Technology*, 2004.

10. Watzenig, D. et al., 'Engine state monitoring and fault diagnosis of large marine diesel engines', *Elektrotechnik & Informationstechnik*, 2009.

11. Li, P. et al, 'The Research of the Intelligent Fault Diagnosis Optimized by ACA for Marine Diesel Engine', *Advancing Computing, Communication, Control and Management*, 2010.

12. Yiang, W. et al., "Wind turbine condition monitoring by the approach of SCADA data analysis', *Renewable Energy*, 2013.

13. Yin, Z. and Hou, J., 'Recent advances on SVM based fault diagnosis and process monitoring in complicated industrial processes', *Neurocomputing*, 2016.

14. Widodo, A. and Yang, B. S., 'Support vector machine in machine condition monitoring and fault diagnosis' *Mechanical Systems and Signal Processing*, 2007.

15. Raptodimos, Y. et al., 'Collection & Analysis of Data for Ship Condition Monitoring aiming at Enhanced Reliability & Safety' *The 26th International Conference on Ocean and Polar Engineering (ISOPE)*, 2016.

16. Foley, D. H., 'Consideration of Sample and Feature Size', *IEEE Transactions on Information Theory*, 1972.

17. Hua, J. et al., 'Optimal number of features as a function of sample size for various classification rules', *Bioinformatics*, 2005.

18. Skittides, C. and Früh, W.-G., 'Wind forecasting using Principal Component Analysis', *Renewable Energy*, 2014.

19. Wang, Y. et al., 'Reducing sensor complexity for monitoring wind turbine performance using principal component analysis', *Renewable Energy*, 2016.

20. Gkerekos, C. et al., 'Ship Machinery Condition Monitoring using Vibration Data through Supervised Learning', *International Conference on Maritime Safety and Operations, Glasgow, UK*, 2016.

21. Maimon, O. and Rokach, L., 'Data Mining and Knowledge Discovery Handbook', *Springer*, 2010.

22. Vapnik, V. N,. 'Statistical Learning Theory', *Wiley-Interscience*, 1989.

23. Kotsiantis, S. B., et al., 'Data Preprocessing for Supervised Learning'. *International Journal of Computer Science*, 2006.

24. Lakshminarayan, K., et al., 'Imputation of missing data using machine learning techniques', *The Second International Conference on Knowledge Discovery and Data Mining, Portland, Oregon*, 1996.

9. AUTHORS BIOGRAPHY

Christos Gkerekos is a PhD Researcher at the department of Naval Architecture, Ocean and Marine Engineering (NAOME) of University of Strathclyde. His research focuses on the use of self-learning algorithms for the condition monitoring of marine machinery.

Iraklis Lazakis is a Senior Lecturer at the department of NAOME of University of Strathclyde. His research interests include ship systems condition monitoring, systems maintenance and reliability, ship operations, risk analysis tools and methodologies, system criticality assessment, shipyard manufacturing and productivity, and asset management.

Gerasimos Theotokatos is a Lecturer at the department of NAOME of University of Strathclyde. His research interests include ship propulsion plant modelling, performance and emissions prediction of marine Diesel engines, the development and testing of propulsion plant system control schemes, the analysis of engine and its turbocharging system under normal and adverse operating conditions, and the experimental investigation of performance and emission parameters of internal combustion engines.

FAULT TREE ANALYSIS AND ARTIFICIAL NEURAL NETWORK MODELLING FOR ESTABLISHING A PREDICTIVE SHIP MACHINERY MAINTENANCE METHODOLOGY

Y Raptodimos and **I Lazakis,** University of Strathclyde, UK

SUMMARY

A dynamic fault tree model for a ship main engine is developed in order to analyse and identify critical systems/components of the main engine. The identified most critical systems are then used as input in an artificial neural network. An autoregressive dynamic time series neural network modelling approach is examined in a container ship case study, in order to monitor and predict future values of selected physical parameters of the most critical ship machinery equipment obtained from the fault tree analysis. The case study results of the combination of the fault tree analysis and artificial neural network model demonstrated promising prospects for establishing a dense methodology for ship machinery predictive maintenance by successfully identifying critical ship machinery systems and accurately forecasting the performance of machinery parameters.

1. INTRODUCTION

Maintenance deals with systems that are subject to deterioration and failure with usage and age. For systems on board ships, it is extremely important to avoid failures during actual operation because it can be dangerous or disastrous in terms of performance, safety and economic losses. The performance of the vessel generally deteriorates with time as a result of fouling or degradation of machinery systems and components. Unwanted failures result in economic impact in form of higher maintenance costs and lower machine reliability and availability. With reduced manning levels and the ever increasing competition, ship maintenance has become one of the major challenges in the marine industry. Technological advances and high cost of ownership have resulted in considerable interest in advanced maintenance techniques. As a consequence, the maritime industry is seeking for increased reliability, maximum uptime and optimal operational efficiency, as well as ensuring safe and sustainable environmental performance in harsh environments.

Optimisation of maintenance is challenging due to highly restrictive and harsh operating conditions of ships in addition to the high level of uncertainty accompanied by these operating conditions. Compared to other industries, data gathering and processing is not always possible as similar equipment in diverse conditions may have different failure patterns. Data is not collected in standard ways which would aid in successful decision making [1]. An additional issue is the constant appearance of new equipment, which makes historical records obsolete and lays other aspects on the replacement decisions.

Efforts have being made to transform corrective/preventive maintenance techniques into predictive ones. Condition monitoring is considered as a major part of predictive maintenance. It assesses the operational health of equipment, in order to provide early warning of potential failure such that preventive maintenance action may be taken. Predictive maintenance consists in deciding whether or not to maintain a system according to its state [2]. Moreover, predictive maintenance focuses on failure prediction, occurring through follow-ups with a specific systematic on parameters and equipment conditions. This type of maintenance did not emerge as a replacement for corrective and preventive maintenance, but as an additional tool, which seeks to minimize maintenance costs and losses in equipment through the monitoring of specific parameters [3].

As part of the above, this paper focuses in combining a Fault Tree model and then an Artificial Neural Network (ANN) for forecasting selected machinery parameters. Fault Tree Analysis (FTA) is a top-down approach which uses failure rates, mean time between failures and minimal cut sets to evaluate the reliability and availability of the examined system [4]. FTA can be applied both in a qualitative and quantitative way. The objective of a Fault Tree is to evaluate the probability of occurrence of the top event. Moreover, Fault Trees are also used to display the causes and consequences of events, identify system critical components and evaluate changes in design amongst other things. Fault Tree diagrams provide important information regarding the likelihood of a failure occurring and the means by which this failure could occur. They can be constructed at any point of a design stage and the FTA results can help improve system safety.

Raza and Liyanage [5] stated that there has been an increasing demand for testing and implementing intelligent techniques as a subsidiary to existing monitoring programs and that Artificial Neural Networks (ANNs) have emerged as one of the most promising techniques in this regard. The equipment condition and the fault developing trend are often highly nonlinear and time-series based. ANNs can be used due to their potential ability in nonlinear time-series trend prediction. Therefore this paper initially proposes the analysis of a Fault Tree model for the main engine of a ship in order to identify the most critical systems/items of the main engine. Physical parameters of the identified systems are then used as input in the ANN model in order to monitor and predict their future values.

This paper is organized as follows: Section 2 briefly describes the research background, containing information regarding predictive maintenance, FTA and ANNs. Section 3 presents and defines the overall methodology. A case study is presented in Section 4 and results are presented in Section 5, followed by the concluding remarks contained in the last section.

2. BACKGROUND

2.1 MAINTENANCE

Ships are part of the marine transportation system and are crucial assets of the supply chain. In this respect, maintenance tasks affect the reliability and availability standards of the shipping industry and are an important factor in the lifecycle of a ship that can minimize down-time and reduce operating costs [9]. The importance of maintenance is demonstrated by the fact that it is the only shipboard activity to have one whole element assigned to it [6]. Also, due to the impact of shipping on the environment and the importance of the safe operation of ships; ship owners and operators pursue to adopt a maintenance plan and procedures that will reduce costs, promote the lifecycle integrity and enhance the energy efficiency of the ship.

Initially, corrective maintenance was applied to ships followed by successful preventive maintenance actions due to International Safety Management (ISM) code [6] and regulations and was then followed by predictive maintenance advances [7]. Predictive maintenance did not emerge as a replacement for corrective and preventive maintenance, but as an additional tool, which seeks to minimize, through the monitoring of specific parameters, maintenance costs and losses in equipment [3]. Therefore, predictive maintenance differs from preventive maintenance by concentrating maintenance on the actual condition of the machinery rather on some predefined schedule dictated by predefined time intervals or system operating hours. Consequently, predictive maintenance is used to define required maintenance tasks based on quantified material and equipment condition. It uses modern measurement and signal processing methods to accurately predict and diagnose items/equipment condition during operation [8].

2.2 FAULT TREE ANALYSIS

Reliability assessment tools include Reliability Block Diagrams (RBD), FTA, Failure Modes Effects Analysis (FMEA), Failure Modes Effects and Criticality Analysis (FMECA), Markov analysis and Bayesian Belief Networks (BBN). FTA is one of the basic methods of assessing reliability. The FTA method allows detailed examination of the system operation principles during the design, operation and accident investigations.

Fault Tree diagrams are a graphical design technique following a top-down approach. It uses a graphic model of the pathways within a system that can lead to a projected, undesirable event or failure. The pathways interconnect contributory events and conditions, using standard logic symbols and the basic constructs in a fault tree diagram are gates and events. The fault tree analysis module is based on sets of rules and logic symbols from probability theory and Boolean algebra. Gates represent logic operators that link the various branches of the fault tree together and determine whether the top event can occur or not. Basic events can be defined as the lower level events in each fault tree branch.

FTA uses failure rates, mean time between failures and minimal cut sets to evaluate the reliability and availability of the system. Moreover, fault trees are also used to display the causes and consequences of events, identify system critical components and evaluate changes in design amongst other things.

Lazakis et al. [7] presented a predictive maintenance strategy utilizing FMECA and FTA by considering the existing ship maintenance regime as an overall strategy including technological advances and decision support system by combining existing ship operational and maintenance tasks with the FTA and FMECA tool.

An innovative ship maintenance strategy is also presented by Turan et al. [4] based on criticality and reliability assessment while utilising the FTA tool with time-dependent dynamic gates in order to accurately present the interrelation of the components for a diving support vessel.

Lazakis and Olcer [9] introduced a Reliability and Criticality Based Maintenance (RCBM) strategy by utilizing a fuzzy multiple attributive group decision-making technique, which is further enhanced with the employment of Analytical Hierarchy Process (AHP). The outcome of this study indicated that preventive maintenance is still the preferred maintenance approach by ship operators, closely followed by predictive maintenance; hence, avoiding the ship corrective maintenance framework and increasing overall ship reliability and availability.

2.3 ARTIFICIAL NEURAL NETWORKS

A neural network can be defined according to Principe et al. [10] as distributed, adaptive, generally nonlinear learning machines built from many different processing elements that receive connections from other processing elements and/or itself. Several distinguishing features of ANNs [11] make them attractive for the development of prognostic tools. First of all, opposed to the traditional model-based methods, ANNs are data-driven and self-adaptive methods, meaning that there are few a priori assumptions about the models under study. They learn from past examples and capture subtle functional relationships among the data even if the underlying relationships are hard to describe or unknown. ANNs do not rely on prior principles or statistics models and can significantly simplify the model synthesized process. They can readily address modelling problems that are analytically difficult and for which conventional approaches are not practical, including complex physical processes having nonlinear, high-order, and time-varying dynamics. Secondly, ANNs have good generalisation capabilities and are universal functional approximators. They are also nonlinear. Real word failure models are generally non-linear. However, these models are still limited in that they are based on a little knowledge of underlying law.

The application fields of neural networks can be categorised with respect to different criteria, such as industrial application, type of reliability problem, life cycle phase in which the algorithms are predominantly applicable and the type of learning problem. The building block of a neural network is a neuron [12], which has several inputs a_n, each of these inputs are multiplied by weights w_{ij} and then added up. Weights are adaptive coefficients within the network that determine the intensity of the input signal [13]. Often a bias is added b_j, which is the node's internal threshold. The result is the neuron activation z as shown in the following equation:

$$z = \sum_{i=1}^{n} a_i w_{ij} + b_j$$

Time series modelling and forecasting has fundamental importance to various practical domains. The main aim of time series modelling is to carefully collect and rigorously study the past observations of a time series to generate future values for the series. Time series forecasting can be termed as the act of predicting the future by understanding the past.

Although linear models possess many advantages in implementation and interpretation, they have serious limitations in that they cannot capture nonlinear relationships in the data which are common in many complex real world problems. Neural nets have the potential to represent any complex, nonlinear underlying mapping that may govern changes in a time series [14]. Zhang et al. [15] showed that neural networks are valuable tools for modelling and forecasting nonlinear time series while traditional linear methods are not as competent for this task. The lack of systematic approaches to neural network model building is probably the primary cause of inconsistencies in reported findings.

With the increased availability of monitoring data on the condition of systems and equipment, neural networks are increasingly applied in the field of fault detection [16], fault diagnostics [17] and for predicting the residual useful life [18].

3. METHODOLOGY

The methodology suggested in this paper initially generates a generic model approach, utilising the capabilities of fault tree modelling. Subsequently, the ANN model is applied for the transition of the generic model into a specific model. The aim of the FTA is to calculate and identify the most critical subsystems/components of the top main system. Once the most critical items of the FTA have been acquired, physical parameters associated with these items are used as input in the ANN in order to conduct a time series analysis aiming in predicting their future values. Figure 1 provides an overview of the methodology approach.

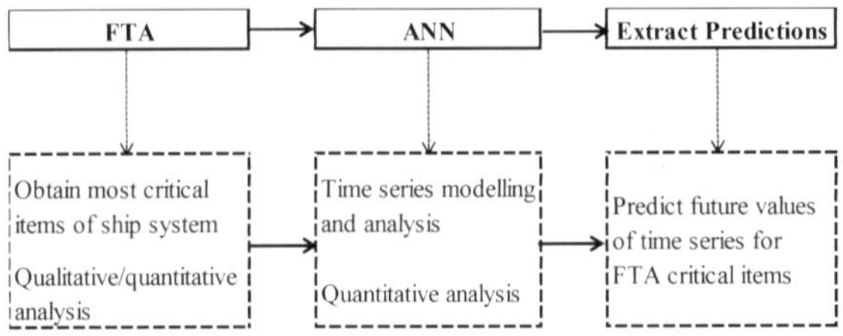

Figure 1: Methodology

3.1 FTA MODEL

A very important part of FTA is the system definition. The basis of the system definition is the fault tree diagram which defines all interconnections and components of the system. Also, reliability parameters should be identified and the system definition must identify important assumptions regarding the system and the conditions that indicate that the components of the system have failed. The top event should also be clearly defined as if a top event is not concisely defined then the fault tree can possibly become too large and complex, resulting in an unfocused system analysis.

One of the advantages of FTA is that it is an event-oriented method. In addition to considering failures in hardware, fault trees can take into account the undesirable events that can possibly occur due to software failures, environmental influences and human errors during the stages of maintenance and system operation. On the other hand, one of the biggest disadvantages of FTA is that all events leading to the top event must be foreseen and the contributors to the occurrence of these undesired events must be anticipated. In addition, FTA requires analyst skill and can be time-consuming, thus costly.

The following steps are usually performed when constructing a fault tree [19]:

- Definition of the FTA scope

- Identification of the top event.
- Identification of the first level events.
- Connection of the first level events with the top event by using gates.
- Identification of the second level events.
- Connection of the second level events with first level by using gates.
- Repetition of the above steps for all subsequent event levels.

3.1(a) Fault Tree Minimal Cut Sets

The FTA can be conducted in a qualitative or quantitative manner, depending on the type of data available. If no data is available, a fault tree can be analysed qualitatively by using minimal cut sets. Qualitative analysis is used to identify what combinations of events cause the top event to occur.

A cut set is a set of basic events, which if they all occur, will result in the top event of the fault tree occurring. A minimal cut set is a combination (intersection) of primary events sufficient for the top event. The combination is a "*minimal*" combination in that all the failures are needed for the top event to occur; if one of the failures in the cut set does not occur, then the top event will not occur (by this combination). To determine the minimal cut sets of a fault tree, the tree is first translated to its equivalent Boolean equations. One of the main purposes of representing a fault tree in terms of Boolean equations is that these equations can then be used to determine the associated minimal cut sets.

The minimal cut set expression for the top event can be written in the general form according to [19]:

$$T = M_1 + M_2 + \ldots + M_k$$

where T is the top event and M_i are the minimal cut sets, each of them consisting of a combination of specific component failures. The general n-component minimal cut can be expressed as:

$$M_i = X_1 \bullet X_2 \bullet \ldots \bullet X_n$$

where $X_1, X_2, \ldots, X_n$ are basic component failures.

If data such as mean time between failures, failure rates, probabilities are available, the fault tree can use quantitative calculation methods and also reliability importance measures [4] such as Birnbaum, Criticality and Fussell-Vesely to identify and calculate most critical items.

3.1(b) Fault Tree Gates

Logic operators known as gates determine how events are generated. A basic event represents the lowest level of a fault tree. Gates are used to represent the failure logic paths between various levels of the tree. Gates can be either static or dynamic. A static gate indicates that the order of the inputs of a gate do not matter, therefore are not sequence-dependent as in dynamic gates. On the other hand, in dynamic gates, the order of the occurrence of input events is vital for determining the output. If dynamic gates are used, then the fault tree becomes a dynamic fault tree. The most common static gates include the AND, OR and Voting gates. An AND-gate indicates that the output occurs if and only if all the input events occur. The OR-gate is used to specify that the output occurs if and only if at least one of the input events occurs. The Voting-gate occurs when at least m out of n input events occurs. Dynamic fault tree gates include the Sequence Enforcing-gate, Priority AND-gate, Spare-gate and Functional Dependency-gate amongst other. The Priority AND-gate indicates that the output will occur if and only if all the input events occur in a particular order from left to right. Thus, items need to fail in temporal order from left to right to trigger the event. Similarly, the Sequence Enforcing-gate forces events to occur in a particular order from left to right. This implies that the left-most event must occur first and that an event connected to such a gate will be initiated immediately after the occurrence of its immediate left event.

3.2 ANN ARCHITECTURE

Data has to be pre-processed prior to using it in the Artificial Neural Network. Also, the neural network architecture has to be established in order to design a network capable of modelling a time series problem and accurately predicting future values of that time series. Figure 2 demonstrates the methodology implementation followed for the neural network structure.

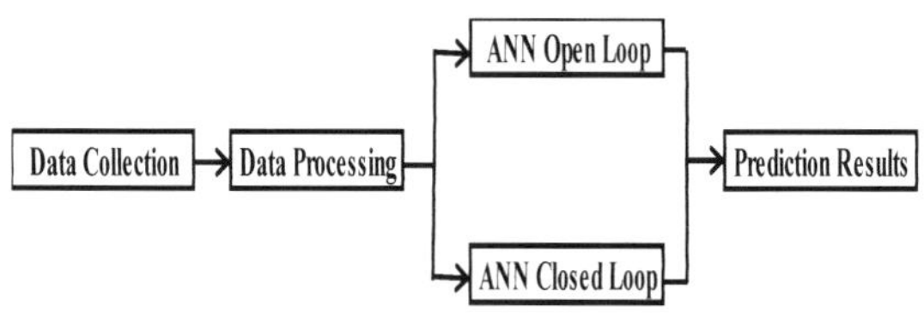

Figure 2: ANN Methodology

Once the data has been processed, the neural network is firstly modelled in open loop and is then converted to closed loop for multistep-ahead predictions.

3.2(a) Data Preparation & Processing

Before the data can be analysed in the neural network, it has to be correctly processed in order to achieve a correct analysis and improve the efficiency of network training.
A time series is a sequential set of data points, measured typically over successive times. It is mathematically defined as a set of vectors y(t), t = 0 ,1 , 2 ,...,d where t represents the time elapsed [20]. The variable y(t) is treated as a random variable. The measurements taken during an event in a time series are arranged in a proper chronological order. The future values of a time series

y(t) are predicted only from the past values of that series. This form of prediction is called nonlinear autoregressive and can be written as:

$$y(t) = f(y(t-1), ..., y(t-d))$$

Where y_t is the observation at time t; and d is the dimension of the input vector or number of past observations used to predict the future; and f is a non-linear function. The data is prepared by shifting time by the minimum amount to fill input states and layer states for network open loop and closed loop feedback modes. This allows the time series data to be trained with the dynamic neural network. Finally, data is divided into two subsets in the network for training and testing purposes. The training set is used for computing the gradient and updating the network weights and biases and the test data is used to measure how well the network generalizes overall.

3.2(b) Neural Network Modelling

An artificial neural network consists of interconnection of neurons. The neurons are usually assembled in layers [12]. Each layer has a number of simple, neuron processing elements called nodes or neurons that interact with each other by using numerically weighted connections [21]. Generally a neural network consists of n layers of neurons of which two are input and output layers, respectively. The former is the first and the only layer which receives and transmits external signals while the latter is the last and the one that sends out the results of the computations. The n-2 inner ones are called hidden layers which extract, in relays, relevant features or patterns from received signals.

The interconnectivity defines the topology of the ANN [5]. The network topology describes the arrangement of the neural network. Successful ANN modelling is based upon the number of neurons, number of hidden layers, values of the weights and biases, type of the activation function, structure of the network, training styles and algorithms as well as data structure. However, the best structure is the one which can predict behaviour of the system as accurately as possible. A crucial step in the building of a neural network model is the determination of the number of processing elements and hidden layers in the network. Hidden nodes are used to capture the nonlinear structures in a time-series. Since no theoretical basis exists to guide the selection, in practice the number of hidden nodes is often chosen through experimentation or by trial-and-error.

ANNs learn the relation between inputs and outputs of the system through an iterative process called training [22]. Neural networks are trained for input data and the output is computed. The error obtained by comparing outputs with a desired response is used to modify the weights with a specific training algorithm. This procedure is performed using training data set until a convergence criterion is met. Neural networks have different learning algorithms for training. The choice of a particular learning algorithm is influenced by the learning tasks a neural network has to perform. The training performance is evaluated using the following performance measures, namely the Mean Square Error (MSE) average sum of square errors and Correlation Coefficient (R) given by the following equations respectively [23]:

$$MSE = \frac{\sum_{j=0}^{P}\sum_{i=0}^{N}\left(d_{ij} - y_{ij}\right)^2}{NP}$$

$$R = \frac{\dfrac{\Sigma\left(x_i - x_{mean}\right)\left(d_i - d_{mean}\right)}{N}}{\left[\left(\dfrac{\Sigma\left(d_i - d_{mean}\right)^2}{N}\right)\left(\dfrac{\Sigma\left(x_i - x_{mean}\right)^2}{N}\right)\right]^{0.5}}$$

where P = number of output processing elements; N = number of exemplars in the data set; y_{ij} = network output for exemplars i at processing element j ; and d_{ij} = desired output for exemplars i at processing element j.

A nonlinear autoregressive dynamic neural network is used for the prediction. A hyperbolic tangent transfer function in the hidden layer and linear transfer function in the output layer are employed, capable of approximating any function with a finite number of discontinuities. The system is firstly modelled as an open loop system to train the network accurately up to the present with all of the data in order to achieve correct predictions; and is then transformed to closed-loop for calculating multistep-ahead predictions. During training, the network weights and biases are updated after all of the inputs and target values have been presented to the network. The network is autoregressive as the only inputs are lagged target values. The neural network is trained using the Bayesian regularization backpropagation algorithm. The term backpropagation refers to the process by which derivatives of network error, with respect to the network weights and biases, can be computed. Bayesian regularization algorithm provides better generalization performance and is most suitable for small data sets compared to other training algorithms. The performance of the network is evaluated using the MSE performance measure and Correlation Coefficient R.

The open loop network is a feed-forward back propagation network. Then, for the multistep-ahead predictions, the open loop network is converted to a Recurrent Neural Network (RNN) closed-loop system, by creating a feedback connection from the output to the network input, thus making the network dynamic. The first two time steps of the input are used as input delay states in order to model the dynamic system.

RNNs can store sequential information in the form of historical data and can be used in forecasting. For example, in an RNN, the input nodes are taken as the value of the current condition X_t and values of the previous time-series condition (X_{t-1}, X_{t-2}, X_{t-3},..., X_{t-d} and X_n). The value of the output X_{t+1} can provide a one-step-ahead prediction of a time-series condition, which is a function of the current value X_t and time-lagged values of the previous condition (X_{t-1}, X_{t-2}, X_{t-3}, ... , X_{t-d} and X_n). The predicted value X_{t+1} of a time series, one-step ahead in the future is given by the following equation:

$$X_{t+1} = F\left(X_t, X_{t-1}, X_{t-2}, \ldots, X_{t-l}, \ldots, X_n\right)$$

Where, l is the time lag, X_{t+1} is the predicted value, X_t is the current value or condition and X_{t-d} is the values of previous condition lagged by time d. The closed-loop network is show in Figure 3.

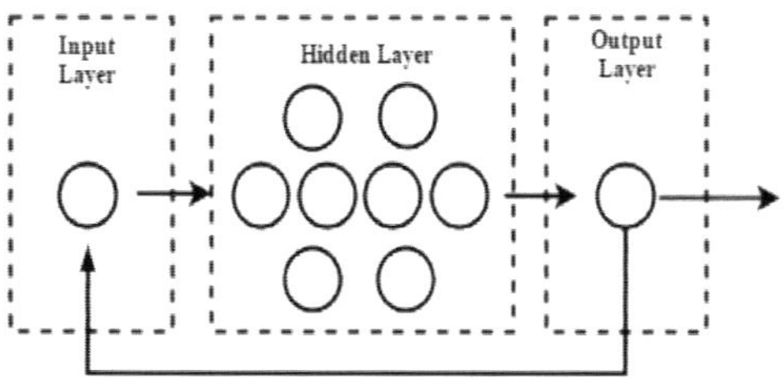

Figure 3: Closed-loop dynamic neural network

4. CASE STUDY

The described methodology is applied for the case study of a Panamax container ship. A Fault Tree model is constructed for the system of an eight cylinder two stroke marine diesel engine. Once the Fault Tree has been constructed and analysed, the most critical subsystems/components of the top main system are identified. Then, physical parameters (e.g. temperature, pressure) associated with the critical items of the Fault Tree are used as input in the ANN in order to conduct a time series analysis aiming in predicting their future values.

4.1 FAULT TREE MODELLING BOUNDARY

As with any modelling technique, the boundaries of a FTA must be defined prior to the construction of the fault tree. In general, defining the boundaries of the analysis involves defining what is in the analysis and what is out of the analysis. Figure 4 displays the boundary condition for the main engine.

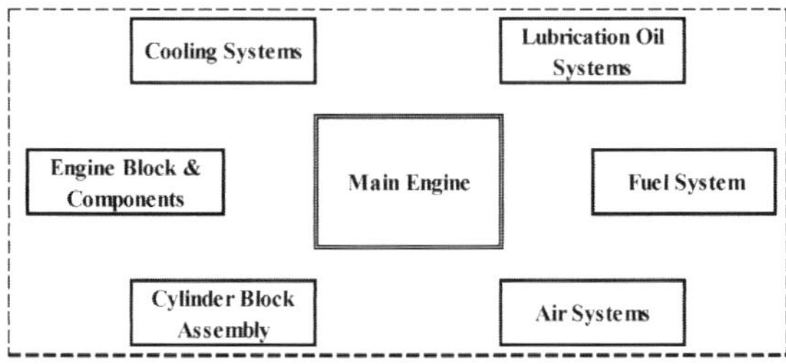

Figure 4: Boundary condition of main engine including sub-systems

As observed in Figure 4, the main engine is divided into six categories which include namely the cooling systems, lubrication oil systems, fuel system, air systems, cylinder block assembly and engine block and components. The cooling system is further divided into the jacket water cooling and central cooling system. The jacket water cooling system consists of the jacket fresh water cooling pump and jacket water cooling and the central cooling system of the sea chest strainer, sea water pipes and central cooler. The lubrication oil system includes the lube oil filter, pump, valves and lube oil cooler. Furthermore the fuel system resembles the lube oil system with the addition of the fuel injectors. The air systems are further separated into the main air system and scavenging air system. In the cylinder block assembly system, the system has been separated into the cylinder system which includes the cylinder head and liner and the piston assembly including the piston crown, rings, stuffing box and connecting rod. Finally, the engine block and components group contains components of the main engine such as the crankshaft, crankcase, camshaft and various bearings. In total, 39 basic events were created in the fault tree representing the components of the various main engine sub-systems as illustrated in Table 1.

Table 1: List of components used for fault tree

Lube Oil Filter	Air Cooler, Piping
Main Lube Oil Pump	Air Cooler
Lube System Valves	Scavenge Air Port
Lube Oil Cooler	Scavenge Air Receiver
J.F.W Cooling Pump	Scavenge Air Manifold
Jacket Water Cooling	Air Receiver
Sea Chest Strainer	Cylinder Head
Sea Water Pipes	Cylinder Liner
Central Cooler	Piston Crown
Fuel Piping System	Piston Ring
Fuel Oil Filter	Piston Rod Stuffing Box
Fuel Pumps	Piston Connecting Rod
Fuel Valves	Camshaft Bearing
Fuel Injector	Thrust Bearing
Main Air Compressor	Main Bearings
Air Distributor	Crankshaft
Air Starting Valves	Crankcase
Air Filter	Camshaft
Auxiliary Blower	Exhaust Valves

For each system and associated component, physical parameters such as temperature, pressure are measured in order to monitor their condition. Key parameters in performance observations include amongst others:

- Engine speed
- Barometric pressure
- Compression pressure
- Fuel pump index
- Exhaust gas temperatures and pressures
- Scavenge air temperature and pressure
- Air and cooling water temperatures prior and after scavenge air cooler.

Table 2 illustrates a sample of physical parameters monitored for the systems and components included and modelled in the Fault Tree analysis. In total, 20 physical parameters are monitored associated with the 6 systems used in the Fault Tree of the main engine as seen in Figure 5.

Table 2: Physical parameters for main engine systems

System	Parameter	Physical Measurement	
		Temperature	Pressure
Cylinder	Scavenging Air	•	
	Exhaust Gas Temperature Outlet	•	
	Jacket Fresh Water Cooling Inlet		•
Piston	Piston Cooling Lube Oil Inlet		•
Lube Oil Cooler	Sea Water Inlet	•	
	Sea Water Outlet	•	
Thrust Bearing	Thrust Bearing Lube Oil Outlet		•

Based on the FTA and utilising Table 2, the selected physical parameters are then used as input for the ANN.

4.2 ARTIFICIAL NEURAL NETWORK

The data composed for analysis of the selected physical parameters was collected through an on board measurement campaign as presented in Raptodimos et al. [24]. The neural network uses a univariate time series data set. The data monitored on board the container ship case study, represents 30 continuous per hour records of physical parameters regarding systems such as the ones shown in Table 2. The ANN constructed consists of one hidden layer with 8 hidden nodes as illustrated in Figure 3. The ANN receives these values as input as a univariate time series data and attempts to predict the upcoming values as output.

5. RESULTS

Results are firstly shown for the FTA of the main engine. Finally, based on the Fault Tree Analysis, results are shown for the Artificial Neural Network.

5.1 FTA RESULTS

A four level fault tree for the ships main engine is constructed including 39 basic events and 14 gates as described in Table 1. Qualitative analysis is performed in order to obtain the minimal cut sets of the fault tree, which provide insight into weak points of complex systems. Figure 5, displays a fragment of the main engine fault tree.

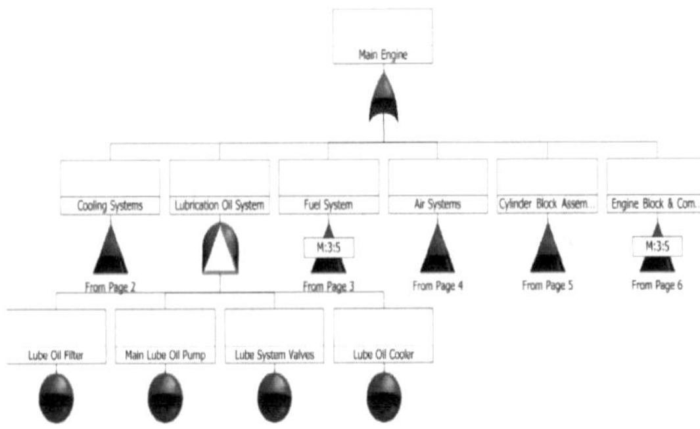

Figure 5: Main engine fault tree diagram

The cooling systems are modelled using an AND gate assuming both the jacket water cooling and central cooling system have to fail. The lubrication oil system has been modelled using a Priority-AND gate, assuming as seen in Figure 5, that the lube oil filter has to occur before the pump, vales and then lube oil cooler occur. By using dynamic gates, the fault tree becomes a dynamic fault tree. The fuel system has been designed using a Voting gate of 3 out of 5 systems and the air system has been modelled as an AND gate assuming that both the main air system and scavenge air system have to occur. The engine block and components gate is composed of a Voting gate assuming that any three of the bearings, crankshaft, crankcase, camshaft or exhaust valves have to occur. Table 3 displays the top five minimal cut sets obtained from the main engine FTA.

Table 3: Top five fault tree minimal cut sets

1	Cylinder Head	Piston Crown	Piston Ring
2	Cylinder Head	Piston Crown	Piston Rod Stuffing Box
3	Cylinder Liner	Piston Crown	Piston Ring
4	Crankcase	Crankshaft	Camshaft
5	Fuel Oil Filter	Fuel Pumps	Fuel Injector

All the minimal cut sets are of third order. As observed, the first three cut sets involve systems/components that were constructed in the cylinder block assembly gate illustrated in Figure 6.

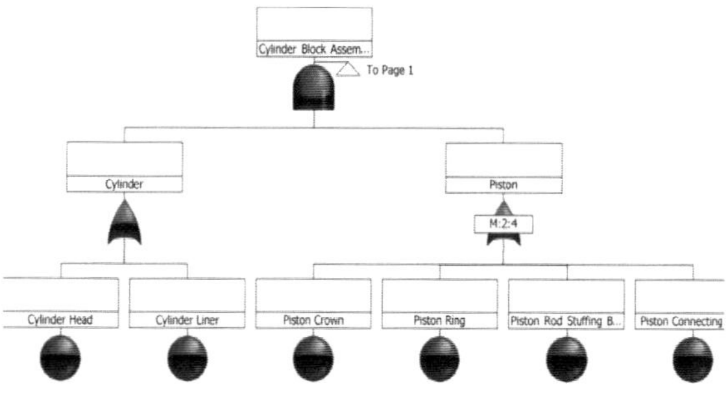

Figure 6: Cylinder block assembly sub-system

Figure 6 shows the cylinder block assembly system modelled with an AND gate, the cylinder OR gate and the piston sub-system modelled with a 2 out of 4 Voting gate.

Since the minimal cut sets have been calculated and involve systems and components inside the cylinder, based on Table 2, physical parameters such as the cylinder exhaust gas temperature can be used as input in the ANN in order to further monitor and examine the performance of the specific system.

5.2 ANN RESULTS

The ANN constructed attempts to predict the upcoming five cylinder exhaust gas temperature values as output, for cylinder 1. The results obtained from the network for predicting the future upcoming five values in time are illustrated in Figure 7.

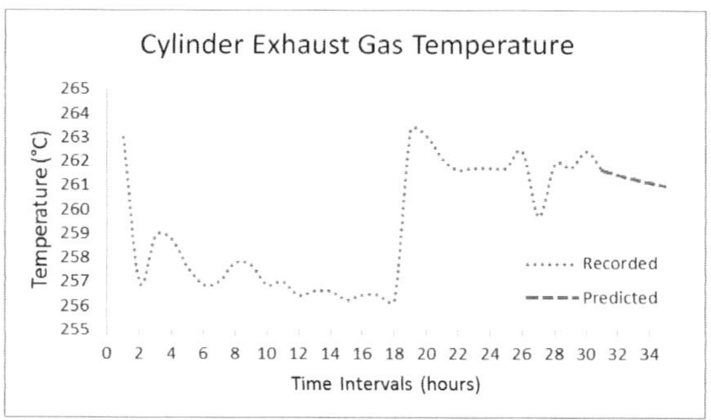

Figure 7: Cylinder 1 exhaust gas temperature prediction graph

Figure 7 illustrates the graph of the recorded values and predicted ones against time. The recorded temperatures are within the range of 254 degrees to 283 degrees Celsius. Variations in temperature, especially the rise of the exhaust gas temperature at some points is observed and is mainly caused by an increase in the engine and fuel load. This is due to the engine governor regulating the engine speed, as the vessel is also sailing at a constant speed of 10 knots. Moreover, these variations could be the result of the specific cylinder condition.

5.2(a) ANN Validation

The results of the ANN are examined and validated using two approaches. The first approach uses the network performance in terms of the regression plots R for the training and testing data as shown in Figure 8 and the error autocorrelation as shown in Figure 9.

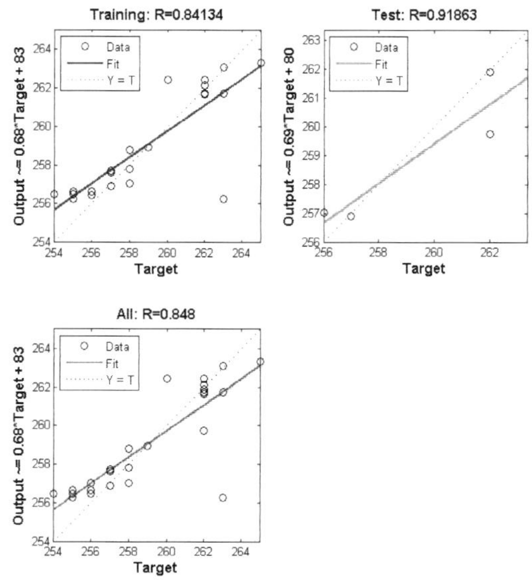

Figure 8: Regression results for training and test data of neural network

The correlation coefficient R, regression plot, is a good measure of how well the network has fitted the data. The regression plot shows the actual network outputs plotted in terms of the associated target values. Regression values measure the correlation between outputs and targets. A correlation coefficient R value of 1 implies a perfect fit of outputs exactly equal to targets. Figure 8 displays the network outputs with respect to targets for training and test sets. Bayesian regularization does not use a validation set but includes this in the training set. The training data indicate a good fit as does the test results showing values of R greater than 84% and 91% respectively. For all data sets, the fit is very good.

Error autocorrelation is used to validate the network performance. The error autocorrelation function defines how the forecast errors are interrelated in time. For a faultless prediction model, there should be one non-zero value that should occur at zero lag implying that the forecast errors are entirely uncorrelated with each other. Therefore, if the network has been trained well then besides the centre line which shows the mean squared error, all other lines should fall within the confidence limits as successfully shown in Figure 9.

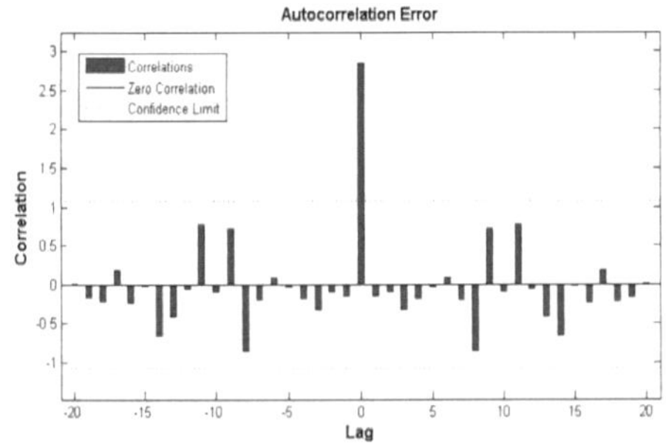

Figure 9: Autocorrelation of error

In the second network validation approach, a sensitivity analysis is conducted [25]. The network predicted values for the temperature are compared with the actual values recorded on board the vessel as seen in Table 4 in order to validate the network prediction accuracy.

Table 4: ANN prediction results versus actual results

Prediction	Recorded Onboard	ANN Prediction	Error
hours	°C	°C	
+1	263.00	261.66	0.51%
+2	260.00	261.45	0.56%
+3	262.00	261.29	0.27%
+4	262.00	261.14	0.33%
+5	263.00	261.00	0.76%

As seen from Table 4, the error difference between the values indicate that the predicted values are extremely close to the actual monitored temperature values recorded on board the vessel, thus verifying the performance and accuracy of the trained network.

6. CONCLUSIONS

Marine automation, electrical and propulsion systems, sensors, robotics, advanced materials, big data analytics, are a few of the categories that can describe the concept behind smart ships. The question of how much data, which data, and how often this should be collected and how has also risen; as although companies adopt condition based maintenance schemes, there seems to be an issue in processing, analysing and utilising the recorded operational data. Intelligent ships will enable owners to make more rapid operating decisions, by analysing real time data, providing real-time information regarding the condition of on board equipment. Thus, this will lead to the evolution of maintenance from fixed intervals, towards tailored predictive maintenance applications, which will optimise maintenance and operation planning and will also boost performance and safety. Therefore, it is clear that such developments have the potential of transforming the design, construction and operation of commercial ships.

This paper proposed the combination of a Fault Tree model with ANN in order to create a generic model in terms of obtaining the most critical systems and components of a main engine through qualitative analysis using minimal cut sets. A minimal cut set is the smallest set of basic events which result in the occurrence of the top event. The set is minimal in the sense that if any of the events do not occur, then the top event will not occur by this combination of basic events. Physical parameters of the most critical systems and components identified from the fault tree are used as input in an artificial neural network for time series analysis and prediction.

The data used for the network represents cylinder exhaust gas temperatures while the vessel was in transient operation. Since no faulty data or failures occurred during the on board measurement, the obtained data does not cover the whole operational range of the system. The case study provided accurate results for predicting upcoming temperature measurements based on previous monitored values. The data monitored represents continuous per hour temperature values. The network predicted values for the temperatures are then compared with the actual values recorded on board the vessel which indicated that the network is capable for time series analysis and has good predictive capabilities.

An important issue in ANN model building is how large the training and/or test sample sizes should be. In the ANN literature, large sample size for training is often suggested for sufficient learning and to ease the overfitting effect in training a neural network. However, Kang [26] found that neural network models do not necessarily require large data sets to perform well, as also demonstrated in this paper, in which the Bayesian regularization algorithm provided sufficient generalization performance capabilities for a small data set. However, the question of how much data is sufficient still exists and ANN models are still limited in that they are based on a little knowledge of underlying law.

In conclusion, the predictive results obtained can be utilised within a maintenance and condition monitoring framework in order to assess the performance of ship machinery equipment based on current and real time information and can be used for prognostic and diagnostic purposes and applications. Finally, it is possible to extend the neural network capabilities by investigating additional analysis in terms of also including temperature ranges with various engine loads and other parameters, which will increase the network generalisation and accuracy.

7. ACKNOWLEDGEMENTS

The work in this paper is partially funded by INCASS project. INCASS has received research funding from the European Union's Seventh Framework Programme under grant agreement No. 605200. This publication reflects only the author's views and European Union is not liable for any use that may be made of the information contained herein.

8. REFERENCES

1. Dekker, R., *Applications of maintenance optimization models: a review and analysis.* Reliability Engineering & System Safety, 1996. 51(3): p. 229-240.

2. Garg, A. and S.G. Deshmukh, *Maintenance management: literature review and directions.* Journal of Quality in Maintenance Engineering, 2006. 12(3): p. 205-238.

3. de Faria Jr, H., J.G.S. Costa, and J.L.M. Olivas, *A review of monitoring methods for predictive maintenance of electric power transformers based on dissolved gas analysis.* Renewable and Sustainable Energy Reviews, 2015. 46: p. 201-209.

4. Turan, O., et al., *Investigating the reliability and criticality of the maintenance characteristics of a diving support vessel.* Quality and Reliability Engineering International, 2011. 27(7): p. 931-946.

5. Raza, J. and J.P. Liyanage, *Application of intelligent technique to identify hidden abnormalities in a system: A case study from oil export pumps from an offshore oil production facility.* Journal of Quality in Maintenance Engineering, 2009. 15(2): p. 221-235.

6. IMO, *International Safety Management (ISM) Code/Guidelines on implementation of the ISM Code.* 1997: London.

7. Lazakis, I., O. Turan, and S. Aksu, *Increasing ship operational reliability through the implementation of a holistic maintenance management strategy.* Ships and Offshore Structures, 2010. 5(4): p. 337-357.

8. Sharma, A., G.S. Yadava, and S.G. Deshmukh, *A literature review and future perspectives on maintenance optimization.* Journal of Quality in Maintenance Engineering, 2011. 17(1): p. 5-25.

9. Lazakis, I. and A.I. Olcer, *Selection of the best maintenance approach in the maritime industry under fuzzy multiple attributive group decision-making environment.* Journal of Engineering for the Maritime Environment, 2015: p. 13.

10. Principe, J.C., N.R. Euliano, and W.C. Lefebvre, *Neural and Adaptive Systems: Fundamentals through Simulations with CD-ROM.* 1999: John Wiley \& Sons, Inc. 672.

11. Zhang, G., B. Eddy Patuwo, and M. Y. Hu, *Forecasting with artificial neural networks: The state of the art.* International Journal of Forecasting, 1998. 14(1): p. 35-62.

12. Barad, S.G., et al., *Neural network approach for a combined performance and mechanical health monitoring of a gas turbine engine.* Mechanical Systems and Signal Processing, 2012. 27: p. 729-742.

13. Nasr, M.S., et al., *Application of Artificial Neural Network (ANN) for the prediction of EL-AGAMY wastewater treatment plant performance-EGYPT.* Alexandria Engineering Journal, 2012. 51(1): p. 37-43.

14. Tang, Z. and P.A. Fishwick, *Feedforward Neural Nets as Models for Time Series Forecasting.* ORSA Journal on Computing, 1993. 5(4): p. 374-385.

15. Zhang, G.P., B.E. Patuwo, and M.Y. Hu, *A simulation study of artificial neural networks for nonlinear time-series forecasting.* Comput. Oper. Res., 2001. 28(4): p. 381-396.

16. Tan, W.L., et al., *Optimum parameters for fault detection and diagnosis system of batch reaction using multiple neural networks.* Journal of Loss Prevention in the Process Industries, 2012. 25(1): p. 138-141.

17. Tamilselvan, P. and P. Wang, *Failure diagnosis using deep belief learning based health state classification.* Reliability Engineering & System Safety, 2013. 115: p. 124-135.

18. Tian, Z., L. Wong, and N. Safaei, *A neural network approach for remaining useful life prediction utilizing both failure and suspension histories.* Mechanical Systems and Signal Processing, 2010. 24(5): p. 1542-1555.

19. Stamatelatos, M. and W. Vesely, *Fault Tree Handbook with Aerospace Applications*, NASA, Editor. 2002: Washington, DC.

20. Hipel, K.W. and A.I. McLeod, *Time series modelling of water resources and environmental systems.* Vol. 45. 1994: Elsevier.

21. Peng, Y., M. Dong, and M.J. Zuo, *Current status of machine prognostics in condition-based maintenance: a review.* The International Journal of Advanced Manufacturing Technology, 2010. 50(1): p. 297-313.

22. Asgari, H., X. Chen, and R. Sainudiin, *Applications of artificial neural networks (ANNs) to rotating equipment.* 2011.

23. Oladokin, V.O., O.E. Charles-Owaba, and C.S. Nwaouzru, *An application of artificial neural network to maintenance management.* Journal of Industrial Engineering International, 2006. 2(3): p. 19-26.

24. Raptodimos, Y., et al. *Ship sensors data collection and analysis for condition monitoring of ship structures and machinery systems.* in *Smart Ships 2016 Conference.* 2016. London: Royal Institution of Naval Architects.

25. Baio, G. and A.P. Dawid, *Probabilistic sensitivity analysis in health economics.* Statistical methods in medical research, 2015. 24(6): p. 615-634.

26. Kang, S.Y., *An investigation of the use of feedforward neural networks for forecasting.* 1992, Kent State University.

9. AUTHORS BIOGRAPHY

Mr Yiannis Raptodimos is a PhD researcher at the University of Strathclyde, Glasgow, UK. His research investigates the development of intelligent ship maintenance tools and strategies. He is also part of the INCASS EU-FP7 funded project. He completed his MEng in Naval Architecture & Marine Engineering in 2014 also at the University of Strathclyde. His previous experience includes industrial experience with regards to bulk carrier dry-docks and machinery inspections.

Dr Iraklis Lazakis is a senior lecturer in the Department of NAOME at the University of Strathclyde. He has 10 years of industrial experience with regards to ship operations involving maintenance and repair and surveys. He has developed his expertise on reliability and criticality based maintenance in order to improve ship safety and operations. He is currently leading INCASS FP7 project and has also participated and contributed in industry, UK and EU funded research related projects in the fields of maintenance, inspection, condition monitoring, data analytics, decision support systems, reliability and risk analysis.

SHIPBOARD POWER SYSTEMS RECONFIGURATION: A COMPARED ANALYSIS OF STATE-OF-THE-ART APPROACHES

L Agnello, M Cossentino, G De Simone and L Sabatucci, National Research Council of Italy ICAR Institute, Palermo, IT

SUMMARY

The Shipboard Power System (SPS) supplies power to navigation, communication, operation and critical systems. The capability of facing single or multiple faults is a mandatory issue for any vessel.

This paper reports a systematic comparison on SPS reconfiguration methods, where most recent contributions to the field have been classified according to taxonomy of criteria, such as: reconfiguration techniques, reconfiguration sub-problems, and characteristics of the electrical layer.

The reconfiguration procedure should be timely in restoring power in faulted areas of the ship, also to avoid subsequent cascade failures; the reconfiguration sub-problems involve priority definition among loads and operations, strongly depending from the electrical layers and the fault diagnosis methodology; moreover, reconfiguration techniques include several different control architectures and load priority schemas.

Literature results encompass several case studies, and employed methods have been deeply analysed.
Finally, some limits of the current state of the art have been identified.

1. INTRODUCTION

As technology evolves, shipboard electronic components demand high performance from the electric power system. The shipboard power system (SPS) supplies energy for communication, navigation, and operation systems. A robust SPS must be able to supply consistent power despite a variety of changing loads and network conditions. A power system consists of various components such as generators, cables, switchboards, loads, circuit breakers, bus transfer switches, and fuses.

The components are continuously updated even during navigation, and therefore power requirements and load demand may change. For instance, a power peak and/or a component failure may produce nearly instantaneous changes in load demand and system topology. In order to meet all of these changing situations, a shipboard system may employ various levels of reconfiguration of the electrical layer via computer-controlled switches across a variety of time scales. This problem of reconfiguration for power systems has been a topic of research for around three decades.

A few reviews studying the problem of the SPS reconfiguration may be found in literature. In [31] authors include a comparison between reconfiguration techniques applied to the terrestrial and SPS domain. They include an analysis of the SPS characteristics; studying and reviewing integrated protection, power distribution, and typical loads on-board. In another review [32] authors surveyed different formulations of the reconfiguration problem and the different techniques used in its solution. They compare the problem of SPS reconfiguration to that of large-scale systems, exploring the issue of optimal reconfiguration from a variety of perspectives. The authors highlight the reconfiguration research field has been mainly focused on finding an optimal switch configuration without considering the system dynamics that results from the switching actions and neglecting the transient behaviour resulting from the reconfiguration actions.

The purpose of this paper is to survey the existing literature in a reasoned manner for highlighting reconfiguration techniques according to intrinsic electrical and problem features.

2. MOTIVATION OF THE RESEARCH

The reconfiguration of an electric network in an SPS is a critical operation requested both to restore power supply, and to meet particular navigation requirements: it is possible, in fact, to change the network topology in order to isolate a fault, to transfer energy to vital loads in critical situations, or to optimize the electrical and electronic systems for the purpose of improving energy efficiency. At present, it is possible to implement several software reconfiguration methodologies.

Software systems are able to intelligently and effectively manage the electrical/electronic hardware on-board, thus allowing a sophisticated and real-time perception of the situation and a ready management of breakdowns, emergencies, energy peaks, etc. Various software methodologies are applicable also considering the underlying electrical architecture and the power supply type.

A typical problem that can be addressed with such reconfiguration methodologies is depicted in [5]. An SPS zonal architecture is composed of AC generators, switchboards, circuit breakers, loads and other electrical equipment, as shown in Figure 1.

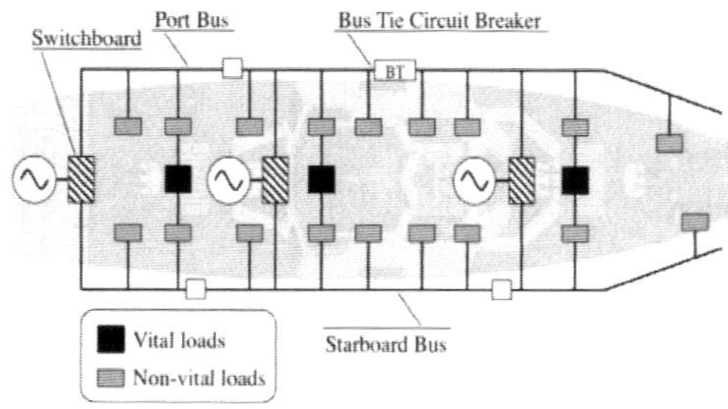

Figure 1: A typical zonal architecture of SPS from [9]

Loads are divided into vital and non-vital categories. Here, an approach using Multi-Agents System (MAS) is used.
The proposed paper studies reconfiguration issue with the aim to identify the relevant variables of the problem and highlighting possible correlations among them.

3. THE PROPOSED COMPARISON APPROACH

The needs motivating the proposed review have been identified by performing a domain analysis, a problem identification and, finally, the formalization of the resulting research questions.

After a fault is encountered, the reconfiguration of the SPS becomes a critical activity. It is desirable to restore power feeding in the shortest time in order to ensure system survivability. This depends on specific features of the system and the occurred problem. The current type and the electrical architecture are deemed to be significant for the reconfiguration strategy.

In the SPS, loads provide various services to the ship. According to the importance of the services being provided, loads in the SPS can be classified into non-vital, semi-vital, and vital, following an increasing priority order. A common approach to the SPS reconfiguration is to supply power to high priority loads and to shed certain loads when needed.

The need for this systematic review is to understand the state-of-the-art about issues affecting SPS reconfiguration, specifically focusing on electrical features, reconfiguration algorithms, and reconfiguration sub-problems.

In particular, this paper considers a couple of research questions:

- RQ1. Is there a relationship between the electrical feature of the system and the technique used for the SPS reconfiguration?

Table 1: Summary of data extracted by systematic review

	Electrical Features		Reconfiguration Sub-Problem			Reconfiguration Technique	
Authors	**Current Type**	**Electrical Network Topology**	**Fault Scenarios**	**Priority Load Reconfiguration**	**Load Shedding**	**Algorithm/Method**	**Control Type**
Xue et al. [1]	AC	ZONAL	Single Failure	✓	✓	Particle Swarm Optimization	n/a
Momoh et al. [2]	n/a	ZONAL	Single Failure	✓	✓	Multi-Agents System	Decentralised
Solanki et al. [3]	AC	RADIAL	Single Failure	n/a	n/a	Multi-Agents System	Decentralised
Huang, Srivastava et al. [4]	AC	RING	n/a	n/a	n/a	Multi-Agents System	Decentralised
Belkacemi et al. [5]	DC	ZONAL	Single Failure	✓	✓	Multi-Agents System	Decentralised
Bose et al. [6]	AC + DC	ZONAL	Multiple Contemporary Failures	✓	✓	Non-Convex vs. Convex Approximation	Centralised
Ouyang et al. [7]	DC	ZONAL	Multiple Contemporary Failures	✓	n/a	Nonlinear Optimization	n/a
Ma et al. [8]	n/a	RADIAL	Multiple Contemporary Failures	✓	n/a	Differential Evolution	n/a
Butler-Purry [10]	AC	RADIAL	Single Failure + Faults Prediction	✓	✓	Multi-Agents System	Centralised
Shariatzadeh, Srivastava et al. [11]	AC	ZONAL	Single Failure + Multiple Contemporary Failures	✓	✓	Genetic Algorithm vs. Particle Swarm Optimization	Centralised
Padamati, Srivastava et al. [12]	DC	ZONAL	Single Failure + Multiple Contemporary Failures	✓	✓	Genetic Algorithm	n/a
Davey et al. [14]	n/a	MESHED GRID	Single Failure of Different Kind	✓	✓	Branch-and-bound optimization	n/a
Srivastava and Butler-Purry [15]	AC	RADIAL	Single Failure + Cascade Failures + Faults Prediction	✓	✓	Probability-based	n/a
Das et al. [16]	DC	ZONAL	Single Failure + Multiple Contemporary Failures	✓	n/a	Q-learning	Centralised
Pal et al. [17]	DC	ZONAL	Single Failure + Multiple Contemporary Failures	✓	n/a	Q-learning + Markov Decision Process	n/a
Bose et al. [18]	DC	ZONAL	n/a	✓	n/a	Delay Analysis Method	Centralised
Srivastava and Butler-Purry [20]	AC	RADIAL	Multiple Contemporary Failures	✓	✓	Rule-based Expert-system	n/a
Feliachi et al. [21]	DC	ZONAL	Single Failure	✓	n/a	Multi-Agents System	Decentralised

- RQ2. How does the technique affect specific sub-problems of the SPS reconfiguration?

4. RESULTS

In this section, a detailed description of the compared papers is reported. In order to answer the two research questions, the most significant characteristics of each paper have been reported, dividing them according to the proposed comparison design, as it can be deducted from Table 1.

A large collection of papers has been selected from various journals and conference proceedings, from 2006 to 2016. After that, this collection has been filtered for identifying a highly focused sample. Title, abstract, methodology section and conclusions of the surveyed papers have been analysed, as it is recommended in [27] and [28].

In the following, a detailed description of the electrical features, problem features, and reconfiguration characteristics of the surveyed papers is reported.

4.1 ELECTRICAL FEATURES

Electrical networks in vessels exhibit very different features. It is possible to classify them according to different classes; the most frequently discussed are: current type (AC/DC), and network topologies. In the following subsections, these will be further discussed.

4.1 (a) Current Type

As depicted in Table 1, several different shipboard power systems have been identified: 7 of them ([1], [3], [4], [10], [11], [15], [20]) refer to the adoption of AC current type, while other 7 ([5], [7], [12], [16-18], [21]) refers to DC current type; only in [6] a mixed AC-DC power system has been found. Both of them have advantages and disadvantages.

For instance, Alternate Current (AC) powered systems have smaller section cables, but bigger and heavier components, as told in [26]; moreover:

- AC generators should work using fixed speed, thus neglecting fuel consumptions efficiency,
- AC transformers are heavy and space-wasting,
- AC current quality is problematic, it is heavily affected by reactive power and harmonic issues,
- Military applications with a support for advanced electrical equipment and weapons are characterized by high-power pulsed loads and they can bring to severe damages to the system,
- AC systems are very diffused in terrestrial applications, also because of their ability to transport energy for long distances with a reduced loss of power. In a ship this may be an irrelevant factor because distances are relatively small.

Besides, Direct Current (DC) powered systems are composed of smaller components, that may save a lot of weight on-board. In particular [6], [16] and [18] refers to a particular kind of DC electrical network, called Voltage DC (MVDC), which specifics for power systems on ships design guidelines are discussed in [33]. Moreover they exhibit:

- compact power converters,
- easier connections,
- no reactive power and harmonic issues.

These features can bring to faults reduction and easier reconfiguration procedures. The main disadvantage of DC systems is that voltage shifts are more difficult to be realized than in AC systems where transformers do that with minimal losses.

4.1 (b) Electrical Network Topologies

The conducted survey highlights two main kinds of electrical network topologies: radial and zonal (see [29]).

In a radial topology, loads are often linked from a single connection point thus reducing ships' cabling. This topology doesn't seem to offer the possibility to segment the system into distinct zones, but it needs shorter cables. Radial topologies are used in 5 ([3], [8], [10], [15], [20]) of the surveyed papers.

The zonal electric distribution systems (ZEDs) seem to be the most used one, mainly for their reliability. In fact, the zonal architecture provides construction cost savings while it inherently offers more operational flexibility [9]. In ZEDs, two parallel and distinct bus rails run along the port and starboard sides of the ship. Components are then positioned across these buses, dividing power distribution in zones. It this way, the electrical network can be configured as a ring bus and it may also use a radial distribution methodology from the switchboards. Moreover, a single fault can be easily isolated and the reconfiguration procedure can be more effective. In [30] the zonal topology outperforms ring bus in terms of service interruption rates. Table 1 reports that 11 papers out of the 18 surveyed ones refer to zonal topologies ([1], [2], [5-7], [11], [12], [16-18], [21]).

Finally, two of the papers exhibit a ring topology or a meshed topology. The ring bus topology is a kind of network spreading around the ship and forming a single main power bus loop. In [4] authors use the ring topology in order to highlight a specific problem of redundant information accumulation (RIA), solving the problem transforming the ring topology into a radial one. A meshed grid has been found in [14]. This is a more complex topology that uses redundancy for power rerouting purposes.

All the discussed power topologies are feasible for shipboard power systems. However, there is always a trade-off between reliability and complexity. Complex topologies require more accurate reconfiguration strategies that have to be taken into account at design stage; instead, simpler topologies may be easily repaired, but they may also be subject to major breakdowns. The trade-off indeed mostly depends on the shipboard specific requirements.

4.2 RECONFIGURATION SUB-PROBLEMS

The reconfiguration sub-problems identified in the surveyed papers are: number of failures, priority-based load reconfiguration, and load shedding. They will be discussed in details in the following.

4.2 (a) Fault Scenarios

A reconfiguration strategy is necessary when a fault (or a series of faults), happens in an unknown instant. In particular, authors of [10] and [15] discuss the reconfiguration strategy that has to be adopted before an expected fault (for instance, before an attacker hits the targeted vessel). Other surveyed papers are related to reconfiguration strategies to be adopted after the occurrence of an unexpected fault.

In [1-3], [5] and [10], reconfiguration and self-healing procedures are related to the simplest case, that is the single fault. Once an anomaly has been detected, the first reconfiguration step is to identify the faulted zone. After that, the reconfiguration approach may disconnect power to the faulted region, and/or it may try to repair that by using specific techniques.

Cases that need more sophisticated strategies occur when multiple faults are detected. In papers [6-8] and [20], the authors take into account the occurrence of multiple contemporary faults. In [15], authors both study the case of single and subsequent (cascade) faults.

Finally, in 4 of the surveyed papers ([11], [12], [16], and [17]), authors discuss both single faults and multiple contemporary faults.

4.2 (b) Priority Load Reconfiguration

An SPS includes several electrical equipments: cables, transformers, generators, loads, and many others. Obviously, not all the loads have the same importance for vessel survivability. For example, a military aircraft carrier can have sensible and vital loads such as weapon systems and operatory rooms. In the case of faults (i.e. due to an enemy attack) the most important electrical equipment in terms of survivability, like engines, countermeasure systems, and so on, should be restored as soon as possible.

The majority of surveyed papers take into account reconfiguration algorithms with priority-load aspects. In particular, authors of [11] and [12] use an objective function in order to maximize the power to unaffected loads and to redirect power towards high priority loads. In this sense vital and semi-vital loads are more relevant than non-vital ones. MAS-based methods ([2], [5], [21]) too, use priorities among agents.

4.2 (c) Load Shedding

In maritime applications, load-shedding policies may have the same importance (or even more) than in the terrestrial counterpart. Load shedding can be acted for different reasons: energy consumption reduction, power outages, and equipment protection. Most of the revised papers investigate load shedding-related problems.

For example, in [2] the proposed system shed the loads using a priority list. In [5] vital loads are non-sheddable loads that affect the survivability of the ship: the load-shedding scheme never intentionally interrupts power to vital loads; instead, non-vital loads are sheddable and they can be immediately disconnected without adversely affecting ship operations, survivability, or life.

In [11], a series of 8 different scenarios are shown, with the corresponding load shedding policies.

4.3 RECONFIGURATION TECHNIQUE

The reconfiguration techniques found in the surveyed papers are mainly based on different algorithms/methods and different logical control types. They will be further detailed in the following subsections.

4.3 (a) Algorithm/Method

Most recent approaches found in literature deal with the reconfiguration issue addressing it by using different methodologies. According to literature, the most used reconfiguration methodologies seem to belong to two main categories: Multi-Agent Systems (MAS), and others methodologies (mostly mathematical methods and knowledge-based methods).

Multi-Agent Systems: Mas-based methodologies are mostly used in shipboard reconfiguration, since each agent can be modelled as a proactive software entity that controls a single electrical component and communicates with other agents. An agent is an autonomous entity that has intelligence, reactivity, and adaptivity [22]. Basically, it is an information processor that executes actions based on communications among other agents and that actively reacts to surrounding environment. Multi-agent architectures are therefore highly taken into account when dealing with the reconfiguration issue. An agent can sense a fault, and besides repairing it, it can communicate to the other agents the new emergency situation. In this way, a global strategy that can reconfigure the electrical layer is possible distributing workloads and responsibilities among agents. Moreover, it is possible to use special hierarchical architectures that

allow system users (i.e. senior officers) to control operations coordinating low-level agents, responding at run-time to sudden crises.
In surveyed papers, this approach is adopted by [2-5], [10], and [21].

The agents try to accomplish fault isolation and power restoration actions, determining which switch in the system needs to be closed and which needs to be opened, restoring the vital load rapidly, minimizing the number of switching operations. Reconfiguration behavior of agent systems performs during the period of reconfiguration requirement.

All the MAS-based approaches show similar characteristics, in particular:

- Decentralised control: the control logic in MAS is decentralised, avoiding problem such as single point of failures, and system topology dependency. This is possible often restricting communications from one agent to their neighbours. This kind of control achieves information awareness like centralised control, and broader functionalities such as decentralised control. This can be obtained using a communication network that should be as reliable as possible (i.e. see [2] and [4]).
- Priority-based load restore: as depicted in [2] and [5], the priority of loads is taken in account and agents restore power primarily to vital loads (engines, weapons), and after to loads of secondary importance; moreover, in [21] priority is taken into account also for energy management purposes.
- Heterogeneous architectures: agents can be applied in various kinds of electrical topologies, such as zonal ([2], [5], [21]), ring ([4]), and radial ([3]).

It is worth noting that MAS-based approaches deal only with single faults scenarios; dynamic behaviour where multiple contemporary faults happen is not present in literature contributions.

Agent-based prototypes are often realized using Matlab and Java Agent DEvelopment Framework (JADE) [13], that is a software framework for the development of intelligent agents implemented in Java.

In this sense, a significant contribution of MAS-based reconfiguration can be found in [5]. The method achieves reconfiguration, restoration and load shedding goals, detecting and isolating faults, reconfiguring and restoring power to healthy loads and performing load shedding when necessary. The proposed system has been further implemented in microcontroller hardware and deployed on a power system test-bed for real-time testing and development purposes. The simulation and experimental results show that agents are capable of performing very fast reconfiguration and load shedding schemes.

Non-MAS methodologies: Among the non-MAS methodologies, the most used algorithms are based on mathematical and knowledge-based methods.

Mathematical methods are related to optimization and approximations algorithms.
In [1], authors use a discrete Particle Swarm Optimization (PSO) algorithm that achieves a rapid and optimal reconfiguration scheme for SPS. The method employs graph theory to generate an emergency reconfiguration, which can quickly restore the system with relaxed operation constraints; then, the particle swarm optimization (PSO) algorithm is used to improve this reconfiguration scheme while considering multiple objectives and stringent constraints. Authors also used a 32-bus electric shipboard test-bed that tests the performances. The model can execute a fast restoration that can satisfy the real-time application requirement, it can also consider multiple objectives for reconfiguration, including minimum power losses, the numbers of switching operations, the numbers of loops in the system and load shedding amount. The graph theory is applied for modifying the network topology in order to accelerate the solution.

In [6] authors consider a reconfiguration method based on low complexity Convex Approximation that is effective in finding optimal solutions. The cumulative distribution function (CDF) of the power delivered to loads is presented to showcase system robustness against random fault scenarios. The method finds an optimal reconfiguration trade-off between power delivery maximization and number of switching actions minimization. Moreover, a separate analysis observes the intermediate dynamic switch states while the reconfiguration is in progress to capture the trade-off more prominently.

The work in [7] presented a Mixed-Integer Nonlinear Programming optimization method to achieve the multiple objectives of reconfiguration, maximizing the restored load and minimizing the number of the switch operations. The DC-zonal SPS line and generator faults are considered, and the reconfiguration method is implemented by means of a software modelling system for mathematical optimization (GAMS) [19], which proves the correctness and feasibility of the proposed method.

In [8], authors use an improved Differential Evolution algorithm, combining chaotic initialization population, improving Pareto elitist selection strategy and adaptive mutation and crossover operator, that obtains an optimal solution. Obtained results are compared with other algorithms (NSGA-II [23], CADA [24], DEMO [25]), demonstrating a better efficiency than other algorithm in solving SPS reconfiguration.

It is possible to find methods that use knowledge in addition to intelligent and smart techniques.

In [11], an intelligent reconfiguration methodology, using a Genetic Algorithm (GA) and a Particle-Swarm Optimization algorithm (PSO), is used to find a post-fault optimal solution for the reconfiguration problem, to maintain the real power balance of the rest of the post-fault isolated system, and to reduce the effect of the fault with consideration of islanding (load shedding) process. The same authors in [12] apply the genetic algorithm for supply restoration and optimal load shedding.

In [14], the authors consider the dynamic impedance measurement as a key element in SPS monitoring. The way to analyse a dynamic shipboard power system is to use equivalent series and parallel impedances to represent power transmission and useful load power, respectively. The impedances are computed and updated in real-time using current and voltage measurements on a trunk line feeding. An optimization algorithm (binary branch and bound, for instance) is then applied determining the best configuration that meets critical power demand and minimizes losses.

In [15], authors present a predictive method for the reconfiguration of an SPS that utilizes electrical and geographical data and advanced techniques that determines the pre-hit reconfiguration actions for vital and non-vital loads, respectively. This probabilistic approach calculates the expected probability of damage (EPOD) for each electrical component on a ship. Further, a heuristic method uses the EPOD to determine control actions to reconfigure the ship's electrical network to reduce the damage to the electrical system.

Authors of [16] and [17] propose a method for dynamic optimal reconfiguration. Optimal reconfiguration is viewed as a learning problem, rather than one of explicit optimization. Reinforcement learning is used as the learning approach. The specific algorithm used here is Q-learning, a very popular reinforcement-learning algorithm, based on dynamic programming. This off-policy method theoretically converges to the optimal reconfiguration. As Q-learning is an online approach, it can be deployed within a real SPS, where the algorithm can not only fine-tune itself with time but also re-adapt in the presence of non-stationary SPS environment. The proposed method also provides the correct sequence of switching operations that would lead to optimal reconfiguration in the shortest possible time.

Finally [18] studies a cross-layer end-to-end delay analysis method for real-time power system reconfiguration after the occurrence of faults. Several centralised sensor network topologies and their impacts on the delay distribution are analysed using a real-time analysis (RTA) framework. Real-time QoS guarantees with current and potential communication technology implementations (FDDI/Ethernet/Gigabit Ethernet) for shipboard are also compared. Specifically, cyber/physical system delays are modelled, including information aggregation, queuing, transmission, communication, and the computation associated with fault isolations and reconfiguration of switch status. One of the main metrics to quantify the performance of the cyber-physical system (CPS) is the real-time delay (a component of QoS). The proposed approach is based on probability theory, real-time queuing theory, and their application to CPS.

4.3 (b) Control Type

Another relevant aspect of the surveyed papers takes into account the control layer strategy. In large-scale shipboard grids, hierarchical control is a good choice because different layers can be controlled independently. Coordination methods are then related to communication techniques, mostly centralised and decentralised.

A centralised control for reconfiguration purposes can be obtained using a central controller, in a master/slave way. This way of control has the advantage that the control algorithm runs on a single node, but the difficulty increases with the increasing number of controlled components. Moreover, a failure on the main controller can lead to a global failure. This kind of control technique is used by authors of [6], [10], [16], and [18]. In particular, authors of [18] focus on communication delays in on-board sensor networks, comparing several centralised architectures: centralised network control (CNC); CNC with backbone; and centralised-cluster network control.

Decentralised control makes coordination independent from the main controller and can overcome single-points failures by distributing computational load among control objects. On the contrary, the accuracy of distributed measurements can impact the effectiveness of the SPS reconfiguration procedure. In [2-5] and [21] authors use the decentralised control technique.

Specifically, in [2] authors restrict agent communications to their immediate neighbours, making the system less dependent on the topology of the system and thus reducing the communication burden within the system. In [3] the locality of information makes the reconfiguration strategy topologically independent and it can works autonomously. In [4], authors deliberately highlights how reconfiguration procedures that require global information are centralised, and they overcome this drawback avoiding single points of failures with decentralised control.

Only in [10], authors implement the architecture of the self-healing system using a centralised monitoring and control framework, but using a multi-agent system with decentralised information and asynchronous computation.

Finally, in [11] authors use both centralised (GA) and decentralised (PSO) techniques.

5. DISCUSSION

In this section a reasoned discussion of the conducted survey is proposed.

5.1 ELECTRICAL FEATURES DISCUSSION

As mentioned before, electrical current types (DC and AC) have some advantage and disadvantage, especially when dealing with variables such as equipment weight, quality of current, and so on. Reconfiguration strategies don't seem to be affected by the existing on-board electrical current type. Moreover, electrical topologies are converging on zonal architectures, mostly because faults isolation and reconfiguration techniques may be more effective. Also, as can be depicted from Table 1, this kind of topology is always used by DC current-based on-board SPS. This answers to the first proposed research question (see 3.1 (b)).

5.2 RECONFIGURATION SUB-PROBLEMS DISCUSSION

As regarding reconfiguration sub-problems, number of failures management seems a key aspect. The most effective contributions in this sense are [10] and [15], that discuss faults prediction. Moreover, [15] takes into account both single fault and cascade failures. In addiction to this, all the MAS methodologies seem to take into account only single failures, lacking multiple-faults scenario.

Almost all of the surveyed papers take into account loads weighting and priority in reconfiguration schema. For instance, in [6] loads are explicitly weighted according to non-vital, semi-vital, and vital loads. Vital loads are restored in a manner such that those loads are serviced optimally and according to their priority. This means that vital and semi-vital loads are restored before non-vital loads. In [12] the used genetic algorithm considers loads multiplied by a weighting factor, so the vital and semi-vital load contributions are greater than the largest non-vital load contribution.

Load shedding aspects are mostly highlighted in [5], where vital loads are non-sheddable loads that affect the survivability of the ship or life. Power to these loads is not intentionally interrupted as part of a load-shedding scheme. The scheme is initiated when a generator can no longer supply all the electrical zones. In [15] the probability-based reconfiguration performs load shedding in a preventive manner, avoiding major damages and blackouts after a predicted fault.

In order to answer the second research question (see 3.1 (b)), it is possible to note that the number of failures management seems not correlated with a specific reconfiguration methodology, but they are addressed using approaches becoming from different categories. Also for load shedding and load priorities, reconfiguration strategy seems not correlated with a specific approach.

5.3 RECONFIGURATION TECHNIQUE DISCUSSION

It is worth nothing that most of the surveyed reconfiguration techniques are MAS-based. This is mostly for the granularity level obtained controlling electrical equipment through agents. In addition to that, all MAS methods provide decentralised control that is aware of single-point failures and does not request global information.
Only in [4] the authors consider the reconfiguration strategy from the information point of view, focusing on a specific problem of information inconsistency. They do not take into account the sub-problems of Section 3.

For non MAS-based methods, the most relevant reconfiguration techniques seem to be the Q-learning and PSO methodologies.

The Q-learning can be deployed within a real SPS because it is an online approach, where the algorithm cannot only fine-tune itself with time but also re-adapt in the presence of non-stationary SPS environment [16]. In [17] two Q-learning methods (∈-greedy and simulated annealing) are compared, where the learning curve of the former seems to reach faster the optimal reconfiguration solution.

It is worth mentioning, the PSO in [1] is able to establish an optimal reconfiguration scheme solving non-linear problem with multiple objectives, thanks to less computational complexity and computing time. In [11] the PSO is compared to a Genetic Algorithm (GA), where both simulation and real-time results indicate that GA is slower than PSO algorithm; nevertheless, it requires less number of tuneable parameters if compared to PSO.

Often the rate of convergence and the simplicity of calculations have been used as the only benchmarks of real-time implementability of reconfiguration algorithms. Safe implementations of a reconfiguration algorithm require the analysis of power systems dynamics during and after reconfiguration. The current body of literature lacks a rigorous examination of dynamic aspects of reconfiguration.

Surveyed papers lack of simulation data, so the repeatability of the experiment is quite difficult. Simulation data are only present in [3], [4], and [21]. Also, temporal aspects such as agents' reaction time, or the execution time of the reconfiguration procedure are neglected. Only authors of [5] report an estimation of the execution time of the algorithm of less than 20 ms., while authors of [21] report agent' steady state reaching after tens of seconds.

6. CONCLUSIONS AND FUTURE WORKS

In this paper, a comparative analysis of the state of the art approaches concerning smart and automatic reconfiguration on shipboard has been proposed. The review here proposed has taken into account three parameters:

- electrical features of reconfigured power systems,
- qualitative features of the problem domain,
- employed reconfiguration technique.

The SPS electrical features are related to the kind of used electrical current (direct or alternate), and electrical architecture topology (zonal, radial, ring, etc.).

Problem domain features embrace the reconfiguration approach adopted when facing failures. In particular, aspects like number of managed failures, priority load reconfiguration, and load shedding have been highlighted from surveyed reconfiguration techniques. Some papers take into account only single failures, while others consider multiple faults that can happen simultaneously, or subsequently, increasing the instability of the whole shipboard SPS. Once the fault has been identified, reconfiguration procedures can consider loads as equally important, or they can weigh them differently according to some priority schemas. Also the load-shedding procedure has been highlighted, because it can be important to protect some sensible load.

Another feature regards the control logic of the reconfiguration technique: a centralised control is less complex to implement and it is particularly effective, but it has the weakness that a problem with the single control component can compromise the managing of the entire SPS; a decentralised control overcomes this weakness, at the cost of an increased complexity in the design and implementation.

In conclusion, MAS-based approaches seem to be the most effective when dealing with single failures of heterogeneous kind; when multiple failures occur, best reconfiguration methodologies use soft-computing methods like Genetic algorithms, Differential Evolution algorithms and Particle Swarm Optimization. Also Q-Learning methods seems effective in this sense.
Future work will be related to a deeper exploration of MAS approaches, filling lacks of the management of multiple faults, and investigating communication issues among agents.

7. REFERENCES

1. XUE, W., FU, Y., 'A two-step method for reconfiguration of shipboard power system', *In 2011 IEEE Electric Ship Technologies Symposium (pp. 167-172) IEEE*, 2011.

2. MOMOH, J.A., XIA, Y., ALFRED, K.C., 'Dynamic reconfiguration for shipboard power system using multi-agent system', *In Power and Energy Society General Meeting-Conversion and Delivery of Electrical Energy in the 21st Century, 2008 IEEE p.1-4*, 2008.

3. SOLANKI, J.M., SCHULZ, N.N., GAO, W., 'Reconfiguration for restoration of power systems using a multi-agent system', *In Proceedings of the 37th Annual North American Power Symposium p. 390-395*, 2005.

4. HUANG, K., CARTES, D.A., SRIVASTAVA, S.K., 'A multiagent-based algorithm for ring-structured shipboard power system reconfiguration', *IEEE Transactions on Systems, Man, and Cybernetics, Part C (Applications and Reviews) 37.5: 1016-1021*, 2007.

5. BELKACEMI, R., FELIACHI, A., 'Design and deployment of a Multi-Agent System on a Hardware Prototype', *In IEEE PES General Meeting. IEEE p. 1-7*, 2010.

6. BOSE, S., PAL, S., NATARAJAN, B., SCOGLIO, C.M., DAS, S., SCHULZ, N.N., 'Analysis of optimal reconfiguration of shipboard power systems', *IEEE Transactions on Power Systems 27.1: 189-197'*, 2012.

7. OUYANG, L., LI, Y., TAN, Y., XIAO, J., CAO, Y., 'Reconfiguration optimization of DC zonal distribution network of shipboard power system', *In Transportation Electrification Asia-Pacific (ITEC Asia-Pacific), 2016 IEEE Conference and Expo. IEEE p. 444-448*, 2016.

8. MA, L., ZHANG, J., REN, G., 'An improved selection strategy differential evolution algorithm for reconfiguration of shipboard power system', *In 2015 5th International Conference on Information Science and Technology (ICIST) IEEE p. 131-135*, 2015.

9. PETRY, C.R., RUMBURG, J.W., 'Zonal electrical distribution systems: an affordable architecture for the future', *Naval Engineers Journal, 105.3: 45-51*, 1993.

10. BUTLER-PURRY, K.L., 'Multi-agent technology for self-healing shipboard power systems', *In Proceedings of the 13th International Conference on Intelligent Systems Application to Power Systems, IEEE*. 2005.

11. SHARIATZADEH, F., KUMAR, N., SRIVASTAVA, A., 'Optimal Control Algorithms for Reconfiguration of Shipboard

Microgrid Distribution System using Intelligent Techniques', *IEEE Transactions on Industry Applications*, 2016.

12. PADAMATI, K.R., SCHULZ, N.N., SRIVASTAVA, A.K., 'Application of genetic algorithm for reconfiguration of shipboard power system', *In Power Symposium, 2007. 39th North American. IEEE p. 159-163*, 2007.

13. BELLIFEMINE, F.L., CAIRE, G., GREENWOOD, D., 'Developing multi-agent systems with JADE', *John Wiley & Sons*, 2007.

14. DAVEY, K. R., HEBNER, R. E., 'Reconfiguration of shipboard power systems', *CEM Publications*, 2015.

15. SRIVASTAVA, S. K., 'Probability-based predictive self-healing reconfiguration for shipboard power systems', *IET Generation, Transmission & Distribution 1.3: 405-413*, 2007.

16. DAS, S., BOSE, S., PAL, S., SCHULZ, N. N., SCOGLIO, C.M., NATARAJAN, B., 'Dynamic reconfiguration of shipboard power systems using reinforcement learning', *IEEE Transactions on Power Systems, 28(2), 669-676*, 2013.

17. PAL, S., BOSE, S., DAS, S., SCOGLIO, C.M., NATARAJAN, B., SCHULZ, N. N., 'Shipboard power system reconfiguration using reinforcement learning', *In North American Power Symposium (NAPS), 2010. IEEE p. 1-7*, 2010.

18. BOSE, S., NATARAJAN, B., SCOGLIO, C.M., SCHULZ, N. N., GRUENBACHER D.M., DAS, S., 'Shipboard Power Systems Reconfiguration—A Cyber-Physical Framework For Response Time Analysis', *IEEE Transactions on Industrial Informatics 10.1: 439-449*, 2014.

19. BUSSIECK, M.R., MEERAUS, A., 'General algebraic modeling system (GAMS)', *In Modeling languages in mathematical optimization, Springer US p. 137-157*, 2004.

20. SRIVASTAVA, S., BUTLER-BURRY, K.L., 'Expert-system method for automatic reconfiguration for restoration of shipboard power systems', *IEE PROCEEDINGS GENERATION TRANSMISSION AND DISTRIBUTION, 153.3: 253*, 2006.

21. FELIACHI, A., SCHODER, K., GANESH, S., LAI, H.J., ' Distributed control agents approach to energy management in electric shipboard power system', *In 2006 IEEE Power Engineering Society General Meeting p. 6 pp. IEEE*, 2006.

22. WOOLDRIDGE, M., 'An introduction to multiagent systems', *John Wiley & Sons*, 2009.

23. WANG, J., 'Multiobjective Optimal Network Reconfiguration of Shipboard Power System Based on Non-Dominated Sorting Genetic Algorithm-II', *Power System Technology 11: 012*, 2012.

24. ABBASS, H.A., SARKER, R., NEWTON, C., 'PDE: a Pareto-frontier differential evolution approach for multi-objective optimization problems', *In Evolutionary Computation, 2001. Proceedings of the 2001 Congress on. IEEE p. 971-978*, 2001.

25. SHI, J., HUANG, M., 'Application of improved binary differential evolution algorithm in distribution network reconfiguration', *Mechanical & Electrical Engineering Magazine 12: 019*, 2008.

26. JIN, Z., SULLIGOI, G., CUZNER, R., MENG, L., VASQUEZ, J.C., GUERRERO, J.M., 'Next-Generation Shipboard DC Power System: Introduction Smart Grid and dc Microgrid Technologies into Maritime Electrical Networks', *IEEE Electrification Magazine 4.2: 45-57*, 2016.

27. KITCHENHAM, B.A., CHARTERS, S., 'Guidelines for performing systematic literature reviews in software engineering', *In Technical report, Ver. 2.3 EBSE Technical Report. EBSE.* 2007.

28. BRERETON, P., KITCHENHAM, B.A., BUDGEN, D., TURNER, M., KHALIL, M., 'Lessons from applying the systematic literature review process within the software engineering domain', *Journal of systems and software 80.4: 571-583*, 2007.

29. TIDD, C.N., 'Hardware Model of a Shipboard Zonal Electrical Distribution System (ZEDS): Alternating Current/Direct Current (AC/DC)', *Massachusetts Institute of Technology, Cambridge Deptartment Of Electrical Engineering And Computer Science*, 2010.

30. DUBEY, A., SANTOSO, S., 'A Two-Level Topology Design Framework for Reliable Shipboard Power Systems', *Inventions 1.3: 14*, 2016.

31. DAHALAN, W.M., MOKHLIS, H.B., 'Techniques of network reconfiguration for service restoration in shipboard power system',

A review. Australian Journal of Basic Applied Science 4.11: 55565563, 2010.

32. NAGARAJ, K.C., CARROLL, J., ROSENWINKEL, T., ARAPOSTATHIS, A., GRADY, M., POWERS, E.J., 'Perspectives on power system reconfiguration for shipboard applications', *In 2007 IEEE Electric Ship Technologies Symposium p. 188-195*, 2007.

33. IEEE Std. 1709 'IEEE Recommended Practice for 1 kV to 35 kV Medium-Voltage DC Power Systems on Ships', *IEEE Industry Applications Society*, 2010.

8. AUTHORS BIOGRAPHY

Luca Agnello holds the current position of post-doctoral fellowship in the ECOS Lab of the ICAR-CNR (Italian National Research Council), Palermo, Italy. He is responsible for soft-computing techniques applied to various field of computer science research. His previous experience includes soft computing techniques applied to medical imaging, neural networks, and big-data mining.

Massimo Cossentino got his master degree in Electronics Engineering and his PhD in Computer Science Engineering from the University of Palermo. From 2001 he is a research scientist of the Italian National Research Council at the Palermo Dept. of the ICAR (Institute for High Performance Computing and Networks) Institute.
He has been an invited associate professor at the Laboratoire Systèmes et Transports (SET) of the Université de Technologie Belfort-Montbélliard (France). In France he also got his "Habilitation a Diriger des Recherce" from the Université Paul Sabatier of Tolouse (France).

His research interests are in the field of Software Engineering and more specifically: self-adaptive and self-organizing systems, software design methodologies and composition of ad-hoc design processes.

He is the author of more than one hundred and seventy scientific papers published in international journals, conference/workshops proceedings, and books. He is currently leading the Engineering Complex and Smart Systems lab of the ICAR-CNR Institute.

Giada De Simone currently has a research grant at ICAR-CNR, Palermo, Italy. She holds a degree in physics, and her previous studies involve biophysics neural networks.

Luca Sabatucci is a research scientist in the ECOS Lab of the ICAR-CNR (Italian National Research Council), Palermo, Italy. Principal interests are in the areas of Self-Adaptive Systems, Requirements Engineering, and Software Patterns. He is the author of about sixty papers and participated in the organization and program committees of several international conferences and workshops.

DIGITIZATION: INFORMATION TECHNOLOGY (IT) AND OPERATIONAL TECHNOLOGY (OT)

J Beel and **K Jones**, Cisco Systems, Inc., USA

SUMMARY

Digitization leverages networked connection of people, process, data and things to deliver new insights and unprecedented operational visibility.[1] IoT, key to digitization, is the networked connection of devices.

IT delivers information processing and communications; OT is used to monitor and control devices. Previously, IT and OT operated independently because OT fell under line of business managers vice IT. OT devices are becoming more connected as operational environments are automated and networked. IT and OT are converging to allow operational control, monitoring and analysis via networks.

Convergence demands advanced cybersecurity that protects both IT and OT while ensuring availability of critical operational systems. Cybersecurity must be threat-based and address a continuum of events that spans before, during and after attacks. Network segmentation, next generation encryption, intrusion prevention, advanced malware protection and identity services are among technologies that can help protect data and networks before, during, and after an attack.

1. INTRODUCTION

Gartner estimates that the number of connected things will grow to 25 billion by 2020 and Cisco sees the Internet of Things (IoT) growing even faster and estimates that 50 billion devices will be connected by 2020.[2] Gartner views the IoT as a powerful force for business transformation with disruptive impact across all industries.[3] The world will continue to become increasingly connected and through the network, Cisco sees creation of unprecedented value and powerful new types of data and insight. Cisco sees the networked connection of People, Process, Data and Things (sometimes referred to as the Internet of Everything) as key to a new digitized world (Figure 1).[4]

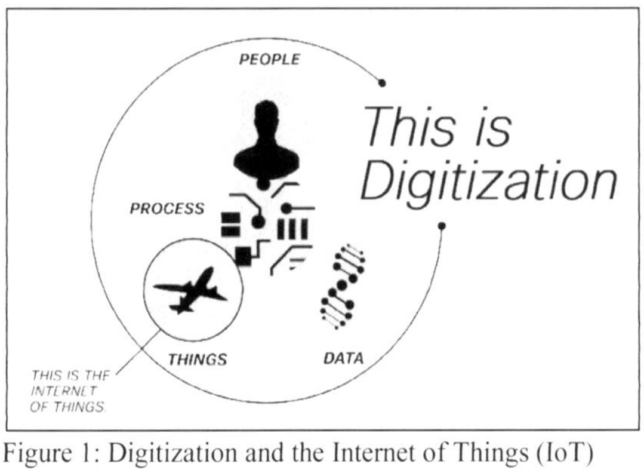

Figure 1: Digitization and the Internet of Things (IoT)

Pratt and Whitney's PW1000G Family of Engines incorporates more than 5,000 sensors in an engine which can generate data at a rate of 10 GB/s. One 12-hour flight for a single twin-engine aircraft can produce up to 844 TB of data.[5] PW1000G engines use this massive amount of data to create less noise and pollution and reduce fuel consumption by 10% to 15%. General Electric (GE) uses large numbers of sensors in manufacturing equipment to digitize operations. GE expects digitization to improve manufacturing efficiency by 40% through data analytics coupled with the IoT. Aircraft such as the Boeing 787 Dreamliner utilize Ethernet-based aircraft data networks that move data at rates up to 12.5 MB/s, blazingly fast compared to older aircraft avionics systems that move data at rates less than 12.5 KB/s. Analysis of engine sensor data is just one example of potential value created by digitization. IoT data can be used to enhance customer experience, optimize fuel efficiency, detect conditions that predict impending problems that will require maintenance, automatically order parts for delivery when and where maintenance will be performed, etc. Aircraft engine manufacturers are a great example of delivering digitized operations and maintenance by leveraging IoT. IoT has taken root and accelerated due to a number of inflection points, which occurred simultaneously and created a perfect storm for this evolution. These factors include[6]:

- Standardized Internet Protocol (IP) across industries
- Network capacity growth
- Explosion of software applications
- Cloud Computing
- Edge Computing
- Big data and data analytics
- Nanotechnology and the miniaturization of sensors and other end devices

The IP network has become prevalent throughout industry. Enterprises made investments to move toward a standard IP network and networks scaled to support increasingly larger enterprises. Cloud computing has also made significant changes to how applications are delivered and how data is stored. Cloud computing can offer economies of scale and speed delivery of applications.[7] Many enterprises, small and large, have adopted or are currently in the process of adopting some form of cloud computing, whether it is via a private, public, or hybrid cloud. Public cloud Infrastructure as a

Service is projected to reach $38B in 2016 and grow to $173B in 2026.[8] Another factor which is becoming increasingly important is edge or fog computing which brings processing closer to where data is being generated.[9] Fog computing can capture value from devices connected to the IoT network while reducing network bandwidth required to support increased automation and analytics.

Tapping into the massive amount of data generated by the rapidly increasing number of devices connecting to the network is how value is created from digitization. Analytics will allow enterprises to harvest data created by IOT and transform it into information and knowledge for appropriate usage. For example, if airline engines can become more efficient and deliver a 2% fuel savings, annual fuel costs can be reduced by $4B.

Nanotechnology and the miniaturization of sensors and edge devices are enabling the IoT. Wearable devices are starting to be seen everywhere along with smaller and smaller sensors, which will all be ultimately connected to the larger network. Sensor technology has accelerated the growth of IoT and enables digitization in maritime industries.

2. DIGITIZATION IN MARITIME APPLIATIONS

Ships carry about 80% of global trade by volume and 70% by value thereby, playing a key role in the global economy.[10] The Maritime Industry faces numerous challenges and some of the areas where improvement is desired are: improved safety, increased capacity, decreased fuel consumption, improved environmental compliance, logistics visibility, more effective maintenance, etc.[11] Digitization through the IoT can enable the maritime industry to reduce the cost of doing business and enable the following capabilities and more:

- Internal and external communications
- Fuel efficiency
- Energy production and distribution
- Efficient fleet management
- Condition-based maintenance
- Asset management and visibility
- Safer navigation and collision avoidance
- Inventory management and visibility
- Port infrastructure management
- Water production and control
- Emergency response and management

Maritime IT systems have evolved tremendously and are critical for route planning, asset tracking, maintenance planning, etc. Digitization now allows real-time asset tracking and cargo management. For example, refrigerated containers can be monitored continuously to minimize losses due to equipment malfunctions/failures or human error. Ericsson offers a Maritime ICT Cloud to connect vessels at sea with shore-based operations, maintenance providers, customer support, partners, port authorities, etc.[12] This cloud solution includes satellite connectivity as well application hosting and support to allow ships' systems to integrate with shore-based fleet management systems and maximize use of real-time data to monitor critical functions, optimize fuel usage, set and maintain efficient courses and enhance crew safety.

Digitization allows new applications at sea that can greatly enhance operations. For example, shipboard IT networks can deliver highly reliable and efficient unified internet protocol communications both for internal and off-ship communications. Real-time voice, video and chat allow ship crews to optimize manning, equipment monitoring, emergency response and management, safety and collaboration both on and off-ship. Unified communications and collaboration can benefit a range of areas from routine maintenance to personal welfare to highly sensitive business or critical military operations.

The shipboard network can also bring Power-over-Ethernet (PoE) powered lighting that can be monitored and managed via the network, thereby helping to decrease energy costs and enhance safety. Other services, such as outdoor lighting sensors and video analytics, can be connected to the network to provide efficient pier operations management and lighting to improve port operations safety and security. Inventory management can leverage radio frequency identification tags on assets and communicating the location of assets via a wireless network. Sensors in critical machinery can enable condition-based maintenance to minimize maintenance cost and optimize system availability. These services can provide efficient fleet management, whereby decisions can be made faster as data is processed and analyzed on the network in near real-time.

The world's militaries operate numerous ships; The Center for Strategic and International Studies (CSIS) suggests that broader military use of the IoT requires improved connectivity, analytics and interoperability. CSIS also recommends scalable security techniques that can be applied to commercial off-the-shelf (COTS) devices to enable greater use of IoT. Much is known about how to provide security for the traditional enterprise IT systems. IT systems in the maritime industry are not much different than those in the enterprise business as they inherit COTS based equipment, processes, policies, security, etc. from much broader application of IT solutions. IT/OT convergence to enable greater digitization of the maritime environment will allow increased operational efficiency, effectiveness and safety. Convergence will demand a different approach to cyber security to ensure security across both IT and OT systems and deliver required safety, availability and integrity of critical operational systems.

3. IT/OT CONVERGENCE

The worlds of IT and OT are converging to allow optimized business processes, enhanced information for better decisions, reduced cost, lower risk and faster project completion.[1] Organizations must manage convergence, alignment and integration of IT and OT. IT networks are commonly used for research, development, chat, email, phone calls, ordering supplies, managing finances and payroll, directing teams, managing projects, performing human resource management, etc. OT, which is a physical-equipment-orientated technology, consists primarily of control networks and industrial control systems used for management of manufacturing machinery, industrial processes, power generation, mining, Heat Ventilation and Air Conditioning (HVAC), etc. IT and OT have traditionally been viewed as two separate network spaces with two separate and different sets of priorities but they converging in many facilities (Figure 2).

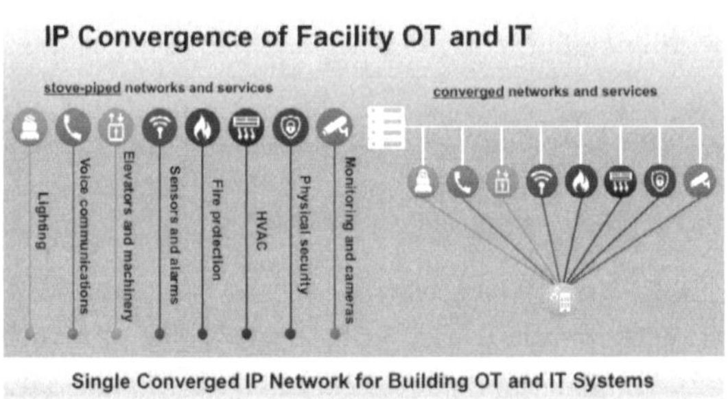

Figure 2: Facility OT and IT Convergence

OT networks frequently require remote access for maintenance and monitoring, are constrained by safety regulations that prevent modification without recertification and often require data sharing with outside organizations. The complexity and cost of replicating Industrial Control Systems (ICS), make offline lab setups for test and evaluation costly. Even with the ability to pre-test security updates, the focus on availability limits deployment of security solutions to standard maintenance windows that are more likely to be limited to a few times a year. When control networks fail, there are risks posed to human lives and environmental safety. Availability and reliability of OT networks are critical and must be maintained at all times to ensure operational performance and safety.[13]

OT systems of the past were typically designed to be independent with no sharing of data or direct connection to other systems or networks. It was generally assumed that any system in a control network was meant to be part of the OT system. Physical separation or "air gaps" between systems and access control by basic username and password were considered sufficient to guarantee security.

But this outdated security approach is showing not just its vulnerability but, its counter-productive nature by allowing sophisticated attacks with potentially dangerous impacts. The STUXNET attack leveraged vulnerabilities on traditional workstations to launch a worm targeting components on Iran's centrifuges and ultimately rendered them physically inoperable. The Target data breach was launched from an HVAC system. Examples like these illustrate the need for an evolved idea of what cyber security encompasses and how OT systems can be protected.

4. CONVERGED CYBERSECURITY

The convergence of IT and OT networks raises several security concerns. The two technology management teams have different priorities. IT teams are primarily focused on confidentiality, or ensuring sensitive information is accessible only by those authorized. Next the focus is on the integrity, or accuracy of the information and availability of the system and data is the final concern. OT teams, however, are focused on availability as the highest priority followed by integrity and confidentiality as lower priorities.

Despite different priorities in IT/OT security approaches, one consistency across both domains is the fact that no single product, technology or methodology can fully secure any network. OT systems require a defense-in-depth network security framework to address both internal and external security threats, just as IT systems do. A balanced OT security framework must address both technical and non-technical elements such as physical policies and procedures. An effective OT security framework should be based on a well-defined set of security policies and procedures, leveraging established IT processes, while balancing the functional requirements of the OT applications. Risk and security assessments should look holistically at the current situation, from technologies to policies, and give a picture of the current security posture and the mitigation techniques available to move to an acceptable risk state.

Converged IT and OT networks require an integrated threat defense with comprehensive visibility and protection capability that spans the continuum of *before*, *during*, and *after* an attack (Figure 3). John Stewart (Chief Security and Trust Officer from Cisco Systems) states that the security threat landscape can be boiled down to three phases: before, during and after.

- *Before* addresses minimizing an attacker's chances of success.
- *During* requires a clear threat containment strategy.
- *After* requires a visibility and forensics capability.[14]

[1] "Gartner Says the Worlds of IT and Operational Technology Are Converging," Christy Pettey and Rob van der Meulen, Gartner, Mar 16, 2011.

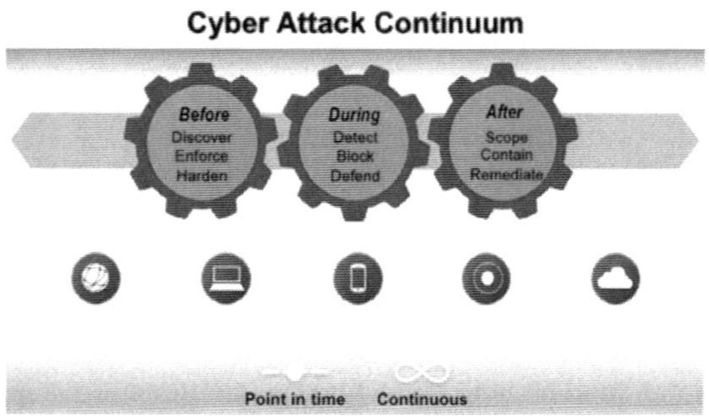

Figure 3: Cyber Attack Continuum

4.1. CONVERGED SECURITY - *BEFORE* THE ATTACK

Before an attack, the focus is on identifying what devices, operating systems, services, applications and users are on the network then translating this information into appropriate policies, access controls and even architectures to minimize the attack surface. Cisco recently analyzed 115,000 Cisco devices on the Internet and discovered that 92 percent of the devices in the sample were running software with known vulnerabilities (26 vulnerabilities per device on average). Also, 31 percent of the devices were "end of sale" and 8 percent were "end of life."[2] Some customers were using versions of software that were more than 6 years old. Currently supported devices operating up-to-date software can greatly improve cybersecurity resiliency and security posture before an attack. Cisco has reduced median time to detect known threats to about 17 hours as compared to 100 to 200 days for the current industry-wide estimated average. Cisco's 2016 Annual Security Report states:

> *"Aging, outdated IT infrastructure is a vulnerability for organizations. As we move closer to the Internet of Things (IoT)-and the internet of Everything (IoE) - it becomes more important for businesses to make sure they are relying on a network infrastructure that is secure, thus ensuring the integrity of the data and communications traversing the network. This is critical to the success of the emerging IoE."*

Segmentation, which breaks the network down into protected physical and logical zones, is a critical step in the *before* stage. Typically in the IT world, the approach has been to create a hardened perimeter defense behind a firewall and allow free flow of traffic in the trusted zone. Microsegmentation breaks down these internal zones behind the hardened perimeter to allow the same physical infrastructure to be used for multiple IT and OT services, while still protecting each segment of the network specific to each service. Through logical segmentation, the attack surface is broken up into segments such that if one micro-area of the network is compromised the rest of the network segments and services are not impacted. Using the Target breach for example, effective segmentation strategy would have ensured that an HVAC system had no ability to communicate with a database containing credit card information thereby, preventing loss of confidential customer data.

Access Control is another aspect to plan and implement in the *before* stage. OT has traditionally focused on basic username and password for security, but convergence onto the broader network requires more sophisticated security controls. Developing and enforcing an enhanced identity and access control policy based on role is a key factor in ensuring protection of converged IT/OT networks. Access should be assigned and enforced based on such factors as identity, location, business function, job role and device type. Relying on previously defined segmentation policy, enforcement should ensure that users, applications and devices are granted access only to authorized segments.

Mobile device management strategies are also important in an OT environment. It is not uncommon for routine maintenance and engineering to be accomplished utilizing portable digital devices, often in the hands of an external user or outside support organization. Such maintenance can create a significant vulnerability. OT networks must treat guests and their devices in a manner similar to bring-your-own-device (BYOD) on an enterprise IT network. OT networks should implement advanced identity management solutions that have the ability to not only manage user access, but also verify device authenticity and security posture before allowing a connection. Additionally, remote access can be secured with Virtual Private Networks (VPNs) that create an encrypted tunnel across the network.

Software-defined segmentation capabilities, such as TrustSec, can simplify and accelerate a consistent security policy across both IT and OT networks. By basing classification on endpoint identity rather than IP addresses, end-to-end policy changes may be implemented centrally and propagated efficiently across the entire network. Modification of users, applications, operating systems, etc. can be changed at the endpoint level and propagated across the network, greatly simplifying access control and segmentation management.

4.2 CONVERGED SECURITY - *DURING* THE ATTACK

While the bulk of security policy, implementation and even funding is dedicated to the *before* stage, mechanisms must be added to monitor, detect and block what does get through. In addition to controlling access, systems can be set up to monitor activity such as who is

[2] Cisco 2016 Annual Security Report, Jan 2016, retrieved from: http://www.cisco.com/c/dam/assets/offers/pdfs/cisco-asr-2016.pdf.

connecting using what device, from where, at what times, and what is being accessed on the network.

In addition to providing a boundary defense, threat focused next-generation firewalls should be implemented to enhance security of converged networks. Traditional network firewalls check for anomalies using IT-based signatures. However, with OT systems connecting to the network, industrial-control signatures must also be incorporated. Next-generation firewalls, which can evoke open source technology to incorporate IT and OT signatures to check for anomalies on the network, are essential. In addition to the standard perimeter security, next-generation firewalls provide protection during an attack by integrating application control, intrusion detection and prevention and malware protection spanning from the network to the endpoint. Firewalls serve a purpose greater than traditional perimeter security because the IoT network becomes an omni-directional threat matrix. Hence, firewalls are required to granularly monitor network traffic and work in conjunction with other deep packet monitoring methods to ensure greater security levels at all points in the network.

Next-Generation Intrusion prevention systems (NGIPS) play another important role in the IoT security framework. Intrusion prevention systems are network control methods that provide network visibility, security intelligence, automation and advanced threat protection. The NGIPS can continually monitor the network and build profiles that provide contextual information to enhance decision-making about intrusion events. In addition to identifying and blocking traffic, a NGIPS can offer malware protection and the ability to sandbox when needed to contain attacks.

4.3 CONVERGED SECURITY – *AFTER* THE ATTACK

A difficult reality is that even with effective policies, controls, defense and monitoring in the *before* and *during* stages, there is still a possibility of a breach. Implementation of tools and processes that offer real-time visibility along with the ability to contain the attack and quickly assess the scope of damage must be part part of any comprehensive security strategy.

One method of creating deeper visibility with existing devices is to use IP-speaking devices as sensors that report back on the network status. Using technologies like NetFlow, every network packet can be tracked as metadata similar to a phone log that shows who talked to whom and for how long. The metadata can then be used to detect network reconnaissance activities, identify compromised control systems unexpectedly attempting to communicate outside their domain, recognize malformed packets as part of an attack or alert network administrators to large data exfiltration. Advanced analytic tools used in conjunction with inherent data provided by network devices offer deep and broad visibility across the attack continuum.

A critical area of consideration in a comprehensive security design is threat intelligence that spans the *before*, *during* and *after* stages of an attack. Attack vectors evolve over time and incorporating products backed by proactive detection and vulnerability research, engine development and research, threat intelligence and a mechanism for communicating current trends can enhance the overall security posture.

Using a defense-in-depth strategy across the attack continuum with greater visibility, and deep packet inspection provides a holistic approach to security that can protect both IT and OT networks.

5. CONCLUSION

Cyber security has many parallels to airport security. The attacker may not be immediately visible, but a series of checks and barriers have been implemented to weed them out. Authentication with a passport or ID is compared against lists of known safe or no-fly rules before access is granted to the gate area. Even then, your destination, your role (passenger, pilot, staff), flight class and the final gate location may limit what areas of the airport you have access to. Your bags will be scanned by security before they can be placed on the plane. Security cameras are in place to monitor suspicious activity that may indicate a possible threat and they also aid in piecing together the timeline and events after an attack. Finally, airport security is never static; as the effectiveness of the security barriers changes (for better or worse), the attackers learn and adapt their methods.

Similarly, cybersecurity is driven by a continuum of what actions can be taken *before* an attack occurs, *during* the attack to ensure continued operations while mitigation takes place, and *after* an attack to detect and defend against threats occurring in the future. Networks may also be operating in more than one phase of the attack continuum at the same time. This attack continuum shows that focusing only on threat detection and blocking, the approach that many take, is insufficient as IT/OT converges.

As OT networks and IT networks converge, cybersecurity is a key factor in determining success. Security needs in a converged IT/OT network will demand adoption of end-to-end architectures that incorporate cyber security focused on a threat-based attack continuum. As the maritime industry pursues digitization to enhance operational effectiveness and gain efficiencies, IT and OT systems will becomes more and more connected and converged. Cyber security will continue to evolve and play a critical role in ensuring that the benefits of such connectivity and convergence can be realized.

6. REFERENCES

1. Matthew Lacey, Helena Lisachuk, Andreas Giannopoulos, Alberto Ogura, "Shipping smarter: IoT opportunities in transport and logistics, The Internet of Things in shipping,", Deloitte University Press, September 15, 2015. Retrieved from https://dupress.deloitte.com/dup-us-en/focus/internet-of-things/iot-in-shipping-industry.html.

2. Telefonica I&D 2011 Cisco Visual Networking Index 2012

3. Gartner Press Release: "Gartner Says 4.9 Billion Connected "Things" Will Be in Use in 2015 Nov 11, 2014.

4. "Digitization with the Internet of Everything, 100 Customer Stories," Cisco Systems, Inc. 2016.

5. Bhoopathi Rapolu, "Internet of Aircraft Things: An Industry Set to Be Transformed," AviationWeek.com, Jan 18, 2016, retrieved from http://aviationweek.com/connected-aerospace/internet-aircraft-things-industry-set-be-transformed.

6. Joseph Beel and Reema Prasad, "Internet of Everything (IoE): Information Technology (IT) and Operational Technology (OT)," American Society of Naval Engineers (ASNE), 2016 Advanced Machinery Technology Symposium (AMTS), 2016.

7. Kevin Kelleher, "Why This Is The Most Important New Technology in Decades," Time.com, May 13, 2015, Retrieved from http://time.com/3856963/2015-the-cloud/.

8. Louis Columbus, "Roundup Of Cloud Computing Forecasts and Market Estimates", Mar 13, 2016. Retrieved from http://www.forbes.com/sites/louiscolumbus/2016/03/13/roundup-of-cloud-computing-forecasts-and-market-estimates-2016/#4a9931a474b0.

9. Ben Dickson, "How fog computing pushes IoT intelligence to the edge," 2 Aug, 2016. Retrieved from: https://techcrunch.com/2016/08/02/how-fog-computing-pushes-iot-intelligence-to-the-edge//

10. Phillip Trcy, "Industrial internet of things maritime use cases." Retrieved from: http://www.rcrwireless.com/20160727/internet-of-things/maritime-industrial-internet-things-tag31-tag99.

11. YoungSoo Cheong, "Connected Smart Ship," Hyundai Heavy Industries and Accenture, Oct 22, 2015. Retrieved from: http://www.thedigitalship.com/conferences/presentations/2015kormarine/3.pdf.

12. "MARITIME ICT CLOUD ENABLES SHIPS TO JOIN THE NETWORKED SOCIETY," Ericsson, Jan 6, 2015. Retrieved from: http://hugin.info/1061/R/1884299/665719.pdf.

13. "Control Networks: Maintain Their Availability During a Cyberattack," Cisco, Nov 2014..

14. John Steward, "Before, During and After: How to Think about Complex Threats," Cisco Blog > Security, July 1, 2014.

7. AUTHORS' BIOGRAPHIES

Joe Beel is a strategic programs manager at Cisco Systems, Inc. He develops and implements strategies to support the U.S. Public Sector focusing on defense. He is a retired naval officer and served as a helicopter pilot and acquisition professional in both the Naval Air Systems Command and Space and Naval Warfare Systems Command.

Kelly Jones is a systems engineer for the U.S. Navy Shipboard Account team at Cisco Systems, Inc. She works with the U.S. Navy and ship integrators to identify requirements, provide technical guidance on the future direction of networking technologies and assists them in implementation, testing and validation of proposed solutions. She served six years in the U.S. Navy as a Cryptologic Maintenance Technician and is a Certified Information Systems Security Professional (CISSP).

SHIP'S CENTRAL COOLING SYSTEM LIVE PERFORMANCE OPTIMISATION AND MODELING

A P Boveri, A Panzera and **F Silvestro,** University of Genoa DITEN, IT
I Crocicchia and **R Lodde,** ABB Marine & Ports, IT

SUMMARY

Central cooling systems play an essential role in ship operations, since guarantee the correct operating mode of vital users such as the main engine and generators. These systems are normally sized according to the design point so that they can deliver the cooling demand even in extreme conditions, while these conditions are very rare events in the ship's life. In order to improve the overall cooling efficiency, a variable frequency drive can be adopted [1-3]. In this work, a physical model of a real Ro/Con (i.e. Roll on roll off Cargo ship) central cooling system is presented with an integrated process control scheme. An experimental campaign has been carried out for six months to collect the overall cooling demand (e.g. seawater and fresh water temperatures, flows and valve position) before and after the introduction of variable speed drive on the seawater pump. An algorithm for the control of the cooling water system is proposed and implemented in the model in order to improve the performances and guarantee the basic features of each component. The estimated annual saving coming from the recoded data and the model are around 40 to 60 percent obtained with a reduction between 10 to 27 percent for the pump speed. The methodology and model are proposed and developed in order to better exploit the data gathered during the on-board measuring period.

1. INTRODUCTION

SINCE the industrial Revolution in the 1700's, the emissions of CO_2 have dramatically increased, resulting in increasing the CO_2 concentration in the atmosphere. In fact, in 2005, this concentration was 35% higher than before the Industrial Revolution [1]. Because of this growth in atmospheric concentration of Greenhouse Gas (GHG), such as CO_2, changes in the energy balance of the Earth's climate system have been discovered. It is therefore likely that these changes are related to the rising problem regarding global warming [2], which has increased the attention of international organization and operators to the environmental impact of using fossil fuels. Although ships are one of the most environmental friendly form of transportation, vessels typically have higher energy consumptions than actually needed. In this context, the International Maritime Organization (IMO) has recently introduced several normative in order to reduce these GHG emissions both for new design ships (i.e. Energy Efficiency Design Index EEDI, Energy Efficiency Operating Index EEOI) and for already existing ones (i.e. Ship Energy Efficiency Management Plan SEEMP). In order to face the challenge of increasingly strict international standards on environmental protection, continuous developments and studies are performed also by shipping industries to reduce the energy and fuel consumptions, together resulting in a reduction of GHG emissions. A promising method to achieve these objectives is to optimise the power balance between the power required by the users and the power generated.

A key concept in ship design is the *"Design Point"*, which is a combination of the main variables that characterize the project (e.g. ship speed, draft, length, deadweight, operative conditions, weather conditions and so on). Conventionally, a ship is designed to operate efficiently close to its Design Point (DP). Actually, ships operate often in conditions far from their DP, where the global efficiency is lower than expected. A good example of a vessel that frequently deviates from its best operative conditions is the container carrier. This kind of ships operate facing huge variations in weather and seas, performing a wide range of speeds and subject to large cargo variations during a round voyage [3] - [5].

At any rate, the main implications of these considerations is that energy efficiency must be viewed from a much broader perspective. In this context, it is not adequate to design ships to operate efficiently only close to their DP. Instead, an optimisation for Multiple Design Point (MDP) can be an efficient solution to improve ship's performances. There are several areas in ships design process where it is possible to operate in order to improve the ship's energy performances. It is possible to achieve energy efficiency improvements by optimising the hull shape in order to minimize the hull resistance and increase the propulsion efficiency, or by optimising the hull form and the ship design in order to reduce the added resistance due to waves and wind. Another area where it is possible to introduce new design concepts and solutions is shipbuilding. In fact, there is the possibility to perform structural optimisations and adopting lightweight materials with high resistance characteristics. Finally, for what concern the shipboard systems design, in many cases (e.g. for cooling system, ventilation, air conditioning etc.) it is possible to adopt for instance Variable Frequency Drives (VFDs) that are one of the most effective technology in order to improve on board energy efficiency. These devices are typically used in cooling system such as seawater cooling system or engine room ventilation system in order to better control the cooling capacity when the vessel is operating far from the extreme conditions. In seawater cooling system, VFDs are introduced in order to optimise the power balance between the power required by the systems and the power supplied by their motors. Furthermore, it is to be noted that for many ships (e.g. such as cargo ships), the pumping systems account for nearly the 70% of the total electrical load. In this perspective, there is the opportunity to reduce pumping system energy consumption by adopting "*Smart Design*", retrofitting and operating strategies. In the following sections, the main characteristics and possible design of a Marine Central Cooling Systems are introduced. Then, the physical model and the actual data recorded on-board the case study ship are presented. Several results are then reported and analysed in order to validate the VFD solution for the case study ship. Finally, some conclusions are drawn.

2. MARINE SEA WATER COOLING SYSTEMS

There are several and different machineries installed on board. These are all designed to work for long periods in adverse conditions. In transportation systems, such as in many other fields, the most common and maximum energy loss is in the form of heat energy. Anyway, this loss has to be reduced by a cooling media to avoid malfunctioning or breakdown in machineries. For this reason, practically more than 35000 merchant ships are equipped with a cooling system on board. A number of centrifugal pumps, which move the fluid in order to obtain a correct heat exchange, manages the operation of such a system.

2.1 DESCRIPTION OF A MARINE CENTRAL COOLING SYSTEM

There are two main approaches to design a marine cooling system: the seawater cooling system and the central cooling system. Due to the aim of this work and the characteristics of the case study ship, only the central cooling system is here presented and studied (Figure 1). This is one of the most used design approach for shipboard cooling systems. In this kind of cooler, fresh water is used in a closed circuit to cool down the engine room machineries. The fresh water returning from the heat exchanger after cooling the machineries is than cooled by seawater in a central heat exchanger.

Otherwise, in this design approach, the seawater is used as a cooling media in large heat exchangers to cool the fresh water of the closed circuit. These are the central coolers of the system. The fresh water system can be a mixed system. In fact, the heat from all the different machineries and other components that need cooling is mixed together in one flow. This is the easiest system to install and regulate, but needs some more cooling area than a split system. On the other hand, in a split system the Low Temperature (LT) circuit is used for low temperature zone; hence its temperature is lower than the High Temperature (HT) circuit. The LT circuit comprises of all auxiliary systems (e.g. auxiliary to propulsion system, to power generation system and so on). The HT circuit mainly comprises of jacket water system of the main engine where the temperature is quite high. Therefore, the ship main and auxiliary systems and machineries are cooled by using fresh water. The main reason for this choice is that seawater causes corrosion. Thus, the heat is extracted from the machineries and heat balance is maintained so that the machineries can operate properly within the most efficient and safe operating parameters.

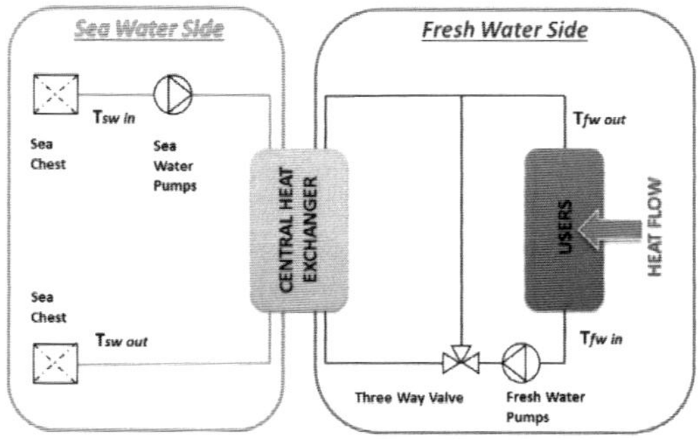

Figure 1 - Typical Central Cooling System Layout

Two or more main sea water pumps installed in the engine room of the ship pump the seawater from the sea chests to the central heat exchanger. By the way, traditionally each pumping system on board ships works basically at a fixed speed under a rated frequency and, where it is required to change the flow rate of the system, the control is performed with a bypass line or throttling valve with a considerable waste of energy. By necessity, efficiency is a lower priority in design. However, several design approaches have been shown to meet all the requirements of a central cooling system facility while minimizing power consumption and cost.

2.2 A VARIABLE FREQUENCY DEVICE INTRODUCED INTO A CENTRAL COLLING SYSTEM

In this work, a practical solution to improve the energy performances of a Sea Water Cooling System (SWCS) is presented considering a SWCS installed on board a case study ship, where a Variable Frequency Drive (VFD) has to be installed in order to control the AC motor that moves the pump.

2.2 (a) General Description of VDFs

AC motors drive many pumps installed on board ships. This kind of motor may change their speed by changing the frequency and voltage in the same ratio. The power electronics converter that converts fixed frequency, voltage power into a variable-frequency, variable-voltage power source, achieves the AC motor speed control. Such converter is called the Variable Frequency Drive [9]. It is widely used with large motor driving pumps or ventilating fans running long hours every day with variable fluid pumping requirement. It is desired to maintain the stator flux constant during the motor operation at all speeds in order to produce full rated torque and to avoid magnetic saturation in the core. Therefore, maintaining a constant rated flux in AC motor requires maintaining the constant rated V/f ratio while reducing the frequency for the entire speed range as reported in equation *(1)*.

$$V_{rms} = V_m / \sqrt{2} = (2\,\pi / \sqrt{2})\, f\, N\, \varphi_m \qquad (1)$$

2.2 (b) How a VDF Works in a Central Cooling System on Board Ships

The VFD, particularly in a fluid pumping plant such as the one installed on board ships, saves energy as described below. The old conventional way of reducing the fluid flow rate was to control the valve position, while the AC motors kept running at full speed (e.g. as shown in Figure 1). This puts the system in conflict; the motor running at full speed while inserting a hydraulic resistance in the fluid flow, resulting in inefficient utilizations of the power. The energy efficient way is to reduce the motor speed in response to reduced flow requirement. The possible savings of introducing a VFD that regulates the seawater pumps flow rate, instead of using only the valve position control, can be from 20% to 30% of the total energy used to move the seawater pumps. The VFD has made significant advances over the last two decades in the industrial and shipboard applications. Figure 2 shows a control diagram for VFD – driven a SWCS on ships. The data flow for the control system is depicted in Figure 3.

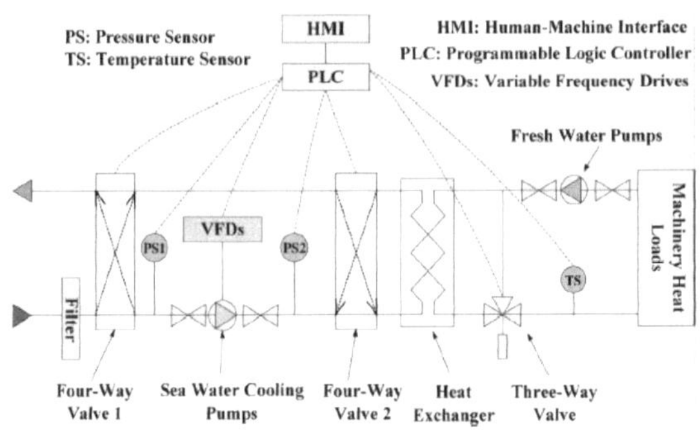

Figure 2 - Control Scheme for a VDF - Driven Sea Water Cooling System [7]

It is possible to note that the Programmable Logical Controller (PLC) receives and analyses the measurement delivered from the sensor for the charged and discharged water temperature of the heat exchanger at the fresh water side, the pressure difference between the inlet and outlet of the pump at the seawater side and control signals. The control signal issued by the PLC is delivered to the VFD to vary the running frequency of the pump to regulate the pumped seawater flow in order to guarantee a constant temperature (i.e. typically close to 36° C) in discharge fresh water according to the measurements from the pressure and temperature sensors. In case of adverse conditions such as seawater high temperature or heavy machinery workload, the system operates at rated speed. On the other hand, in case of seawater low temperature conditions and/or normal machinery workload, the discharge fresh water temperature is still low although the pump operates at the lowest speed. In this context, the three-way valve at the fresh water side can be regulated to vary the fresh water flows in order to reach the required temperature. The seawater pumps are enabled when the discharged fresh water temperature is beyond the allowed ranges due to variations in machineries workload and seawater temperature as well as increasing in head losses of seawater pipe and heat exchanger.

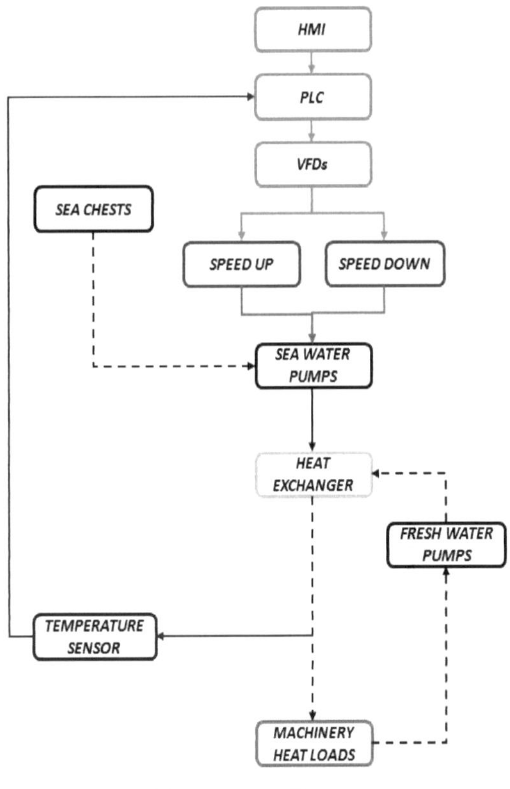

Figure 3 – Control System Data Flow

In the following section, the physical model of the SWCS is presented and studied. Further, a model is developed using *MATHWORKS* Simscape library (i.e. in Simulink), which enables the users to create model of physical systems within the Simulink environment.

3. MODEL AND DESIGN OF A SWCS WITH A VFD, BASED ON ACTUAL DATA

The system under consideration is a "Central cooling System", characterized by a seawater side and a second closed circuit where fresh water flows. Between these two circuits, a central heat exchanger operates in order to decrease the fresh water temperature, as it is shown in

Figure 2. The seawater circuit is composed of two sea chests, which suck the water that is then transferred to a settling-case positioned between the bottom and the double bottom of the ship. From the latter, the water is pumped by two centrifugal pumps and flows through two plate exchangers. These are the central core of the seawater cooling system. The fluid leaking from the exchangers (i.e. which has increased its temperature), is ejected overboard through a drain suction head. The fresh water circuit, instead, is a closed hydraulic circuit composed of two plate heat exchangers. These allow the heat exchange between the fresh water system and the sea water side. Two centrifugal pumps are also installed in addition to a huge network of branched pipes, which reach all utilities on board the ship who need to discharge heat outside in order to correctly operate. A three-way valve is installed in order to recirculate the fluid. This has the purpose to evolve the right amount of water to the exchangers, in order to obtain the correct operating temperature of the machinery and of the main motor as well.

3.1 PRESENTATION OF THE ACTUAL DATA RECORDED ON-BOARD THE CASE STUDY SHIP

The system chosen in order to perform an energy efficiency analysis is installed on board a RO/CON ship (i.e. Roll-On / Roll-off cargo ship, fitted for the transport of containers). The main characteristics of this case study ship are here reported in Table 1.

Table 1 – Case study ship's main characteristics

CHARACTERISTICS OF THE CASE STUDY SHIP	
Ship type	*Roll-On / Roll-off cargo ship, fitted for the transport of containers*
DWT / TEU	*26881 t (Summer) / 800 TEU*
Length	*210.92 m Over All* *196.80 m Between Perpendicular*
Breadth	*32.26 m*
Draft	*9.75 m*
Ship Speed	*Max. 22 knots - Design 18 knots*
Main Engine	*1 x MAN-B&W-Uljanik 8S60MC-C – 12640 kW*
Shaft Generator	*1 x 1400 kW*
Generators	*4 x 1180 kW DD/GG*
Transportable cars	*3890*
Passengers	*12*

The available data regarding the system are measurements of temperatures (seawater in and out, fresh water temperature beyond the users), valve position and pumps flow rate. These measurements account for two operating months and have been collected by *ABB Marine & Ports* through a measurement campaign since November 19 2014 and January 14 2015. These are here presented in Figure 4, where their behaviour over time is shown.

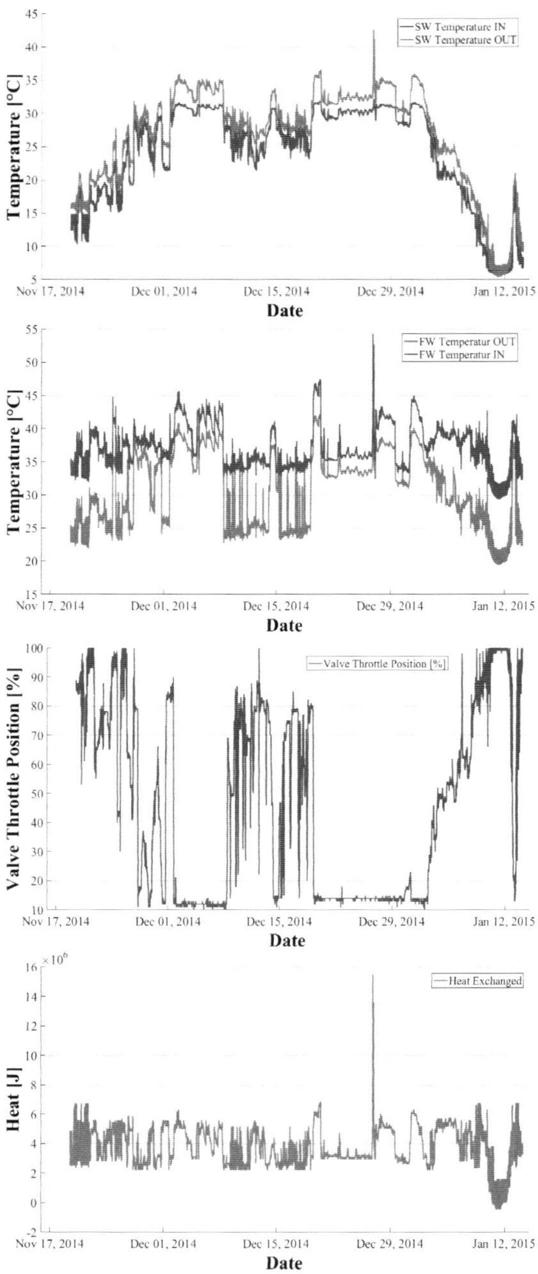

Figure 4 - Actual Data Time Behaviours

From the first chart in Figure 4 it is possible to highlight that the case study ship has recorded very different sea water temperature values (e.g. from 35°C to 5°C). This phenomenon is mainly due to the ship operating areas, which starts from the sub-tropical region through the equator and finally ends close to Artic Circle. In the second diagram reported in Figure 4, the data named *FW temperature Out* correspond to the temperature of the fresh water side before the machineries/users. It is noteworthy that, in most of cases, this temperature is very close to the target value of 36°C. On the other hand, the data named *FW temperature In* identifies the fresh water temperature after the machineries/users (i.e. after the heat exchange between the cold fluid and the hot machineries). Then, in the third chart the percentage position of the three-way valve (i.e. the one positioned in the fresh waterside after the users and before the central exchanger, as it is shown in Figure 1) is reported. In order to fully understand its behaviour, it is important to note that when the valve position is close to zero it means that the valve is completely open and all the possible water flow is evolved in direction of the central heat exchanger. Otherwise, when its position is close to 100% it means that the valve is closed and all the water flows through the bypass line. In this context, the valve behaviour shows that, in case of seawater with high temperature (e.g. close to 35°C) the valve is almost fully open, on the contrary, when the seawater temperature is low (e.g. less than 15°C) the valve is often close. Finally, considering the last diagram, the heat exchanged in the central heater exchanger is report as the results of some calculations performed using the previous data. These calculations and their hypothesis are here reported in the following formulas and respective explanations.

3.2 AN EQUILIBRIUM MODEL FOR EVALATING ENERGY PERFORMANCIED OF A CENTRAL COOLING SYSTEM WITH A VFD

In order to fully understand how a central cooling system with VFDs works, it is important to introduce the mathematical models beside each main component. In this work, the system is divided in four main components: pumps, central heat exchanger, three-way valve and the machinery heat exchanger. For each component, the physical equations that govern the way they work are introduced and commented as well.

3.2 (a) Seawater Pumps Model

Usually the main seawater pumps are installed closed to the sea chest for pumping this water to the heat exchanger for cooling the fresh-water side. Applying the affinity law reported in equation *(2)* it is possible to understand how the presented energy efficiency solution works. In fact, with a reduction close to the 10% for the electrical motor revolutions speed (i.e. the motor that moves the seawater pump) it is possible to obtain a reduction close to 30% in power demand.

$$P_2/P_1=(n_2/n_1)^3 \tag{2}$$

According to the electrical machinery theory and affinity law of centrifugal pumps *(2)*, the running frequency of the motor, considering a variable frequency control can be calculated applying the following equation *(3)*.

$$f_2 = f_1 \cdot (P_2/P_1)^3 \tag{3}$$

Where f_1 is the rated frequency of the motor, P_1 is the rated power, P_2 is the input power that is calculated with *(4)* and f_2 is the motor frequency at this power input.

$$P_2 = \frac{Q \cdot H_{TOT} \cdot \gamma}{\eta_E \cdot 6120} \tag{4}$$

Where Q is the water flow rate required by the system, H_{TOT} is the total suction head (i.e. including pipe loss and pressure drop), γ is the gravity of the seawater and η_E is the overall efficiency of a single pump with VFD.

3.2 (b) Central Heat Exchange model

In marine applications, it is usual to adopt the Plate Heat Exchanger (PHE) in order to exchange heat from a hot side to a cold one. PHEs use a series of stacked metal plates to transfer heat between two fluids with different temperature, and in this case study, are installed with the fluids flowing in a counter-flow arrangement. This system behaviour can be described by some equations with the effectiveness-Number of Transfer Units (NTU) method. This method is mostly used in conditions where the size of a heat exchanger and the inlet temperature are known, and the heat transfer rate and the fluid exit temperature are sought [10]. Due to the collected data, which are available in this work, the size of the heater, the flow rates and the inlet temperatures are known for both side of the system, moreover, for the seawater side, the outlet temperature is known. Therefore, only the outlet temperature in the fresh-water side is unknown and can be calculated using the NTU method. As a first step, it is useful to define *(5)* the heat exchanger efficiency ε, which identifies the rate between the heat actually exchanged $\dot{q}_{sw}$ and the maximum heat exchangeable $\dot{q}_{max}$ by the PHE.

$$\varepsilon = \frac{\dot{q}_{sw}}{\dot{q}_{max}} \tag{5}$$

By using the NTU method is then possible to calculate the fresh water outlet temperature (i.e. which is the unknown parameter) as reported in equations *(6)*.

$$T_{fwout} = T_{swin} + (T_{fwin} - T_{swout}) \cdot e^{-NUT \cdot (1-C)} \tag{6}$$

Where T_{swin} and T_{swout} are respectively the seawater inlet and outlet temperatures, T_{fwin} is the fresh water inlet temperature (i.e. evaluated after the machinery heat exchanger module), *NUT* is calculated as defined in equation *(7)*. *C* is the ratio between the flow-stream capacities of the two as defined in equation *(8)*,*(9)* and *(10)*.

$$NUT = \frac{1}{1\text{-}C} \cdot ln(\frac{1\text{-}C \cdot \varepsilon}{1\text{-}\varepsilon}) \tag{7}$$

$$C = \frac{C_{min}}{C_{max}} \tag{8}$$

$$C_{min} = \dot{m}_1 \cdot c_{pfw} \tag{9}$$

$$C_{max} = \dot{m}_{sw} \cdot c_{psw} \tag{10}$$

Where c_{pfw} and c_{psw} are the specific heats for the fresh water and the seawater, respectively.

3.2 (c) Three-Way Valve Model

A three-way valve is usually controlled in order to either mixing or diverting applications. In this work, it is modeled using the equations *(11)*, *(12)* and *(13)* here reported.

$$\dot{m}_1 = \dot{m}_{fw} \cdot (1\text{-}v) \tag{11}$$

$$\dot{m}_2 = \dot{m}_{fw} - \dot{m}_1 = \dot{m}_{fw} \cdot v \tag{12}$$

$$\dot{m}_{fw} = \dot{m}_1 + \dot{m}_2 \tag{13}$$

Where $\dot{m}_1$ is the seawater mass flow, $\dot{m}_{fw}$ is the fresh water mass flow (i.e. the constant mass flow calculated before the fresh water pumps as reported in Figure 1). Moreover, *v* is the three-way valve position (i.e. with value between 1 and 0 depending respectively if the valve is diverting or mixing the flow rate) and $\dot{m}_2$ is the water flow flowing in the bypass piping. A PID controller is installed on the case study ship on the outlet piping for live time control of the three-way valve. This controller controls and regulates the flow opening degree according to the temperature signals, with a slower control compared with the one chosen to regulate the seawater pumps.

3.2 (d) Machinery Heat Exchange Demand Model

This heat exchange module has been modeled as a sum of heat loads from many machineries instead of a complex model of each main engine detail. The five groups of loads modeled for the case study system are:

- Cooling air recovery system
- Lubricating oil cooling
- Cooling of the motor sleeve
- Cooling auxiliary engines
- Cooling constant loads.

Each of these users need to evolve a certain amount of heat as a function of the conditions under which the ship is operating (e.g. cruising, manoeuvring, port etc.). The first three utilities are in close relation with the operation of the main engine, while the fourth group can be considered as independent of it. In fact, the flow of heat generated by the auxiliary engines depend strongly on the time in which are found to be in operation in the various operating conditions of the ship and not by the main engine operating point. With regard to the constant heat sources, these are consisting primarily of the refrigerator, air conditioning, shaft generator, air ventilation in the main engine control room, as well as for the transformer room. The first three users considered are closely dependent on the main engine's operating point. In fact, the heat exchanged is a function of the power delivered and the number of operating revolutions of the main engine. The main engine installed on-board the case study ship is a MAN B & W 8-S60MC-C8.2-TII, whose load and layout diagrams are shown in Figure 5.

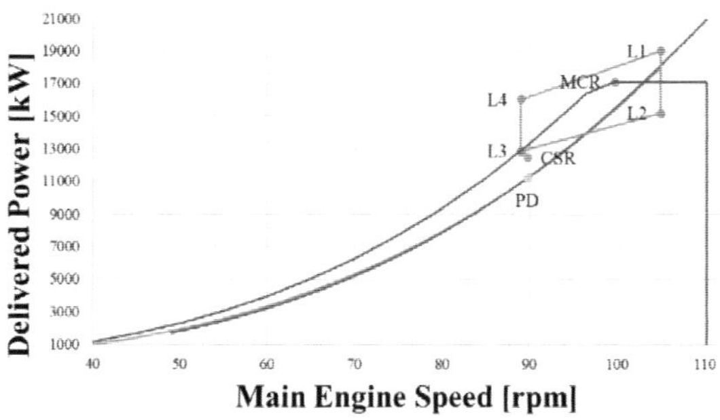

Figure 5 - Main Engine Load, Layout Diagrams and Propeller Load Curve

In the available data, for the case study ship, the power delivered by the main engine is not provided. Instead, the ship speed is available. By using the propeller curve reported in Figure 5 (i.e. that was obtained with the traditional calculation performed in design phase on the propeller and the ship's hydrodynamics characteristics), it was possible to calculate the power delivered in function of the ship's speed. In order to calculate the total heat to exchange (i.e. from the ship's machineries to the fresh-water side) it is possible to apply the following equations *(14)*-*(19)* provided by the main engine project guide. It is also possible to note that the main users are here divided in four class, each depending on the ship operating conditions (e.g. main engine load percentage, electrical generators load in function of the ship operating scenario, etc.):

- The scavenge air heat $\dot{q}_{air}$ is calculated in function of the main engine load conditions (i.e. power and speed) as reported in equations *(14)* and *(15)*.

$$\dot{q}_{air\%} = 2 \cdot P_{\%}^{1.68} \cdot n_{\%}^{-0.83} \quad (14)$$

$$\dot{q}_{air} = \dot{q}_{air\ max} \cdot \dot{q}_{air\%}/100 \quad (15)$$

- The lube oil system heat $\dot{q}_{oil}$ is calculate as a function of the main engine working conditions (i.e. the power delivered at a specific speed) as shown in equations *(16)* and *(17)*.

$$\dot{q}_{oil\%} = 67.301 \cdot log(n_{\%}) + 7.63 \cdot log(P_{\%}) - 245 \quad (16)$$

$$\dot{q}_{oil} = \dot{q}_{oilmax} \cdot \dot{q}_{oil\%}/100 \quad (17)$$

- The main engine's jacket heat $\dot{q}_{H.T.}$ is calculated in a similar way to the previous and the main equations are here reported in *(18)* and *(19)*.

$$\dot{q}_{H.T.\%} = e^{(-0.0811 \cdot log(n_{\%}) + 0.8072 \cdot log(P_{\%}) + 1.2614)} \quad (18)$$

$$\dot{q}_{H.T.} = \dot{q}_{H.T.\ max} \cdot \dot{q}_{H.T.\%}/100 \quad (19)$$

- Finally, the auxiliaries engine heat is calculated in function of the electrical load analysis and the generator's heat characteristic and is a constant depending on each different ship operating scenario.

The final heat load in function of the ship speed has been evaluated and is reported in Figure 6, where each main contribute is highlighted for the navigation condition.

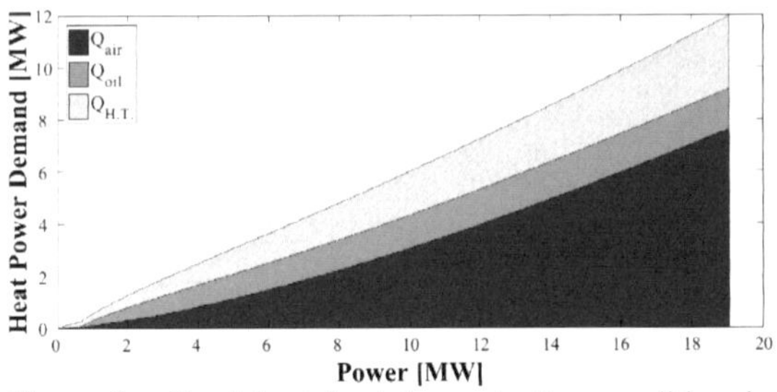

Figure 6 - Total heat load in navigation condition in function of the ship speed

3.3 DEVELOPMENT OF A PHYSICAL MODEL IN SIMULINK FOR SIMULATING DYNAMIC PERFORMANCIES

In order to evaluate the possible energy savings in dynamic simulations, the Simscape physics library (i.e. a recent tool present in MATWORKS Simulink) has been used. This because of the aim of this work, which is to study and analyse the possibility to optimise the energy performances of the case study ship introducing a VFD that operates in real time. The development of the model in Matlab-Simulink-Simscape environment is central for the dynamic analysis of the system and the study of control architectures. Moreover, this simulation platform is well suited to model physical systems with and study their dynamic behaviour. The model simulations using Simscape blocks are performed in an analogous way to the ones performed using Simulink blocks. The difference is that, on the beginning, the compile phase requires a relatively longer time, since it is necessary to wait that the symbolic manipulator solves automatically the DAE system (Differential Algebraic Equation) consisting on all the physical equations of the model. In Figure 7 it is possible to identify the main components of the central cooling system under exam as they have been modeled. These components have been modeled with the physical law previously presented in 3.2.

3.3 (a) Seawater Block

The seawater block presented in the second picture in Figure 7, is composed of the sea chests, the seawater centrifugal pumps and an hydraulic resistance that simulate the actual piping loss. Furthermore, the electrical motor that moves the pumps is modeled with a VFD, where the motor frequency is controlled with a Pulse Width Modulation (PWM) converter.

3.3 (b) Central Heat Exchanger Block

The central heat exchanger, presented in the third picture in Figure 7, is composed of two different components. These are the seawater and fresh water sides, which interact through a thermal connection in order to simulate the heat exchange from the fresh water closed circuit to the seawater side. This heat exchange is regulated as reported in 3.2 (b).

3.3 (c) Three-way Valve Block

The three-way valve shown in the fourth picture in Figure 7 connects the three fresh water circuit lines. This valve divides, as a function of its position, the water flow between the central heat exchanger and the by-pass line. This modulation is realized through a reduction of the pipes area proportional to the percentage of the valve opening position. Finally, there are two resistors, to take into account the loss due to the presence of the elbows, flow meters and temperature for both lines.

3.3 (d) Machineries Heat Power Demand Block

In conclusion, the machineries model is presented in the last picture in Figure 7, where the fresh water temperature is used as input as well as the heat power demand coming from the machineries Then, the heat exchanger is modelled as a single pipe that exchanges heat with the cooling system, increasing the temperature of the fresh water that evolves within. For the movement of the fluid a centrifugal pump has been installed, which

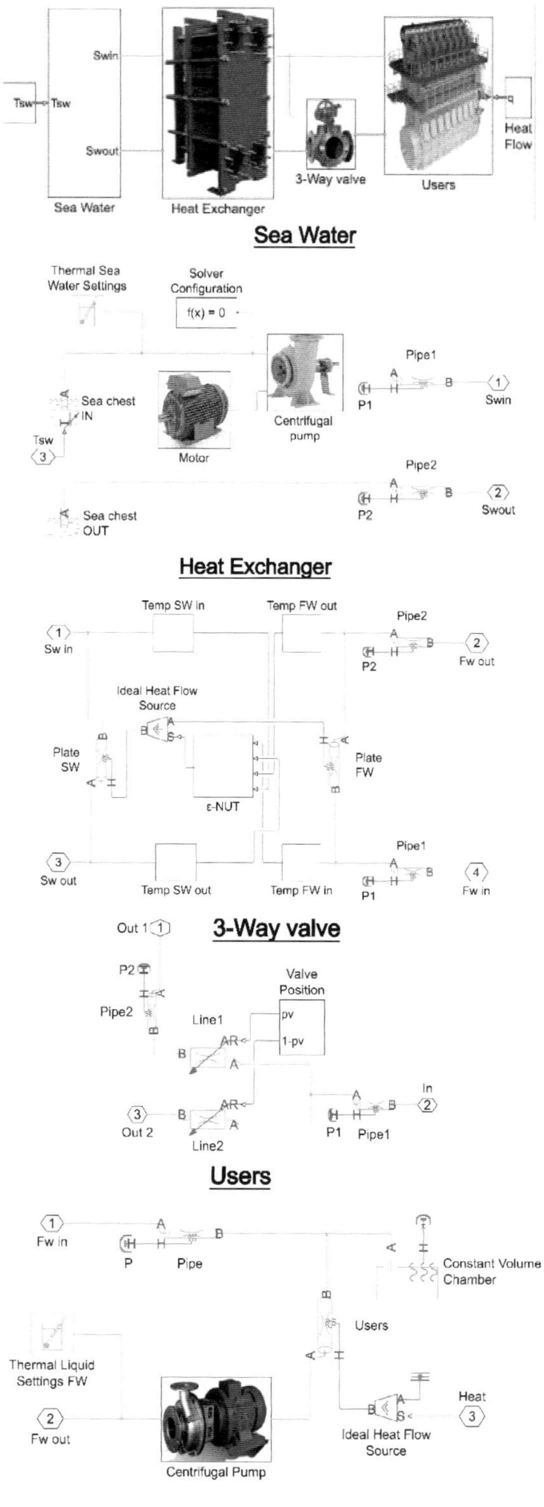

Figure 7 - Simscape Central Cooling System Model and Main Blocks

works at fixed speed and mass flow. It was no possible to install another VFD controlled by a PLC, because many machineries that require to be cooled also require a constant pressure and a constant mass flow, in order to work in safety conditions.

3.4 SIMULATIONS AND RESULTS

Firstly, a series of simulations have been performed in order to verify that the model developed reflects as closely as possible the behaviour of the real one. Once the response of the model to changes on its input parameters has been studied and the characteristic parameters have been defined, it was possible to perform a simulation using as an input the allowable collected data of a typical ship mission (e.g. eight days with the port, manoeuvring and navigation scenarios). This simulation uses as input data, the sea water temperature recorded and the heat exchanged with the utilities, calculated through the formulation presented in 3.2 (d) in function of the ship speed (i.e. which was a known data). Furthermore, the control temperature of the fresh water before the machineries heat exchanger has been set to 28°C. The behaviours of the input variables are here presented in Figure 8.

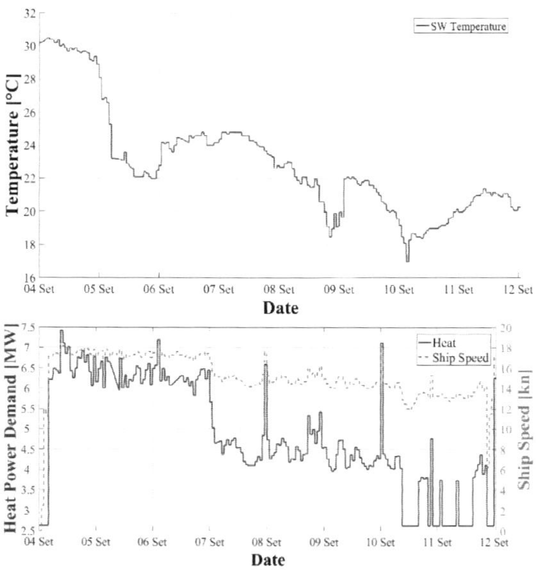

Figure 8 - Input Parameters Behaviour

On the other hand, the simulation variables are the seawater pumps flow rate and the three-way valve position, which are both controlled by a PLC as shown in Figure 2. The simulation results are presented in Figure 9, where is possible to compare the behaviour of the set temperature, the three-way valve positon and the seawater pumps speed recorded on board the case study ship (i.e. without the VFD) with the analogous parameters obtained with the implementation of a VFD on the seawater pumps (i.e. by using the model previously presented). In the proposed results (i.e. first chart in Figure 9), it is possible to note that for the first thirty hours of simulations the set temperature has not been reached. This is mainly because in that time horizon a heavy operating condition has been meet (i.e. the temperature of the seawater in input was approximately close to 30°C, in addition to a heat demand close to 7 MW for the entire simulation). In Figure 9, it is also presented the comparison between the set temperature (i.e. first chart), the three-way valve position (i.e. second chart) and the pumps speed both recorded and simulated (i.e. in the last chart in Figure 9).

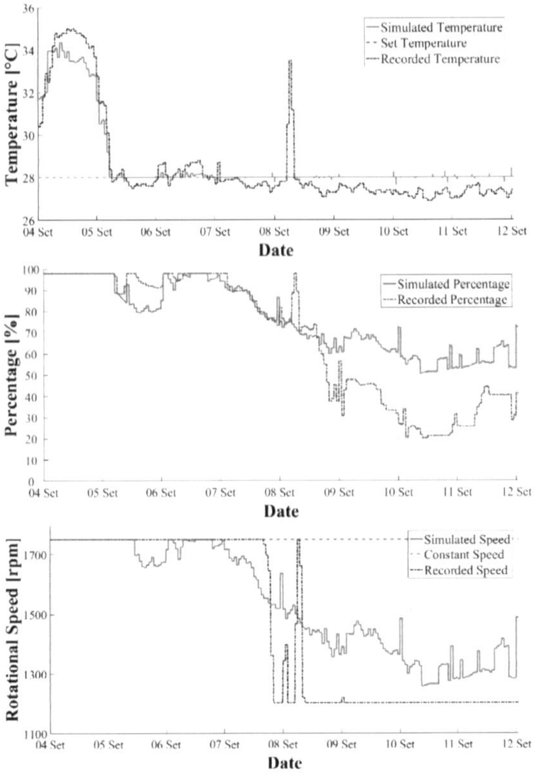

Figure 9 - Simulation Results Presentation

The comparison between the simulation and recorded parameters and variables has highlight the real possibility to operate in a more efficient way, reducing the waste of energy supplied by the electric motors that moves the pumps (Figure 10). In Figure 10, it is possible to note that the required power by the sweater pumps is constant (i.e. close to a value of 150 kW), on the other hand, in the proposed simulation where the VFD is implemented, the required power in variable from 50 kW to 150 kW depending on the operating conditions.

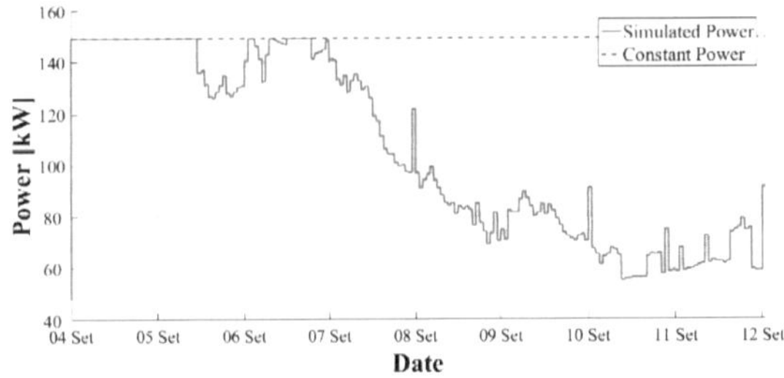

Figure 10 - Power Required by the Seawater Pumps

The main results considering the energy, cost and CO_2 emissions savings are presented and reported in Table 2 considering as time horizon both the mission of 8 days and one operating year for the case study ship. It is also to be noted that these results are provided considering heavy ship's operating conditions such as seawater high temperature and high heat workload from the machineries. Otherwise, considering a normal operating scenario, with seawater temperature close to 20 °C and a heat workload close to 5 MW, an energy saving close to 20-25 % is possible.

Table 2 – Simulation Results

SIMULATION RESULTS (8 DAYS MISSION)	
Energy Required without VFD	*63.9 MWh*
Energy Required with VFD	*55.6 MWh*
Energy Saving	*8.3 MWh*
Energy Saving [%]	*13% (considering the overall system energy)*
Marine Diesel Oil Consumption	*1.67 t*
Economic Saving	*900 $ (MDO at 540$/t)*
Energy Saving (1 operating year)	*420 MWh*
Economic Saving (1 operating year)	***45000 $***
CO_2 emission saved (1 operating year)	***586.5 t***

4. CONCLUSIONS

The analysis presented in this work has highlighted the effective possibility to improve, both in design and in retrofitting phase, the energy efficiency and management cost savings of a typical central cooling system. These improvements can be achieved with the introduction of a Variable Frequency Drive to the seawater pumps and a PLC that work in real time in order to optimise the energy performances of the system. In order to perform this energy analysis, a physical model of this system has been proposed and implemented. Then, a dynamic model has been built in Simscape environment and dynamic simulations have been performed. Their results are here proposed and indicate that for the same ship, there are significant energy savings benefits introducing a VFD. Moreover, considering one of the worst possible operating scenario (i.e. high sea water temperature and high machineries workload), savings close to 15 % of energy and 600 t of CO_2 issued are possible, with an increase up to 25% of energy saved considering normal operating conditions.

5. REFERENCES

1. U.S. ENVIRONMENTAL PROTECTION AGENCY, 2009 U.S. ENVIRONMENTAL PROTECTION AGENCY, *Available at: http://www.epa.gov/climatechange/emissions/co2.html*, 2009.

2. SOLOMON, S., QIN, D., MANNING, M., CHEN, Z., MARQUIS, M., AVERYT, K., M. TIGNOR, "Climate change 2007: The physical science basis contribution of working group to the fourth assessment report of the intergovernmental panel on climate change". Technical Report Summary for Policymakers, *Intergovernmental Panel on Climate Change*, 2007.

3. ABB, BU Marine and Cranes, "Energy efficiency guide", *Available at:* https://library.e.abb.com, Helsinki, April 2013.

4. J. LARKIN ET. AL., "Influence of Design Parameters on the Energy Efficiency Design Index (EEDI)," *SNAME Symposium on Climate Change and Ships*, February 2010.

5. IMO MEPC, "Marginal Abatement Costs and Cost-Effectiveness of Energy-Efficiency Measures", *IMO MEPC 61/INF.18*, 23 July 2010.

6. ABS, "Ship Energy Efficiency Measures, Status and Guidance", *Available at: www.eagle.org.*

7. C. L. SU, W. L. CHUNG, AND K. T. YU, "An energy-savings evaluation method for variable frequency drive applications on ship central cooling systems," *IEEE Trans. on Industry Applications, vol. 50, no. 2, March/April 2014*, pp. 1286-1294.

8. S. V. GIANNOUTSOS AND S. N. MANIAS, "Development of an integrated energy efficiency control system for ship power balance and diesel generator fuel consumption optimisation," *in Proc. 2013 IEEE LAS Annual Meeting.*

9. MUKUND R. PATEL, "Shipboard Electrical Power Systems", *CRC Press*, Ch. 11.

10. C. H. LEE, Z. W. LIU, C. N. CHEN, M. Y. CHO, F. T. LIN AND J. A. JIANG, "Assessment of Energy Savings With Variable Speed Drives in Ship's Cooling Pumps," in *IEEE Transactions on Energy Conversion*, vol. 30, no. 4, pp. 1288-1298, Dec. 2015.

6. AUTHORS BIOGRAPHY

Alessandro Boveri received the degree in naval architecture and marine engineering from the University of Genoa, Genoa, in 2014, where he is currently pursuing the Ph.D. degree in electrical engineering with the Department of Electrical, Electronic, Telecommunication Engineering and Naval Architecture. His current research interests include modeling of shipboard power system, probabilistic and stochastic approaches for load demand and optimisation, energy efficiency solutions for marine applications.

Alessandro Panzera was born in Reggio Calabria, Italy, in 1992. He received the bachelor and master's degree in naval engineering from the University of Genoa respectively in 2013 and 2016. His master thesis was focused on ship's systems performances optimisation. Currently he is with ABB Marine & Ports for an internship in as proposal engineer.

Federico Silvestro (SM '16) received the electrical engineering degree and the Ph.D. degree in electric power systems from the University of Genova in 1998 and 2002, respectively. He is currently Associate Professor at DITEN, University of Genova. He is an author of more than 120 scientific papers. His current research interests include distributed generation and smartgrids, energy saving, dynamic security assessment, and knowledge-based systems applied to power systems. Prof. Silvestro is secretary of the JWG C6.25/B5/CIRED on Control and Automation Systems for Electricity Distribution Networks of the Future.

Irene Crocicchia was born in La Spezia in 1984. She graduated in Environmental Engineering at University of Genoa, Italy. Working in ABB since 2011, as new technology development specialist she focused on water treatment and energy efficiency technologies for the process industry. She then joined ABB Marine and Ports Business Unit where she actually works as bid and proposal engineer for automation solutions dealing with onboard energy efficiency implementations.

Romano Lodde was born in Oristano in 1979. He started to work in 2002 as a stage technician. He then joined ABB Marine and Ports in 2005, where he currently works as a Technical Advisor for on-board systems and automation.

NETWORK-CONNECTED LIGHTING IN THE MARITIME ENVIRONMENT

K Jones and **J Beel**, Cisco Systems, Inc., USA

SUMMARY

Light-Emitting-Diode (LED) lighting adoption is accelerating due to lower energy costs, extended lifespans and reduced maintenance. Lighting devices more frequently contain embedded processors which enable them to sense, detect, communicate and be networked. Low-voltage Ethernet power for lighting reduces cost and routes data to and from network-connected lighting. Lights can become data and sensor nodes which collect and disseminate information such as occupancy, temperature, O2 levels, particulate matter, motion, video, sound, etc. Network-connected Lighting can reduce weight and increase safety aboard ships. Smart+Connected Lighting solutions can also be used to deliver and support services such as asset management, emergency/incident response, transportation coordination/oversight and physical security.

Network-Connected Lighting advantages include:

- Reduced energy consumption
- Networked management
- Increased safety
- Reduced weight
- Enhanced situational awareness
- Improved emergency response
- Sensor networking

Network-connected lighting advantages apply both to shipboard applications and maritime facilities such as ports and shipyards.

1. INTRODUCTION

Adoption of LED lighting products continues to grow within the U.S. but there is substantial room for additional LED market penetration.[1] The U.S. Department of Energy estimates that LED lighting has the potential to reduce U.S. lighting energy consumption by 75% but, to date, less than 6% of the solid-state lighting (SSL) energy savings has been achieved, indicating tremendous untapped potential. LED lighting can generate savings of 50-70 percent independently and 80 percent when coupling with smart controls.[2]

In addition to the promise of reduced energy consumption, the replacement of legacy lighting with LED lighting products offers the potential for delivery of connected lighting systems. LED lighting has the potential to be more controllable than other lighting technologies and can deliver increased automation and functionality or "smart lighting."[3] Smart lighting offers the capability of connecting to networks to deliver Smart+Connected Lighting. Connecting lighting to networks enables lighting systems and luminaires to become multifunction devices. Smart+Connected Lighting systems can connect to larger management systems and networks to deliver greater control of lighting and integration with other infrastructure functions.

1.1 INDOOR LIGHTING OPPORTUNITIES

Information Technology (IT) and facilities Operational Technology (OT) have been converging over the past decade or so and leading to more networked buildings or ships and workspaces as detailed in Figure 1. Internet Protocol (IP) Telephony or Voice over IP (VoIP) enabled replacement of legacy stand-alone phone systems in the early 2000s and IP cameras supporting security and surveillance became more prominent in the late 2000s, with both converging onto the primary network for management control efficiency. Building management control systems have also been migrating to the broader network and are taking advantage of low-voltage Power over Ethernet (PoE) and IP networks to manage systems such as Heating, Ventilation and Air Conditioning (HVAC). We are now at the point where smart lighting is also being connected to the network, just as phones and cameras were connected to the network in the previous decade. The advantages gained from IP convergence of IT and lighting for a building are also directly applicable to use in a ship, either military or civilian.

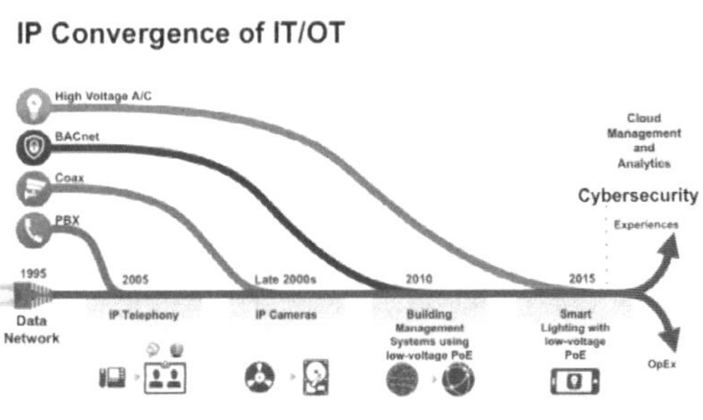

Figure 1: IP Convergence of IT/OT

1.2 OUTDOOR LIGHTING OPPORTUNIES

Outdoor lighting has been a key element of Smart City initiatives around the world. Electric streetlights are essential to any municipal environment but, they are a major energy draw and a target for improved efficiency. Outdoor lighting can deliver safety, security and create an environment suitable for key maritime industry operations such as cargo handling at ports or industrial operations at shipyards. Outdoor lighting can be connected with broader networks and acquire data and deliver information and services to other connected devices and systems. Smart+Connected lighting can scale from very small applications to extremely large implementations across a major city. The city of Jakarta, Indonesia is upgrading approximately 90,000 street luminaires with connected LED lighting to reduce energy costs and improve public services utilizing ellular connectivity to enable networked monitoring and management of connected lighting.[4]

Outdoor LED light fixtures can contain processors and function as sensor-equipped smart devices. The common streetlight can be capable of capturing, analyzing and transmitting data in near real-time. Light poles may follow the path of the phone which not too long ago was used soley to allow two people to talk to each other. Now, phones are powerful computing devices that can deliver many useful applications and can serve as a connection point, camera, sensor, etc. Similarly, light poles can do far more than provide light; they can serve as a multifunctional smart node that can monitor security, weather, traffic, etc.[5] Outdoor lighting can also be used to provide wireless connectivity. Light poles already have electricity and they are usually high above the ground making them ideal for providing wireless access coverage.

Network-connected lighting can be applied to port operations to improve efficiency and effectiveness. Vehicle traffic can be routed to optimize cargo movement/throughput via wayfinding while enhancing safety utilizing communication of real-time conditions and locations to drivers and operations managers. Cargo container and vehicle location and status can become transparent utilizing wireless location services, Radio Frequency Identification (RFID), GPS and asset management applications. Personnel can communicate and collaborate in real-time to enhance decision making based on actual environmental conditions and streaming video.

The City of Hamburg has connected ships to roads to bridges to lights to create a port-wide network. The network gathers data from around the port, creates intelligence and puts it to use in real-time via mobile applications and digital signage. Smarter coordination of vehicle traffic, bridge raises and shipping movement means that more goods are trans-shipped in the port area. Creating a connected port will double port capacity by 2025, without doubling the space required. Additionally, operating costs are expected to fall by 70 percent over 7 years.[6]

2. SHIPBOARD LIGHTING

The U.S. Navy has replaced conventional lights with LED lights on some of its ships. This effort delivered reduced energy consumption, improved working conditions, reduced maintenance and reduced spare light bulb storage requirements.[7] However, this implementation of LED lighting fails to achieve additional efficiencies and enhancements that are possible by connecting the LED lighting to the network. Connecting LED lighting to the network will enable shipboard lighting to also deliver key services, enhance safety and increase mission effectiveness.

2.1 POWER

LED lighting is clearly more energy efficient than conventional lighting in standard configurations however, many LED luminaires are also now capable of being powered by low voltage Power over Ethernet (PoE). PoE is the ability for the Local Area Network (LAN) switching infrastructure to provide power over a copper Ethernet cable to an endpoint or powered device. Space, Weight and Power (SWaP) are important in ship design. PoE reduces the weight of required cabling because only a network cable is required to be routed to each connected LED light and there is no need to route heavier high voltage cables to each light. Ethernet cables also cost less to procure than high voltage cables. Finally, installation costs are also reduced through use of PoE because installation of network data cables does not require a licensed electrician as would installation of standard high voltage cables. Low voltage PoE cables also do not pose any electrical hazard to personnel and contribute to increased safety, which is very critical in the maritime environment. Enhancements have been made to PoE over the past few years. PoE+ (IEEE 802.3at) now delivers 30 watts and UPOE (Universal PoE) now delivers 60 watts per Power Sourcing Equipment (PSE) port which enables more choices in PoE powered lighting.

2.2 NETWORK PLATFORM FOR CONVERGED BUILDING SERVICES

It is advantageous to utilize a foundation for simpler convergence of building services onto a single IP network that can support an ecosystem of solutions. Cisco's Digital Ceiling platform for converged building services is one such solution.[8] A digital ceiling brings the capability to converge disparate building systems such as lighting, heating, cooling, closed-circuit television, metering, security, sensors, etc. on a single IP network as depicted in Figure 2. Convergence on a single network allows richer data collection and correlation to support more comprehensive analytics to enable more efficient building management. As no one company covers all aspects of building management and networking, it's highly desirable to utilize a platform that leverages an ecosystem of industry-leading partners to support Network platform for converged building services implementations.[9]

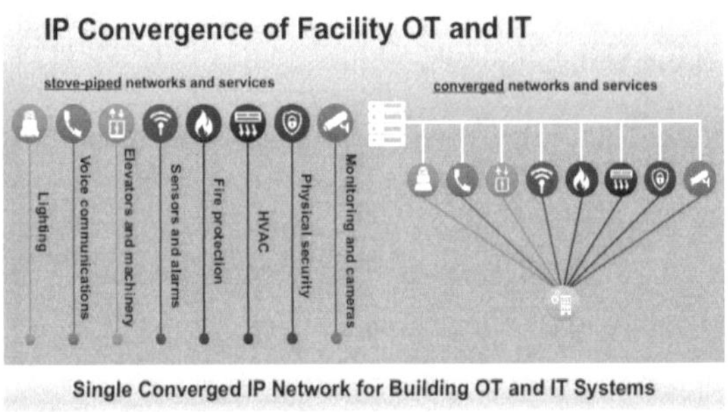

Figure 2: Converged IP Network for Building OT and IT

Some network platforms for converged building services leverage Constrained Application Protocol (CoAP) which is a specialized web transfer protocol for use with constrained nodes and constrained networks within the Internet of Things (IoT). CoAP is designed to be interoperable with other Internet protocols, be efficient for small, inexpensive sensors and be secure. The protocol is designed for machine-to-machine (M2M) applications such as building automation.[10] Just like (Hypertext Transfer Protocol) HTTP, CoAP is based on Representational State Transfer (REST) which provides interoperability between systems and a familiar development environment for many application developers. CoAP has been designed to work with microcontrollers with as little as 10KiB of RAM and 100 KiB of code space, making it ideal for the billions of nodes expected on the IoT. CoAP also utilizes default Datagram Transport Layer Security (DTLS) parameters that are equivalent to 3072-bit RSA public keys.

2.3 SWITCHING

A converged IT/OT network platform to deliver a converged building services solution can leverage enterprise IT switches to distribute power to PoE powered devices. Switches have been improved to provide specific capabilities to optimize use of PoE devices which will enable adding IoT devices without requiring additional high voltage electric power sources. Switches now include dual power supplies to ensure high availability, key to lighting services. Switches also provide converged wired and wireless access eliminating the need for a separate wireless controller to manage a mixed wireless and wired environment which may be very useful in IoT applications.

Small form factor PoE switches enable extension of enterprise-class services well beyond the wiring closet. These compact switches even include fan-less designs that minimize noise making them far less obtrusive in occupied spaces. These new switches are ideally suited for space-constrained environments and for applications where multiple cable runs would present challenges. Some switches enable PoE pass-through which gives the ability for a switch to be powered via uplink from another PoE device with no direct electrical power line required at the downstream switch. Again, this capability reduces required high voltage wiring.

Perpetual PoE (PPoE) provides uninterrupted power to a powered-down device even when the switch is booting. PPoE eliminates the need for a backup power source for switches to ensure constant availability of electricity to LED lights or other critical PoE powered devices. Some switches also include hibernation modes that allow switches to be hibernated in an ultra-low power mode using energy management suite capabilities during periods of non-operation to allow even greater control and optimization of energy consumption. Switches can even leverage Energy-Efficient Ethernet (EEE), IEEE 802.az, to dynamically sense idle periods between bursts of traffic and rapidly switch interfaces to a lower power mode during idle times to reduce power consumption

Enterprise switching brings powerful management and security capabilities to the network platform for converged building services. Technologies like TrustSec increase security on the network with policy-based access control, identity management and role-aware networking. IEEE 802.1AE Media Access Control Security (MACsec) protects against monitoring or altering data by providing by encrypting each packet on the wire using symmetric key cryptography. MACSec works with IEEE 802.1x which can provide dynamic port-based access-control based on identity of users or devices. Simplified management of the distributed switches used in a network platform for converged building services can utilize IT features like Plug and Play (PNP) and Software Defined Networking (SDN) with policy controllers such as Application Policy Infrastructure Controller – Enterprise Module (APIC-EM). Combined, these technologies can be used to automate installation, configuration and deployment of supported switches. These features combine to increase security, improve productivity, cut costs, reduce downtime and enhance the overall user experience.

3. OUTDOOR LIGHTING

Smart+Connected Lighting solutions enable improved city management by enhancing the global transition to outdoor LED lighting fixtures to enable more than just a lighting technology upgrade. Smart+Connected lighting solutions can deliver an open architecture-based light-sensory network that can be deployed with LED luminaries to deliver a gateway for adopting a common platform on which to launch critical smart-city solutions as depicted in Figure 3.

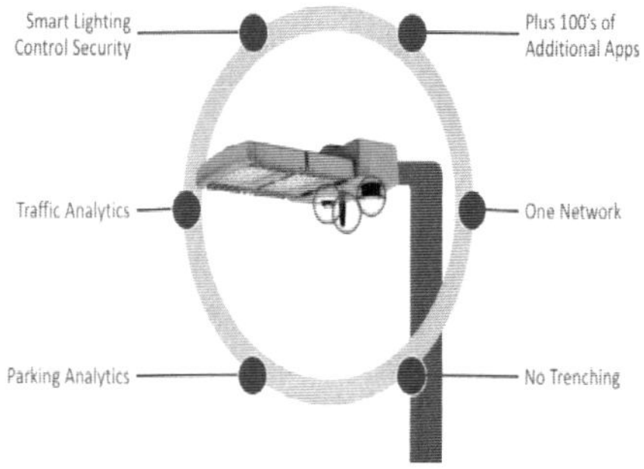

Figure 3: Smart+Connected Lighting Platform

3.1 IP-ENABLED SENSORY NODES

Adapters, such as a Smart+Connected City Multi-Sensor Core Node, converts any LED Light fixture into an IP-enabled sensory node. Adapter nodes leverage standards-based 802.11 a/b/g/n wireless protocol data transports that support a variety of application-specific sensors to deliver multi-service capabilities at each LED luminaire. The light sensory network provides both lighting control and cloud-based IoT services. Nodes are available as internal, or embedded versions, and externally mounted versions. External adapter nodes can connect via standard National Electrical Manufacturers Association (NEMA) photocontrol sockets. Core adapter nodes provide the following capabilities:

- Onboard GPS
- LED luminaire control
- Electrical and sensor data measurement and reporting
- Auxiliary power to other devices
- Network and luminaire status monitoring
- Embedded antennae
- Light dimming
- Fault management.

4. CONCLUSION

LED lighting provides efficiency and durability at a reasonable cost. This is driving relatively rapid replacement of conventional incandescent lighting with LED lighting, mostly to reduce energy costs. Internet of Things (IoT) growth has rapidly accelerated due miniaturization of sensors and processors, network capacity growth, standardization on Internet Protocol as well as other factors. The growth of LED lighting and IoT are well timed to allow the two trends to leverage each other and provide even greater benefit from adoption of LED lighting.

LED lighting coupled with IoT can converge Operational Technology (OT) and Information Technology (IT) networks onto a single IP network. This single network can include a "Network platform for converged building services" which can serve as a foundational component of delivering new capabilities and enabling the networking of people, process, data and things within ships or buildings. A network platform for converged building services aboard ship will reduce weight, enhance safety and help digitize numerous capabilities to enhance operational effectiveness, create new value and enhance customer experiences.

Smart+Connected (S+C) Lighting adoption is accelerating in cities around the world. Modification and/or replacement of outdoor light fixtures with smart nodes to provide network access, sensors, communications, location services, etc. form the foundation of S+C lighting. The same capabilities used to light city streets and create enhanced government and citizen services can also be adopted for use in the maritime industry to enhance operations at piers, ports, shipyards, etc. S+C Lighting can improve safety, security, cargo throughput, loss prevention, asset management and many other essential functions within the industry.

5. REFERENCES

1. Mary Yamade and Kelsey Stober, "Adoption of Light-Emitting Diodes in Common Lighting Applications," Navigant, U.S. Department of Energy, July 2015. Retrieved from: http://energy.gov/eere/ssl/downloads/led-adoption-report.

2. James Crowther, Christoph Herzing and Gordon Philips, "The Time is Right for Connected Public Lighting Within Smart Cities," Cisco and Philips, Oct 2012.

3. James Brodick, "What's Next for Solid-State Lighting?" U.S. Department of Energy, NEMA electroindustry, Feb 2015.

4. "Philips Lighting implements one of the world's largest connected street lighting systems in Jakarta," Nov 15, 2016. Retrieved from: http://www.newsroom.lighting.philips.com/news/2016/20161115-philips-lighting-implements-one-of-the-worlds-largest-connected-street-lighting-systems-in-jakarta.html.

5. "Why the IoT Is Coming to Street Lights," Brian Buntz, 22 Apr, 2016. Retrieved from: http://www.ioti.com/smart-cities/why-iot-coming-street-lights.

6. "Hamburg Gets Smarter," Cisco, retrieved from: http://www.cisco.com/web/tomorrow-starts-here/cars/hamburg/index.html.

7. Admiral Kevin Slates, "SEEING THE LIGHT: SHIPBOARD LEDS," Oct 2, 2014. Retrieved from: http://navylive.dodlive.mil/2014/10/02/seeing-the-light-shipboard-leds/.

8. Zeus Kerravala, "A Network platform for converged building services Is Now a Business Imperative," ZK Research, Aug 2016.

9. Cisco. Retrieved from: http://www.cisco.com/c/en/us/solutions/digital-ceiling/partner-ecosystem.html

10. "CoAP RFC 7252 Constrained Application Protocol." Retrieved from: http://coap.technology/.

6. AUTHORS' BIOGRAPHIES

Kelly Jones is a systems engineer for the U.S. Navy Shipboard Account team at Cisco Systems, Inc. She works with the U.S. Navy and ship integrators to identify requirements, provide technical guidance on the future direction of networking technologies and assists them in implementation, testing and validation of proposed solutions. She served six years in the U.S. Navy as a Cryptologic Maintenance Technician and is a Certified Information Systems Security Professional (CISSP).

Joe Beel is a strategic programs manager at Cisco Systems, Inc. He develops and implements strategies to support the U.S. Public Sector focusing on defense. He is a retired naval officer and served as a helicopter pilot and acquisition professional in both the Naval Air Systems Command and Space and Naval Warfare Systems Command.

ENERGY SAVINGS FOR A SHIP IN IRREGULAR WAVES USING REAL-TIME OPTIMAL CONTROL OF PROPELLER PITCH AND ELECTRIC PROPULSION

H Makino, **Y Hirano**, and **N Umeda**, Osaka University, JAPAN
T Ohtsuka, Kyoto University, JAPAN
K Tanizawa and **H Sekiguchi**, National Maritime Research Institute, JAPAN
H Susaki, Furuno Electric Co. Ltd., JAPAN
J Suzuki and **M Fukazawa**, Kamome Propeller Co. Ltd., JAPAN

SUMMARY

In order to save energy for a ship sailing under various wave conditions, we developed an optimal-control system of propulsion power and propeller pitch. In this system, incoming irregular wave data are estimated using a wave-radar and the Fourier transformation. To respond to that wave situation, the optimal-control signals of an alternating-current induction motor and a specifically designed controllable-pitch propeller are calculated in real time using an efficient algorithm of receding horizon control. The numerical results indicate that, for a short-crested irregular seaway with a significant wave height of 3 m and a mean wave period of 12 s, the energy required to propel a 217 m bulk-carrier with 14.7 knots is reduced by about 2% in head waves and by about 7% in the following waves when the optimal-control system proposed herein is used.

1. INTRODUCTION

Owing to global economic development, shipping, which is the main logistics sector, is required to operate more safely and efficiently. Thus, several approaches have been proposed and investigated, such as ship sensors and smart operational measures. Moreover, several application services are utilized to improve ship safety and efficiency. Under these circumstances, Ballou [1] noted the need for a comprehensive platform for the optimization of deployment, smart navigation planning, and energy management in ships in order to improve energy efficiency. Domínguez [2] developed a new intelligent tool using real-time cloud computing, big data, and analysis cognition for the purpose of improving marine traffic safety. Moreover, Gowri et al. [3] proposed a method for monitoring the ocean situation by integrating images obtained using a synthetic-aperture radar with data from an automatic identification system for the purpose of ship safety management and efficient operation.

Energy conservation in ships is a major issue in the effort toward reducing global warming. In fact, the International Maritime Organization (IMO) adopted a revision of the MARPOL Convention in July, 2011 [4] that requires the CO_2 emissions of all new ships to be reduced by 10%–30%. This regulation presents a new challenge to naval architects and marine engineers, who must develop novel ideas and technologies to satisfy this regulation.

The challenges for future research are to develop and implement new energy-saving methods while satisfying design and operational constraints. For this reason, we propose using a controllable-pitch propeller (CPP) as an energy-saving device. In general, a ship engine is controlled by its governor that maintains a constant engine speed and propeller pitch. However, a ship at sea runs under oscillatory external environment; therefore, its optimal control for realizing minimum energy consumption could be different from the theoretically constant revolution and pitch control. In a previous study, Matsunaga et al. [5] managed to improve fuel efficiency by optimizing the control of the diesel-engine propulsion via sequential quadratic programming [6]. However, this control method is computationally expensive, making it unsuitable for real-time applications. Nevertheless, Kitagawa et al. [7] implemented this method on a full-scale ship by using a pre-calculated control-procedure database, and the incoming wave data were estimated from the measured propeller torque using a time series-based prediction technique. As a result, they obtained energy savings of 0.7% in moderate head seas and 1.9% in moderate following seas. The major limitation of this method, in addition to the prohibitively long a priori calculation time for optimal control, is the slow response of diesel engines and CPP control. Regarding the limitations of the power plant, Kushida et al. [8] performed a numerical study of energy efficiency improvements for alternating-current (AC) electric propulsion systems with direct-current (DC) batteries in head seas. However, wave prediction using a propeller torque is not straightforward, making it unsuitable for wider application.

In order to make this approach applicable, Makino et al. [9] addressed the remaining limitations that hinder the realization of greater energy savings at sea. To overcome the limitation of slow CPP control, direct use of servo motors, which is expected to lead to rapid CPP control, is proposed. To deflect the problem of long computation time, the receding horizon control algorithm [10] was used to rapidly calculate the optimum CPP pitch ratio and the drive frequency for the AC propulsion motor, thereby minimizing fuel consumption under feasible actuator constraints. As a result, about 5% improvement in fuel efficiency in regular waves was confirmed by numerical calculation. Furthermore, fuel efficiency improvement of about 10% was obtained in regular waves through trial calculation by varying the operational weights of the CPP rotation number, the propeller pitch, and the propulsion-motor input frequency.

In this paper, we describe the updated system based on the above modifications and examine how it works in a more realistic ocean environment, i.e., in short-crested irregular waves. For this purpose, we use the wave radar developed by Hirayama et al. [11].

2. STUDY OUTLINE

The concept of the proposed system is shown in Figure 1. It consists of the following steps. First, approaching waves are detected by a wave-radar, and the obtained radar images are analyzed using a three-dimensional (3D) Fast Fourier transformation. Then, the command values of the propeller pitch and the driving frequency of the propulsion motor are calculated so that energy consumption is minimized with actual constraints based on a nonlinear model predictive control (NMPC). Finally, the values are used as input for a CPP, which has quick-response ability and an electric propulsion motor. In order to realize this system, we investigate the following elements.

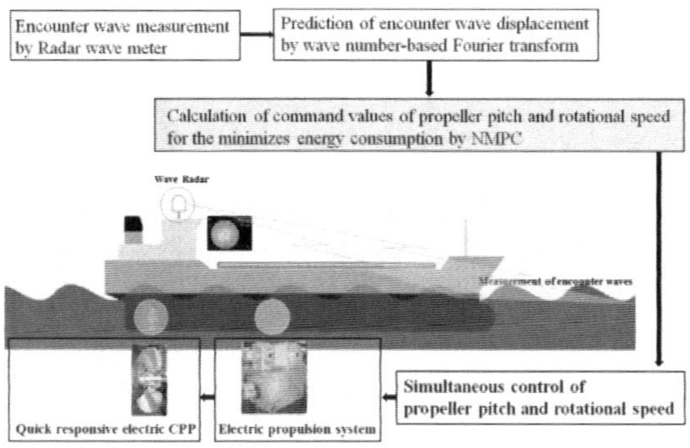

Figure 1: Concept of the novel ship control system.

2.1 DIRECT-DRIVE SERVO CPP

A conventional CPP is driven by a hydraulic system, which means that the CPP response is relatively slow. To improve the CPP response speed, we propose using an electrically driven CPP. To implement this, an electric actuator is installed inside the hub of a new CPP type. The power for the actuator is to be supplied through a slip ring mounted on the shaft. The prototype CPP configuration is determined based on the size of the actuator to be installed in the propeller hub. This new type of CPP is called a direct-drive servo CPP (DDS-CPP), because each actuator is directly driven by a servo motor. Figure 2 shows a schematic diagram of the proposed DDS-CPP system, and Figure 3 shows the actual DDS-CPP [12].

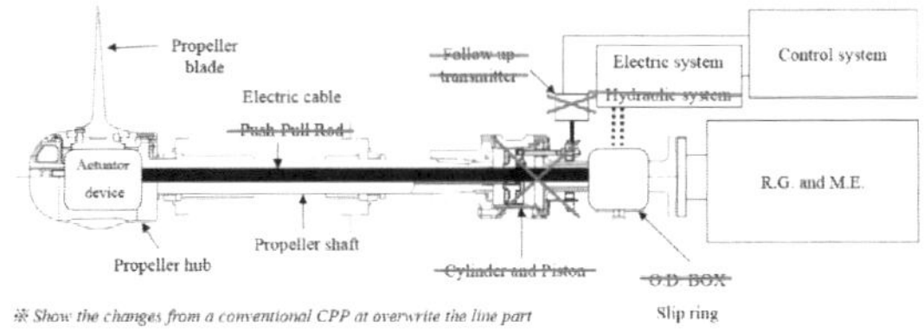

Figure 2: Schematic of the direct-drive-servo controllable-pitch-propeller system.

Figure 3: Direct-drive servo CPP (DDS-CPP).

2.2 ELECTRIC PROPLUSION UNIT

The simplest configuration of the electric propulsion unit is shown in Figure 4. In this experimental system, the electric motor in the electric propulsion unit is a three-phase induction motor; its rotation speed is controlled by an inverter, while the load is controlled by a DC electronic load that is connected to a DC motor via a torque/rotation speed meter. In addition, the inverter and the DC electronic load can be controlled simultaneously using an arbitrary signal generator controlled from a PC.

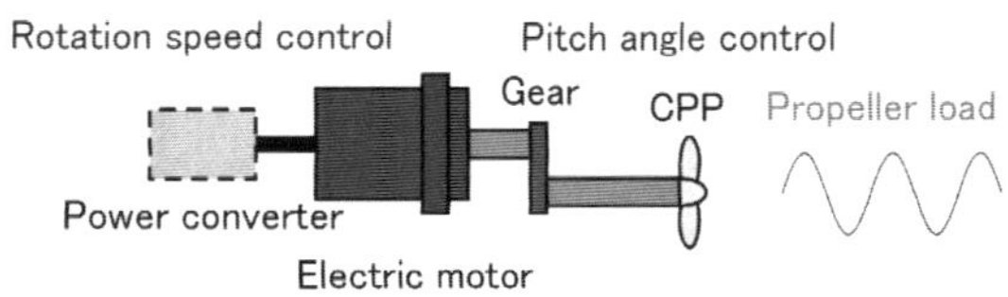

Figure 4: Electric propulsion unit.

2.3 WAVE RADAR

The wave radar used in this system can predict approaching wave trains in real time. Figure 5 shows the X-band radar that acquires wave images; Figure 6 shows an example of the obtained radar images. 3D Fourier-transform spectral analysis is performed on these radar images. By performing the inverse Fourier transform on them, the wave surface after Δt s is obtained. Finally, by changing these continuously, a time series of the wave surface is obtained [11]. The real-time display of the predicted waves is shown in Figure 7. As a result, it is possible to optimally control the CPP for the coming wave trains.

Figure 5: Picture of the wave radar.

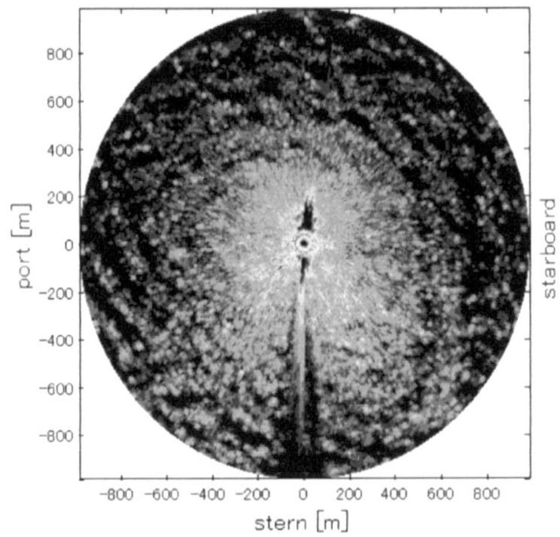

Figure 6: Radar image of the wave.

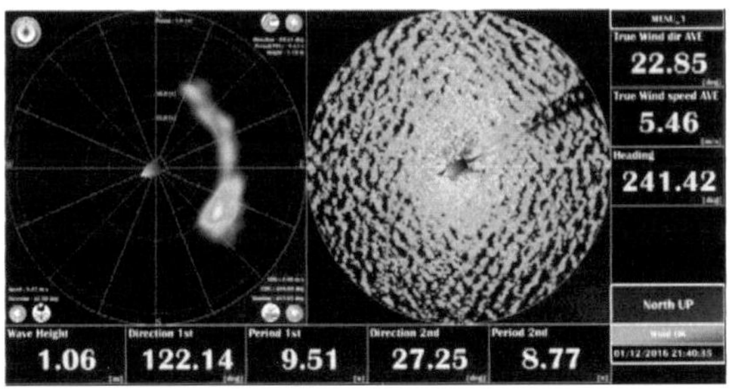

Figure 7: Real-time display of the predicted incoming waves.

By using these factors, we can realize an energy-saving system, if appropriate optimal control can be realized in real time. The optimal-control procedure will be described in the next section.

3. CONCEPT OF CONTROL SYSTEM

In order to control the system, it is necessary to establish a control law. Our goal here is to develop a system that provides real-time optimal control of the propeller-blade angle and the drive frequency, so that the angle can be optimized in anticipation of the impending waves, thereby minimizing energy consumption. To achieve optimal real-time control, we use the NMPC method, because it can anticipate future events and take the appropriate control actions. This constitutes a fundamental advantage of NMPC over proportional-integral-derivative controllers and linear-quadratic-regulator controllers.

The key components of NMPC include a prediction model, a receding horizon, a performance index, and constraint handling. Section 4 discusses in detail the fundamental aspects of the prediction-model equations. The receding horizon of NMPC shows the time, over which the model predicts its behavior, and uses this information for real-time optimized decision-making. This study uses a 20 s receding horizon. The control inputs are the propeller pitch ratio and the drive frequency of the electric propulsion unit. The performance index depends on the amount of electrical energy used and the squared difference between the actual ship and command speeds. The key model constraints are the drive frequency, the propeller-pitch ratio, the rate of change of the propeller-blade angle, and the average ship speed. For incident waves, we evaluate the energy savings obtained through real-time optimal control by comparing its energy requirements with those of fixed-pitch and fixed-drive-frequency systems. Figure 8 shows a schematic diagram of the control system.

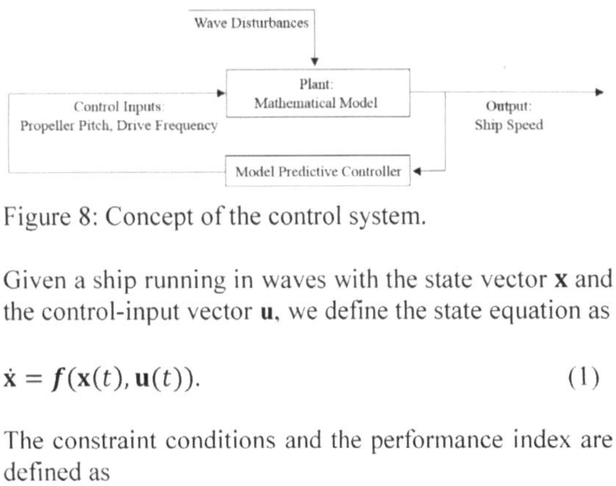

Figure 8: Concept of the control system.

Given a ship running in waves with the state vector **x** and the control-input vector **u**, we define the state equation as

$$\dot{\mathbf{x}} = \boldsymbol{f}(\mathbf{x}(t), \mathbf{u}(t)). \quad (1)$$

The constraint conditions and the performance index are defined as

$$\boldsymbol{C}(\mathbf{x}(t), \mathbf{u}(t)) = 0, \quad (2)$$

$$J = \int_t^{t+T} \boldsymbol{L}(\mathbf{x}(\tau), \mathbf{u}(\tau)) d\tau, \quad (3)$$

where $\boldsymbol{T}$ is the time over which the control system is evaluated (horizon) and τ is the integration variable. The control input at each time t is determined so as to minimize the performance index with a receding horizon. The optimal control $\mathbf{u}_{\mathrm{opt}}(\tau)$ is a function that spans over the horizon and depends on t and $\mathbf{x}(t)$ as $\mathbf{u}_{\mathrm{opt}}(\tau; t, \mathbf{x}(t))$ $(t \leq \tau \leq t + \boldsymbol{T})$. However, the actual input to the system is given only by $\mathbf{u}_{\mathrm{opt}}(\tau)$ at time t, that is, $\mathbf{u}(t) = \mathbf{u}_{\mathrm{opt}}(\tau; t, \mathbf{x}(t))$, which is a state-feedback control law. Next, by discretizing the stationary condition and the performance index $\boldsymbol{J}$, the stationary condition is obtained from the following Euler-Lagrange equation [10]:

Performance index:

$$J = \sum_{i=0}^{N-1} L(\mathbf{x}_i^*(t), \mathbf{u}_i^*(t))\Delta\tau(t), \quad (4)$$

Stationary condition:

$$\mathbf{x}_0^*(t) = \mathbf{x}(t), \quad (5)$$

$$\mathbf{x}_{i+1}^*(t) = \mathbf{x}_i^*(t) + \boldsymbol{f}(\mathbf{x}_i^*(t), \mathbf{u}_i^*(t))\Delta\tau(t), \quad (6)$$

$$\boldsymbol{\lambda}_N^*(t) = \left(\frac{\partial\varphi}{\partial\mathbf{x}}\right)^T (\mathbf{x}_N^*(t)), \quad (7)$$

$$\boldsymbol{\lambda}_i^*(t) = \boldsymbol{\lambda}_{i+1}^*(t) + \left(\frac{\partial H}{\partial\mathbf{x}}\right)^T (\mathbf{x}_i^*(t), \mathbf{u}_i^*(t), \boldsymbol{\lambda}_{i+1}^*(t), \boldsymbol{\mu}_i^*(t))\Delta\tau(t), \quad (8)$$

$$\frac{\partial H}{\partial u}\left(\mathbf{x}_i^*(t), \mathbf{u}_i^*(t), \boldsymbol{\lambda}_{i+1}^*(t), \boldsymbol{\mu}_i^*(t)\right) = 0, \quad (9)$$

$$C\left(\mathbf{x}_i^*(t), \mathbf{u}_i^*(t)\right) = 0. \quad (10)$$

In these equations, N is the number of sample points on the horizon, $\Delta\tau(t)$ is $\boldsymbol{T}(t)/N$, and $\mathbf{x}_i^*(t)$ corresponds to $\mathbf{x}^*(i\Delta\tau, t)$. Because the horizon $\boldsymbol{T}$ depends in general on time t, $\Delta\tau$ depends on time as well. Note that only the τ axis is discretized; the dependence on t remains continuous at this stage of the problem. In other words, the discretized problem is solved for continuous time t, although, in the actual implementation, the time t will also be discretized. Given the initial state of the discretized problem, $\mathbf{x}_0^*(t) = \mathbf{x}(t)$, the control sequence $\{\mathbf{u}_i^*(t)\}_{i=0}^{N-1}$ is optimized at each time t. The actual control input to the system is given by $\mathbf{u}(t) = \mathbf{u}_0^*(t)$. Here we define the Hamiltonian as [10]:

$$\mathbf{H}(\mathbf{x}, \mathbf{u}, \boldsymbol{\lambda}, \boldsymbol{\mu}) = L + \boldsymbol{\lambda}^T \boldsymbol{f} + \boldsymbol{\mu}^T \boldsymbol{C}. \quad (11)$$

The vector $\mathbf{U}(t)$, which consists of a series of control inputs $\mathbf{u}(t)$ and Lagrange multipliers $\boldsymbol{\mu}_i^*(t)$, is defined as

$$\mathbf{U}(t) = \begin{bmatrix} \mathbf{u}_0^*(t) \\ \boldsymbol{\mu}_0^*(t) \\ \vdots \\ \mathbf{u}_{N-1}^*(t) \\ \boldsymbol{\mu}_{N-1}^*(t) \end{bmatrix}. \quad (12)$$

Because the state $\mathbf{x}_i^*(t)$ and the costate $\boldsymbol{\lambda}_i^*(t)$ are functions of $\mathbf{U}(t)$ and $\mathbf{x}_0^*(t) = \mathbf{x}(t)$, Equations (9) and (10) can be written as the following algebraic equations:

$$\boldsymbol{F}(\mathbf{U}(t), \mathbf{x}(t), t) = \begin{bmatrix} \left(\frac{\partial\mathbf{H}}{\partial\mathbf{u}}\right)^T (\mathbf{x}_0^*(t), \mathbf{u}_0^*(t), \boldsymbol{\lambda}_1^*(t), \boldsymbol{\mu}_0^*(t)) \\ \mathbf{C}(\mathbf{x}_0^*(t), \mathbf{u}_0^*(t)) \\ \vdots \\ \left(\frac{\partial\mathbf{H}}{\partial\mathbf{u}}\right)^T (\mathbf{x}_{N-1}^*(t), \mathbf{u}_{N-1}^*(t), \boldsymbol{\lambda}_N^*(t), \boldsymbol{\mu}_{N-1}^*(t)) \\ \boldsymbol{C}(\mathbf{x}_{N-1}^*(t), \mathbf{u}_{N-1}^*(t)) \end{bmatrix} = 0, \quad (13)$$

Equation (13) must hold at all times t. In addition, the initial value of $\boldsymbol{F}$ and its time derivative shall be zero. In other words, if the initial value of $\boldsymbol{F}$ and its time derivative are zero, F is also zero. This is equivalent to satisfying Equations (14) and (15) simultaneously, so that it can be further rewritten as a simultaneous linear equation for $\dot{\mathbf{U}}(t)$ [10]:

$$\frac{d}{dt}\boldsymbol{F}(\mathbf{U}(t), \mathbf{x}(t), t) = -\zeta\boldsymbol{F}(\mathbf{U}(t), \mathbf{x}(t), t) \; (\zeta > 0), \quad (14)$$

$$\boldsymbol{F}(\mathbf{U}(0), \mathbf{x}(0), 0) = 0, \quad (15)$$

$$\frac{\partial\boldsymbol{F}}{\partial\mathbf{U}}\dot{\mathbf{U}}(t) = -\zeta\boldsymbol{F} - \frac{\partial\boldsymbol{F}}{\partial\mathbf{x}}\dot{\mathbf{x}}(t) - \frac{\partial\boldsymbol{F}}{\partial t}. \quad (16)$$

Since a simultaneous linear equation can be more easily solved than an algebraic equation, the time integral of the control input $\dot{\mathbf{U}}(t)$ can be determined in much less computational time. This method is called the continuation/GMRES (generalized minimal residual) method [10].

4. SYSTEM EQUATIONS

4.1 FORMULATION OF STATE EQUATION

4.1 (a) Ship equation for surge direction

An irregular wave is defined as a superposition of regular waves having random phases with different frequencies. As shown in Figure 9, we set the coordinates of the irregular waves as ξ-η, the coordinate of the hull as x-y, and the angle from the ξ direction to the hull as χ.

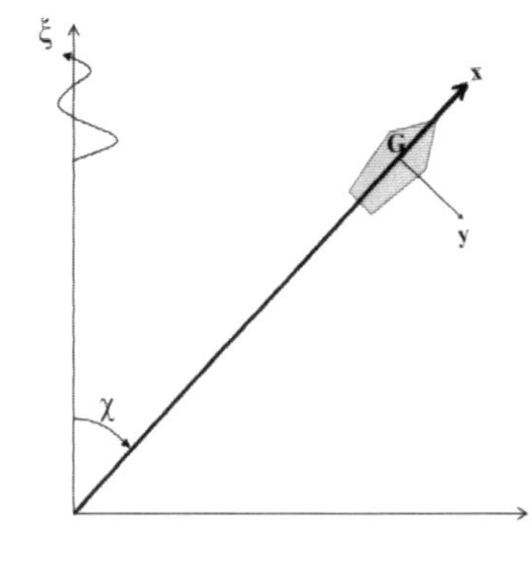

Figure 9: Coordinate systems of irregular waves.

In this case, the surge movement of the ship is expressed as

$$(m + m_x)\dot{u} = T_e - R + X_w + X_{add}, \quad (17)$$

where m is the ship mass, m_x is the added mass, X_w is the wave-exciting forces, T_e is the effective propeller thrust, R is the ship resistance, u is the ship speed, and X_{add} is the added resistance in waves.

$$X_w = \int_0^\infty \int_{-\pi/2}^{\pi/2} |H_x(\omega, x)| \sqrt{2S_w(\omega, \alpha) d\omega d\alpha} \cos(\omega t - k\xi_G \cos\alpha - k\eta_G \sin\alpha + \varepsilon_X(\omega, \chi) + \delta(\omega.\alpha)), \quad (18)$$

$$T_e = (1 - t_p)\rho n^2 D^4 K_T\left(\frac{U_p}{nD}, p\right), \quad (19)$$

$$R = \frac{1}{2}\rho u^2 S_F \left\{C_{F0}\left(\frac{uL}{v}\right) + C_R\left(\frac{u}{\sqrt{Lg}}\right)\right\}, \quad (20)$$

$$U_p = \int_0^\infty \int_{-\pi/2}^{\pi/2} \omega \sqrt{2S_w(\omega, \alpha) d\omega d\alpha} e^{-kZ_p} \cos(\omega t - k\xi_G \cos\alpha - k\eta_G \sin\alpha + kx_p \cos\chi + \delta(\omega, \alpha)), \quad (21)$$

$$X_{add} = 2\int_0^\infty \int_{-\pi/2}^{\pi/2} K_{AW}\left[F_n, \alpha, \frac{\lambda}{L_{pp}}\right] * \left(\frac{4\rho g B_{MAX}^2}{L_{pp}}\right) S_w(\omega, \alpha) d\omega d\alpha, \quad (22)$$

where H_x is the transfer function, ε_X is the phase difference for wave enforcement, t_p is the thrust-deduction fraction, ρ is the wave density, n is the rotational speed of the propeller, D is the propeller diameter, K_T is the thrust coefficient of the propeller, U_p is the propeller inflow velocity, w_p is the effective wake fraction, Z_p is the vertical distance from the calm-water plane, x_p is the distance in the surge direction from the hull centre of gravity to the propeller, S_F is the wetted surface area of the ship, C_{F0} is the frictional resistance of the ship, L_{pp} is the ship length between perpendiculars, v is the kinematic coefficient of viscosity, C_R is the residual resistance of the ship, g is the gravity acceleration, K_{AW} is the resistance-increase coefficient in regular waves, F_n is the Froude number, B_{MAX} is the ship breadth, S_w is the ocean wave spectrum, AE is the ship aft end, FE is the ship fore end, $S(x)$ is the cross-sectional area of the ship, $B(x)$ is the ship width at x from AE, $d(x)$ is the cross-sectional draught of the ship, and p is the propeller pitch.

For the transfer function of resistance increase in waves, as expressed by Equation (22), the calculated value using the slender-body theory of low-frequency approximation is used for the Kochin function in Maruo's formula. Furthermore, for the short wavelength region, the value according to the semi-empirical formula of Fujii and Takahashi is used. Then, K_{AW} in Equation (22), is calculated based on these theory. And, as K_{AW} is discrete data, the integral part is expressed as the summation of these using fixed Froude number. The differences of each parameters α, $\frac{\lambda}{L_{pp}}$ are 5 degree and 0.05 respectively. In addition, H_x and ε_X are defined as follows:

$$H_X(\omega, \chi - \alpha) = Hxc(\omega, \chi) + iHxs(\omega, \chi), \quad (23)$$

$$Hxc(\omega, \chi - \alpha) = -\rho gk\cos\chi \int_{AE}^{FE} C_1(\chi) S(x) e^{-k\left(\frac{d(x)}{2}\right)} \sin(kx\cos\chi)\, dx, \quad (24)$$

$$Hxs(\omega, \chi - \alpha) = \rho gk\cos\chi \int_{AE}^{FE} C_1(\chi) S(x) e^{-k\left(\frac{d(x)}{2}\right)} \cos(kx\cos\chi)\, dx, \quad (25)$$

where

$$C_1(x) = \frac{\sin\left(\frac{B(x)}{2} * k\sin\chi\right)}{\frac{B(x)}{2} * k\sin\chi},$$

$$\varepsilon_X = -Tan^{-1}\left(\frac{H_{XS}}{H_{XC}}\right). \quad (26)$$

4.1 (b) Propeller shaft revolution

Propeller thrust is generated by the rotation of the propeller shaft. Therefore, the equation of shaft rotation is

$$2\pi J_p \frac{dn}{dt} = Q_m[N, f, t] - Q_p[n, U_p, t], \quad (27)$$

where J_p is the inertial moment, Q_m is the motor torque, f is the drive frequency of the propulsion motor, N is the rotational speed of the motor, t is time, n is the rotational speed of the propeller, U_p is the propeller inflow velocity and Q_p is the propeller torque, which is given by:

$$Q_p = \rho n^2 D^5 K_Q\left(\frac{U_p}{nD}, p\right), \quad (28)$$

where K_Q is the torque coefficient of the propeller.

4.1 (c) Controllable-pitch propeller

The time-variant propeller-pitch ratio can be expressed as a first-order lag system written as

$$T_p\dot{p} + p = p^*. \quad (29)$$

Here, T_p is the time constant of the CPP controller, $\dot{p}$ is the propeller-pitch ratio, and p^* is the commanded propeller-pitch ratio. By assuming that the propeller thrust and torque are independent of the rate of change of the propeller pitch, the thrust and torque coefficients, as functions of the pitch ratio, may be given as

$$K_{T}=a_1(p)J^2 + a_2(p)J + a_3(p), \quad (30)$$

$$K_{Q}=b_1(p)J^2 + b_2(p)J + b_3(p). \quad (31)$$

Here, the following relational expression is used to convert the propeller pitch angle into the propeller pitch ratio:

$$p = \frac{P}{D} = 0.7\pi tan\left(\frac{\pi\theta}{180}\right), \quad (32)$$

where J is the forward constant and θ is the propeller-pitch angle.

4.1 (d) Equivalent-circuit model for propulsion motor

By representing a three-phase squirrel-cage induction motor with an L-type equivalent-circuit model, the relationship between the torque and the input frequency, at which a constant control V/f is required, is modeled as

$$Q_m = \frac{90f^2}{\pi N}\left(\frac{V_{\mathrm{Rated}}}{f_{\mathrm{Rated}}}\right)^2 \frac{\frac{r_2'}{s}}{\left(r_1+\frac{r_2'}{s}\right)^2+\left(x_1+x_2'\right)^2}, \quad (33)$$

where f is the drive frequency, N is the rotational speed of the motor, V_{Rated} is the voltage rating, f_{Rated} is the drive-frequency rating, r_1 is the primary-winding resistance, r_2' is the secondary-winding resistance, x_1 is the leakage reactance of the primary-winding resistance, x_2' is the leakage reactance of secondary-winding resistance, and s is the slip of the induction motor.

In addition, taking into account the difference between the time constants of the ship motion and that of the electric circuit, the L-type equivalent circuit is used as a quasi-static model, so any transient response is ignored.

4.2 WAVE SITUATION

In this study, we assume that the wave input data includes the ocean wave spectrum and the phase given by the wave radar. Nevertheless, in this paper, an irregular wave is expressed as the standard spectrum by the Pierson-Moskowitz-type ITTC spectrum (1978), and directional spreading function, besides, the phase created by uniform random numbers from 0 to 2π is used as an alternative.

$$S_w(\omega, \alpha) = S_w(\omega)\mathrm{D}(\alpha), \quad (34)$$

$$S_w(\omega) = \frac{A}{\omega^5}\exp\{-\frac{B}{\omega^4}\}, \quad (35)$$

$$\mathrm{D}(\beta) = \frac{8}{3\pi}cos^4\beta, \quad (36)$$

where

$$\mathrm{A} = 173\frac{H_{1/3}^2}{T_1^4}, \mathrm{B} = \frac{691}{T_1^4},$$

where $S_w(\omega)$ is the ocean wave spectrum, $H_{1/3}$ is the significant wave height, T_1 is the average wave period, and $\mathrm{D}(\beta)$ is the directional distribution function.

Here, regarding the number of divisions of ω, the range from 0.3 to 1.19 is divided into 52 parts, and the range of 1.19 to 2.0 is divided into 12 parts, at unequal intervals based on the effective range of the transfer function. In addition, 36 divisions are made based on the direction distribution α, in the range from −π/2 to π/2, as expressed by Equation (36). The ocean wave directional spectrum expressed by Equation (34) is shown in Figure 10.

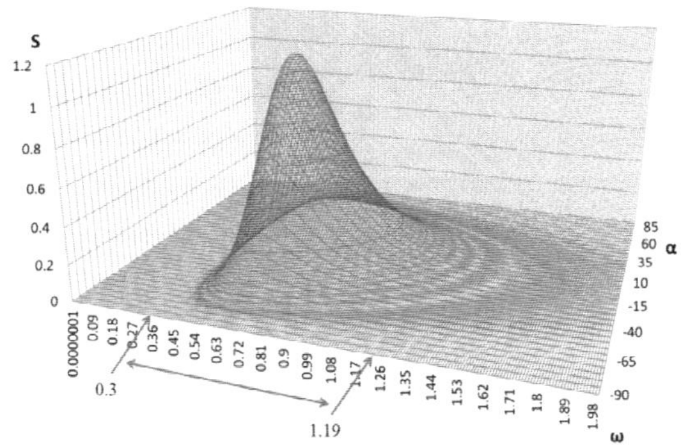

Figure 10: Ocean wave directional spectrum.

4.3 CONSTRAINT CONDITION FOR MODEL-PREDICTIVE CONTROL

The constraints are composed of the following four conditions:

1. The drive frequency of the propulsion motor f must be between 0 and a given maximum:
$$0 \le f \le f_{\max}. \quad (37)$$
2. The propeller-pitch ratio p must be within the available range:
$$p_{\min} \le p \le p_{\max}. \quad (38)$$
3. The absolute value of the rate of change of the propeller-pitch ratio $\dot{p}$ must be less than or equal to a given maximum:
$$-|\dot{p}_{\max}| \le \dot{p} \le |\dot{p}_{\max}|. \quad (39)$$
4. The ship speed must be equal to the specified speed.

The first three conditions are based on inequalities related to the design of the actuators, whereas the fourth condition is an equality condition. We transpose the inequality constraints 1-3 to the equality constraint by using dummy input variables, and the equality constraint 4 is integrated into the performance index.

4.4 PERFORMANCE INDEX FOR MODEL PREDICTIVE CONTROL

The performance index J is assumed to be obtained by integrating the power P_e delivered by the propulsion motor, and the square of the error of the ship speed is added to the performance index instead of constraint 4 for the ship speed. Thus, we obtain:

$$J = \int_t^{t+T} \boldsymbol{L}(\mathbf{x}(\tau), \mathbf{u}(\tau))d\tau, \quad (40)$$

$$\boldsymbol{L}(\mathbf{x}(\tau), \mathbf{u}(\tau)) = w_1 \cdot P_e + w_2 \cdot (u - \hat{u})^2 \cdot H(\hat{u} - u) + \sum_{i=3}^{5} w_i \cdot \mathbf{u}_i(\tau) + w_6 \cdot (f - \hat{f})^2 + w_7 \cdot (p - \hat{p})^2, \quad (41)$$

where w_i $(i = 1\text{-}7)$ are the weights, $\hat{u}$ is the commanded ship speed, $\hat{f}$ is the reference drive frequency of the propulsion motor in calm water, and $\hat{p}$ is the reference propeller-pitch ratio in calm water.

$$P_e = 3I_{\text{out}}^{'\,2} \frac{1-s}{s} r_2^{'}, \quad (42)$$

where

$$I_{\text{out}}^{'} = \frac{V_{\text{out}}}{\sqrt{(r_1 + r_2^{'}/s)^2 + (x_1 + x_2^{'})^2}}, \quad (43)$$

and

$$V_{\text{out}} = \frac{V_{\text{Rated}}}{f_{\text{Rated}}} \cdot f, \quad (44)$$

As for the value of the weights, since the evaluation function mainly consists of the power consumption and the ship speed, the influence of terms related to the dummy inputs and the inequality constraint conditions are set to be as small as possible. Moreover, with respect to the ship speed, the step function H is multiplied, so that it becomes 0 when the ship speed exceeds the commanded ship speed.

5. NUMERICAL CALCULATION FOR IRREGULAR WAVES

5.1 MODEL CHARACTERISTICS OF SUBJECT SHIP

The ship used in this study is a 217m bulk-carrier that is assumed to be equipped with an electric propulsion system consisting of a 9070 kW AC induction motor, invertors, and rechargeable batteries. The gear between the induction motor and the propeller has a ratio of 1:11. The principal particulars of the ship are given in Table 1.

Figure 11 shows the relationship between the ship speed, the drive frequency, and the rotational speed without gear in the equilibrium state in calm water. The sections, where the curves overlap, show that the rotational speed can be changed without changing the drive frequency.

Table 1 Ship principal particulars.

Items	Ship
Length between perpendiculars [m]	217.0
Breadth [m]	32.3
Mean draught [m]	12.2
Block coefficient	0.840
Induction-motor power [kW]	9070
Wake fraction	0.295
Thrust-deduction fraction	0.133
Propeller diameter [m]	7.1
Maximum speed of blade angle [degrees/s]	3.0
Propeller-blade angle range [degrees]	10–26
Moment of inertia of propeller shaft system [$\text{kg} \cdot m^2$]	77670.0
Time constant of CPP controller [s]	1.0
Maximum drive frequency of propulsion motor (f_{max}) [Hz]	59.9
Maximum propeller-pitch ratio (p_{max})	1.0726
Minimum propeller-pitch ratio (p_{min})	0.3878
Absolute value of rate of change of propeller-pitch ratio ($\dot{p}_{max}$)	0.1646
Reference ship speed ($\hat{u}$) [knots]	14.7
Reference drive frequency of propulsion motor in calm water ($\hat{f}$) [Hz]	44.93
Reference propeller-pitch ratio in calm water ($\hat{p}$)	0.8

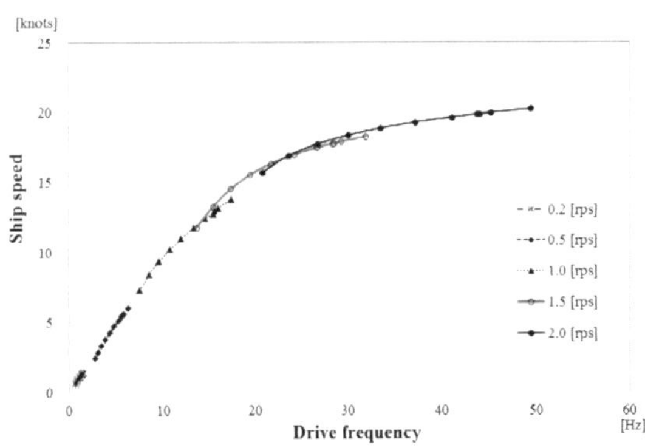

Figure 11: Ship speed vs. drive frequency for various propeller rotational speeds without gear in calm water.

5.2 CALCULATION RESULTS AND DISCUSSION

In order to verify the proposed method, numerical calculations are performed assuming head and following seas. In this numerical calculation, we employ an average wave period of 12 s, as it was calculated for a regular wave with a wavelength per ship length of 1 ($\lambda/L = 1$) in the trial calculation of the previous study [9]. In addition, the significant wave height is set at 3 m, which is the average value at the time of sailing for the North Atlantic Ocean.

Both the case of optimal control and the case of no control are considered in several wave-cycle encounters. In the former case, the propeller-pitch ratio and the propulsion-motor drive frequency change over time. In the latter case, these two parameters remain constant. The design speed of the ship is used as the target ship speed for this control system.

5.2 (a) Numerical calculation for head seas

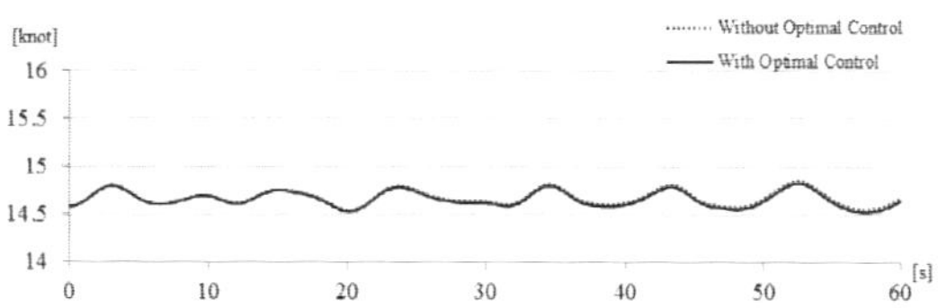

(a) Ship speed.

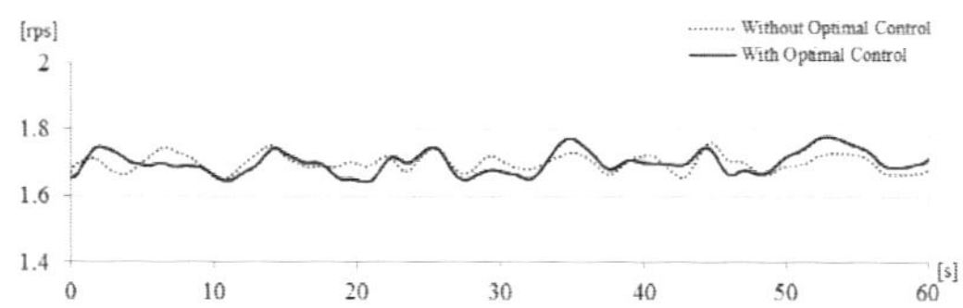

(b) Propeller speed.

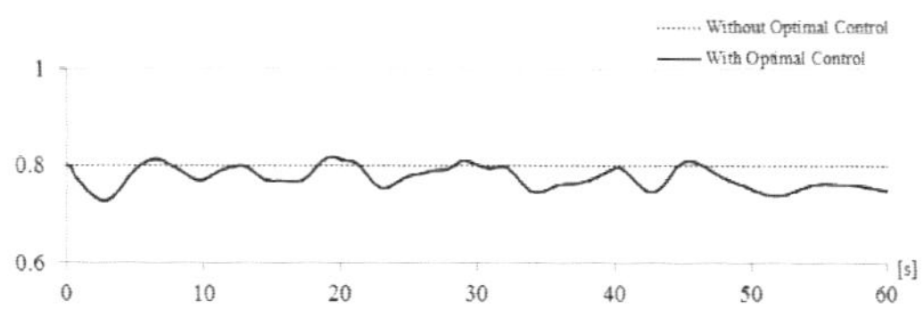

(c) Propeller-pitch ratio.

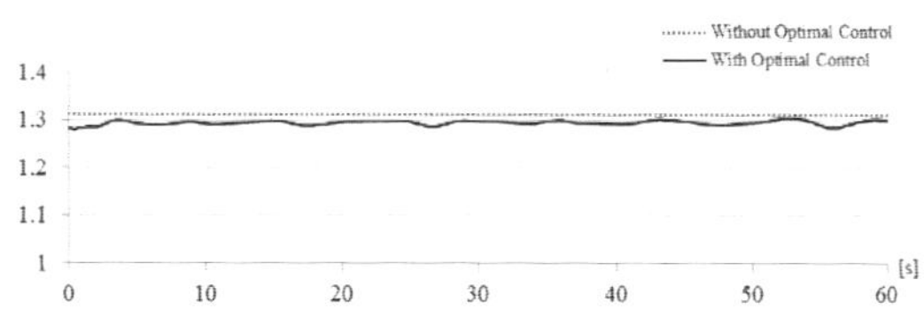

(d) Drive frequency of the propulsion motor.

Figure 12: Simulation results for a ship in head seas: (a)–(d) comparison of situations with and without optimal control of ship speed, propeller speed, propeller-pitch ratio, and drive frequency, respectively.

Figure 12 shows the results of the numerical calculation for head seas and a design speed of 14.7 knots. The different panels in this figure show the ship speed, the propeller speed, the propeller-pitch ratio, and the drive frequency as functions of time.

The difference between the results with control and those without control reveals the effect of optimal control. For the NMPC, we use a receding horizon time of 20 s throughout this study. The ship speed's fluctuation is shown in Figure 12(a) and does not seem to be significantly affected by control. This indicates that the propulsive power is too small to modify the ship speed in the face of the waves' force. This also means that the average ship speed with the control is the same as that without control, which means that the effect of the control can be obtained by directly comparing the case with control to the case without control. Figure 12(b) shows the output result of the propeller speed. It can be confirmed that the output value with optimal control and without control are almost the same trend. In addition, each average value is about 1.7 rps. Figure 12(c) shows a comparison diagram of the output results for the propeller-pitch ratio. The pitch fluctuation due to the optimal control is conspicuous as compared with the without optimal control with fixed pitch, and this result led to better propulsion efficiency. In Figure 12(d) (the output result is non-dimensionalized by the reference frequency), the fluctuation of the drive frequency due to the optimal control is not so large. The mean of the drive frequency with optimal control is less than that without control, so that optimal control saves propulsive energy (in this case, an energy reduction of about 2% is realized with the optimal control).

5.2 (b) Numerical calculation for following seas

Figure 13 shows the results for following seas. The results for the ship speed are shown in Figure 13(a). There is no significant difference between the speed with control and that without control. The difference between maximum ship speed and minimum ship speed is less than 1 knot. Figure 13(b) shows the output result of the propeller speed; the output value with control is smaller than that without control. Without control, the peaks correspond almost exactly to those of the ship speed; thus, the propeller speed increases with the ship speed. On the other hand, with control, the propeller speed peaks occur about 5 s later than the ship speed peaks. As for the fluctuation of the propeller-pitch ratio shown in Figure 13(c), the trough value coincides with the peak of the propeller speed. This is the difference between determining the ship speed by optimally controlling the pitch and the propeller speed according to the waveform, and control without determining the ship speed, only by the propeller speed according to the waveform. The amplitude of the controlled drive frequency shown in Figure 13(d) (the output result is non-dimensionalized by the reference frequency) is small, but its mean is significantly smaller than that without control, which indicates that optimal control saves propulsive energy (in this case, about 7% of the energy required without control is saved).

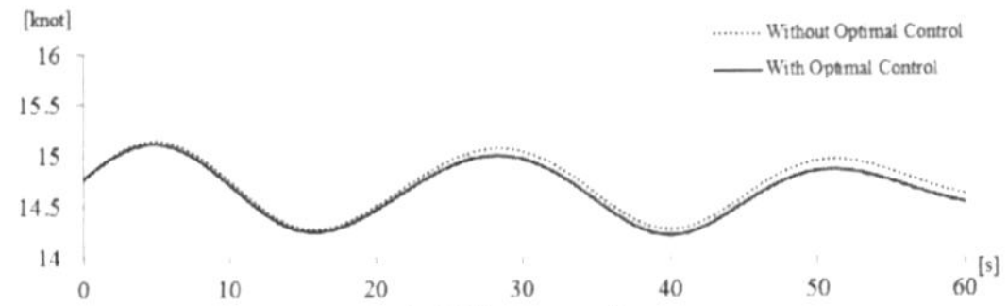

(a) Ship speed.

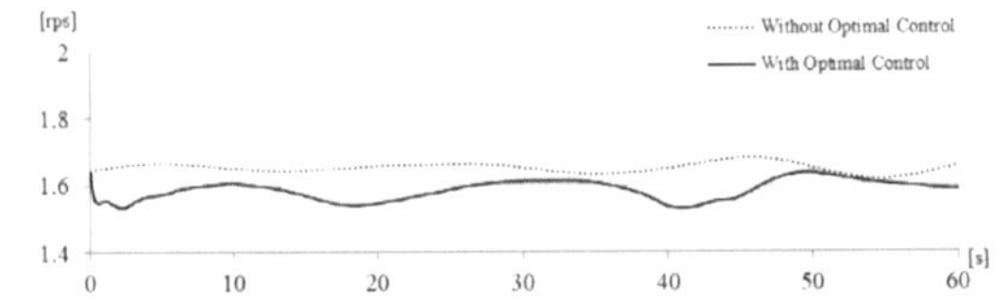

(b) Propeller speed.

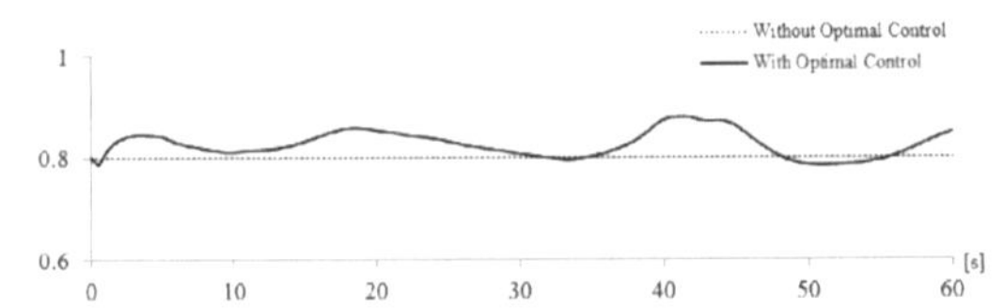

(c) Propeller-pitch ratio.

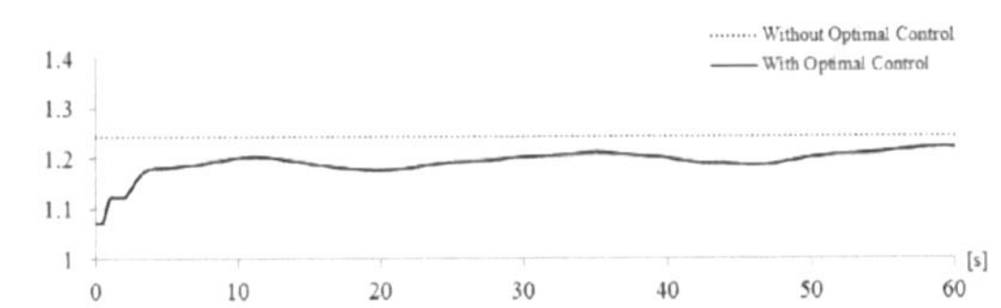

(d) Drive frequency of propulsion motor.

Figure 13: Simulation results for a ship in following seas: (a)–(d) comparison of situations with and without optimal control of ship speed, propeller speed, propeller-pitch ratio, and drive frequency, respectively.

6. CONCLUSIONS

This paper proposes an energy-saving method for ships sailing on wavy seas. The method exploits real-time optimal control of the ship's propulsion system, which consists of a servo-motor-driven CPP and an electric propulsion motor. A numerical simulation assuming conditions of short-crested irregular waves indicates that such optimal control can lead to energy savings of about 2% in head waves and 7% in following waves, respectively. The required computation time is much less than the real time and all constraints are satisfied, so applying this control method in actual operations should be feasible. The key element here is that, if the propeller pitch and the drive frequency of the propulsion motor are varied in anticipation of wave encounters, the propulsive energy required could be reduced. Implementing this method requires real-time optimal control, highly responsive actuators, and a wave radar.

7. ACKNOWLEDGEMENTS

The research was supported by Japan's Ministry of Transport Land, Infrastructure, Transport, and Tourism, Nippon Kaiji Kyokai (ClassNK), and Kamome Propeller Co., Ltd. as a research and development project for next generation marine technology. The authors would like to thank Enago (www.enago.jp) for the English language review.

8. REFERENCES

1. Ballou, P.J., Ship Energy Efficiency Management Requires a Total Solution Approach, Marine Technology Society Journal, Volume 47, Number 1, pp. 83-95, 2013.

2. García-Domínguez, A., Mobile Applications, Cloud and Big Data on Ships and Shore Stations for Increased Safety on Marine Traffic, 2015 IEEE International Conference of Industrial Technology (ICIT), pp. 1532-1537, 2015.

3. Gowri, R. and Vinoth, B., Ship Safety Management Using Mobile Applications with Cloud and Big Data Environment, International Journal of Innovative Research in Computer and Communication Engineering, Volume 4, Number 5, pp. 9762-9769, 2016.

4. The International Convention for the Prevention of Pollution from Ships, MARPOL, International Maritime Organization (IMO), 2011.

5. Matsunaga, Y. and Umeda, N., Energy Efficiency Improvement by Optimal Control of Propeller Pitch and Electric Propulsion for a Ship in Waves, Proceedings of the 11th International Marine Design Conference, University of Strathclyde (Glasgow), pp. 533-543, 2012.

6. Barclay, A., Gill, P.E., Rosen, J.B., SQP Methods and their Application to Numerical Optimal Control, International Series of Numerical Mathematics, Volume 124, pp. 207-222, 1998.

7. Kitagawa, Y., Tanizawa, K., Hirata, K., Ueno, M., Umeda, N., Fukazawa, M., Komiya, S., Validation of Optimal Pitch Control Algorithm for Energy Efficient Propulsion in Waves, Conference Proceedings of the Japan Society of Naval Architects and Ocean Engineers, Volume 16, pp. 355-356, 2013.

8. Kushida, R., Umeda, N., Sekiguchi, H., Matsunaga, Y., Energy Efficiency Improvement by the Use of a Series Hybrid Propulsion System Applied to a Bulk Carrier in Head Waves, Proceedings of the 6th Asia-Pacific Workshop on Marine Hydrodynamics, UTM (Johor), pp. 483-488, 2012.

9. Makino H, Umeda N, Ohtsuka T, Ikejima S, Sekiguchi H, Tanizawa K, Suzuki J, Fukazawa M, 'Energy Saving for Ship Propulsion in Waves Based on Real-Time Optimal Control of Propeller Pitch and Electric Propulsion', Journal of Marine Science and Technology, 2016 (under review).

10. Ohtsuka, T., A Continuation/GMRES Method for Fast Computation of Nonlinear Receding Horizon Control, Automatica, Volume 40, pp. 563-574, 2004.

11. Hirayama, T., Nomiyama, D.H., Nishimura, K., Hirakawa, Y., Takayama, T., Very Short-Term Prediction of Encountering Waves Utilizing Marine Radar, Proceedings of the 18th Ocean Engineering Symposium, The Society of Naval Architects of Japan, Volume 18, CD-ROM, 2005.

12. Fukazawa, K., Odagiri, T., Suzuki, J. et al., Development of Ship Energy Saving Technique by Using Control of CPP Shaft Speed and Pitch Angle," Journal of Japan Institute of Marine Engineering, 2017 (in press).

9. AUTHORS BIOGRAPHY

Hidenari Makino holds the current position of Postdoctoral research fellow at Osaka University supported by Kamome Propeller Co. Ltd. He is responsible for the optimal control of the system in the project described above. He worked at Osaka University and Kobe University as an assistant professor in naval architecture and a post-doctoral research fellow at Kobe University in maritime science, respectively.

Yuichiro Hirano is a MSC candidate at Osaka University in naval architecture and ocean engineering. He is responsible for the numerical works of optimal control in the project described above.

Naoya Umeda, MRINA, is a professor at Osaka University in naval architecture. He is responsible for education and research in the areas of ship stability, manoeuvrability and control as well as tasks as the coordinator of correspondence group of the second generation intact stability criteria at the IMO.

Toshiyuki Ohtsuka is a Professor at the Graduate School of Informatics, Kyoto University. Before moving to Kyoto University, he spent 1995-1999 as an Assistant Professor at the University of Tsukuba, and then he joined Osaka University, where he was a Professor from 2007 to 2013. His research interests include nonlinear control theory and real-time optimization with applications to aerospace engineering and mechanical engineering.

Katsuji Tanizawa is a Senior director of Fluids Engineering Department at National Maritime Research Institute, Tokyo, Japan. He is responsible for all research projects in the departments. He is a director of the Japan Society of Naval Architects and Ocean Engineers as well as the research planning committee.

Hidenori Sekiguchi is a Senior Researcher with National Maritime Research Institute, Tokyo, Japan. He is responsible for the design of electric propulsion unit of the system in the project described above. His research interests include integrated hybrid and electric propulsion systems for ships.

Hironori Susaki is a Head in Charge of Researching Department at Technology Development & Researching Laboratory., Furuno Electric co., Ltd. He is responsible for the wave radar of the system in the project described above.

Junichi Suzuki is a Manager at CPP Design Grp./ Engineering Dept., Kamome Propeller Co., Ltd. He is responsible for the design of CPP of the system in the project described above. His previous experience includes strength analysis of the CPP and optimization of CPP shaft alignment system of merchant ship, etc.

Masaki Fukazawa is a Director/General Manager at Engineering Dept., Kamome Propeller Co., Ltd. He is responsible for the design of FPP, CPP, Side Thruster and K7-Flap Rudder. His previous experience includes development of research of highly skewed CPP and hydrodynamic design of CPP blade, etc.

UNLOCKING THE HIDDEN VALUE OF DYNAMIC POSITIONING SYSTEMS FOR SMART SHIPS

M.R. Carter, Sonardyne International Limited, UK

SUMMARY

Over the last decades dynamic positioning (DP) systems have grown in popularity and applications, by proving highly effective and efficient in keeping vessels safe in a multitude of applications and operations and providing invaluable support to vessel crews and reducing the burden from increasing operational demands and complexities.

The key technology components which every DP system relies upon are the environmental, motion and position reference sensors which provide the crucial measurements for the vessel automation system to function in the first place - whether these are below surface (acoustic), on surface (laser and microwave) or above surface (Global Navigation Satellite Systems or GNSS) position reference systems.

Increased positioning integrity and data analytics based on systems and sensors that are already in service are enabling technologies for smarter ships. Today, they already help to improve performance, reduce costs and help make faster and better operational decisions. The expected benefits range from identifying, diagnosing and predicting operational performance to monitoring vessel as well as fleet efficiencies and effectiveness across all vessel operations. All of these will be necessary parts of a smart ship solution.

1. INTRODUCTION

The combination of an absolute position reference such as acoustics and inertial navigation for high integrity vessel positioning was first suggested at the Offshore Technology Conference, Dallas, TX, US 1975 [1]. Time has passed and technology and systems have matured. GNSS, acoustic, inertial and local surface (laser and microwave) based position references have now been in routine operational use for many years in the oil and gas industry and beyond. Due to the high integrity positioning requirements of applications such as offshore drilling, these positioning systems now achieve high levels of reliability by following key design principles such as independence, redundancy and fault tolerance. As a result it is common for drill ships to hold station for more than three months at a time with minimum user input to guidance, navigation or control.

Fault tolerant, high availability navigation systems used in the oil and gas industry today can already be considered "smart" and are an enabling technology for increased levels of autonomy where there are shared requirements for high integrity navigation. However, with up to 200 people on-board a typical offshore drilling rig they are far from unmanned and still rely on human intervention to take action when automatic fault detection and decision making are defeated.

The development of future system capability is moving towards both higher levels of automation and the enabling of remote access for troubleshooting, support and decision making activities. Fortunately many of the building blocks that achieve these developments exist today on the latest generation of offshore oil and gas vessels.

The following sections provide an overview of systems currently in use by the oil and gas industry to ensure prolonged, uninterrupted DP operations. The aim is to highlight some of the key building blocks for smarter, autonomous ships.

2. CURRENT, STATE OF THE ART DP REFERENCE SYSTEMS

IMO 645 rules require at least three independent position references to be available on a Class 2 or 3 dynamically positioned vessel. Typically, two or more Differential GNSS (DGNSS) systems are used alongside one or more Long and Ultra Short Baseline Acoustic (L/USBL) reference system. In order to claim independence, the DGNSS systems are usually manufactured by different companies and with differential corrections delivered by various means. A conventional DP system is shown in Figure 1.

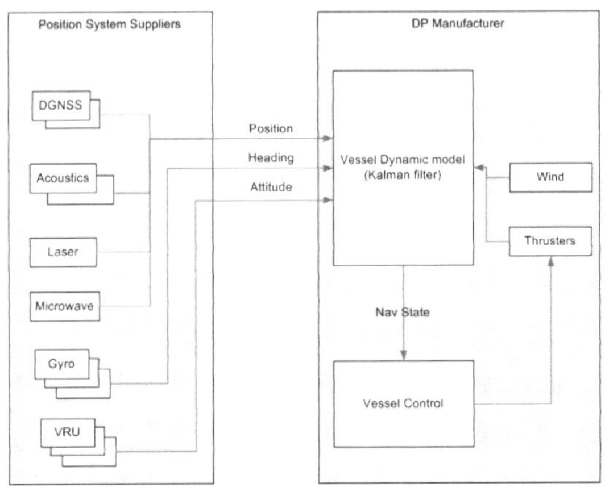

Figure 1: Conventional DP system

2.1 VULNERABILITIES

The use of two or more DGNSS systems in an attempt to meet an acceptable level of independence leaves the ships positioning vulnerable to satellite signal systematic errors and signal outages originating from the GNSS space segment. Also GNSS signals are weak and easily disrupted, sometimes for long periods of time by unintentional or intentional interference (jamming and spoofing) and natural phenomena such as scintillation.
As a common error source, these disruptions can cause all DGNSS receivers to suffer systematic errors and outage. A simultaneous bias in multiple DGNSS references can cause the vessel control system to reject a single L/USBL reference telegram as the multiple DGNSS references will have a higher update rate and combined weighting in the DP model. Subsequent increase in thruster activity to regain position using the acoustic reference, can result in excessive noise and cavitation that degrades its performance.

Acoustic based L/USBL systems are highly repeatable and robust positioning systems. Although, acoustic signals can be interfered with by environmental factors such as aeration clouds, momentary increases in the noise level and physical masking. As a result, no manufacturer will claim position date availability is 100%

Radar, cameras and local position references offer additional location information in some applications. However, the targets they need to identify in their field of view are often either not available or cannot be easily installed offshore.

The consequences of the combined effects and constraints described above have been attributed to a number of station keeping incidents over recent years in the oil and gas industry. The lessons learned in oil and gas highlight the need to improve the robustness of position measurement equipment as levels of autonomy increase and are adopted in the maritime industry in general.

3. INERTIAL NAVIGATION

Inertial navigation is the computation of velocity, position and orientation from measurements of acceleration and rotation. Acceleration (velocity change) is measured by an orthogonal triad of accelerometers. Rotation (angular change) is measured by an equally orthogonal triad of gyroscopes. The composite mechanical structure is called the Inertial Sensor Assembly (ISA). An Inertial Measurement Unit (IMU) is composed of the ISA and a processor that performs various forms of compensation using e.g. factory calibration coefficients. The outputs of an IMU are compensated delta angles and delta velocities typically at a rate of 100-200Hz. An inertial navigation system (INS) is based on the dead-reckoning principle. INS velocity is computed from an initial value by integration of acceleration over time. Similarly a position vector is computed from velocity and an orientation vector (attitude and heading) is computed from integration of measured rotation (angular change). One can imagine how a fixed error (bias) in acceleration when integrated will cause an error in velocity that grows proportionally with time. In turn the velocity error when integrated causes a quadratic increase in the position error. This position error growth is also known as drift.

The drift of a typical commercial airplane inertial system is on the order of 1NMPH (Nautical Mile Per Hour). This is adequate for guidance from one airport to another (e.g. crossing the Atlantic) but hardly useful for navigation of slow moving marine vehicles over periods of weeks and months or where high accuracy positioning is required.

Improved performance can be achieved by Aiding the INS (AINS) with measurements from one or more external sensors as shown in Figure 2; The AINS algorithm is a Kalman filter that determines INS errors from external sensor measurements. The difference between a measurement (observation) and its expected value (based on the INS) is fed into the Kalman filter and is used to refine its state vector and covariance matrix. The state vector is composed of INS errors in position, orientation and velocity and often also key inertial sensor errors such as gyroscope and accelerometer bias. Errors in external sensors can also be modelled for improved performance.

Due to the need for external aiding an Inertial Navigation System is not a completely independent reference. However, the combination of absolute positioning and INS does provide a complementary solution that is proven to increase the integrity of a vessel's navigation system and reduce the number of station keeping incidents as a result. The following important advantages are also realised [2]:

• "Ride Through Capability"- resilience to short term outages in the acoustic reference or GNSS due to noise increase and aeration.

• Faster update rate (1Hz – 5Hz) and high precision improves DP weighting that increases the probability of detecting GNSS disruptions

When INS is combined with an acoustic reference the ability to detect anomalies in the DGNSS system is considerably enhanced and enhanced system can manage prolonged periods where DGNSS is unavailable.

Sonardyne's DP-INS implements this principle using a Lodestar IMU configured for Acoustically Aided INS (AAINS) operation. Lodestar Attitude Heading Reference System (AHRS) and Lodestar AAINS share an identical hardware platform custom built for tightly integrated marine applications. DP-INS optimally combines the complementary characteristics of the latest

Wideband acoustic signals with high integrity inertial measurements. It uses a tightly coupled integration of range and bearing from at least one transponder to aid the INS and control integration drift.

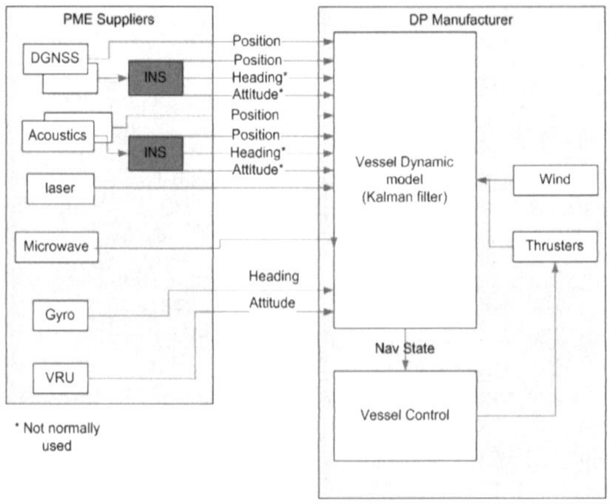

Figure 2: High integrity Aided INS system

This approach makes optimum use of Sonardyne acoustics to exploit and enhance the positive characteristics of the inertial sensors. The tightly coupled integration also ensures almost zero latency between the raw acoustic and IMU observations. The resulting output is resilient to acoustic disruptions and maintains independence from GNSS.

4. UNLOCKING THE VALUE FOR SMART AND AUTONOMOUS SHIPS

GNSS, acoustic and local position references, combined with INS have made it possible to position vessels to acceptable levels of reliability in the offshore oil and gas industry. However there are still a number of station keeping incidents per year where position references are cited as a contributory factor and there are a number of initiatives amongst oil companies, marine societies and other organisations to improve reliability as a result. When the lessons from the offshore oil and gas industry are applied to vessel navigation in general and specifically the trends towards increased automation and ultimately unmanned vessel operations, it is clear that more robust vessel positioning is required. This is particularly true where position excursions cannot be tolerated such as when navigating narrow channels, fairways and shipping lanes or where traffic-separation measures are in place.

Much of the data required to support high integrity navigation for increased autonomy is available in sensors on-board vessels today. The challenge is to achieve the correct level of integration using enough data in a useable format from a disparate set of sensors. Based on lessons learned from the offshore oil and gas industry there are some simple metrics that are often ignored that can address this challenge and make vessel positioning more robust.

4.1 ACCURACY

Accuracy is the most commonly used term to measure navigation system performance and is the easiest to understand. It is a measure of the error in position or the deviation of the reported position from the true position. However, real-time positioning systems have no knowledge of the true position so accuracy is reported as a statistical quantity associated with the distribution of the position error. This is more correctly termed "precision" but the terms are often used interchangeably in marine sensor descriptions. Various terms are used to express the accuracy, each with a confidence level. Sonardyne DP-INS reports a one dimensional root mean square error (1DRMS) or radial error at the 68% confidence level.

It is well-known that an inertial navigation solution will drift with time unless it is constrained by aiding data [2]. In the event of loss of aiding data the reported 1DRMS error will increase with time even though the precision over a few seconds will remain very good. For this reason, it is important that vessel control systems make use of the reported position error rather than internally computing the error from position alone with limited hysteresis. The correct use of reported sensor accuracy parameters allows the control system to implement logic to handle degraded accuracy and raise appropriate alarms on-board and / or remotely.

4.2 INTEGRITY

Although accuracy is easy to understand it has some limitations as a metric for measuring navigation system performance in a marine context. In some scenarios a position accuracy of a few meters is often acceptable (e.g. when calculated based on riser angle tolerances in offshore drilling) and the latest Sonardyne DP-INS systems are well within the accuracy requirements of today's specifications of typically between 0.2% and 0.5% water depth.

A second metric, as important to mission critical applications but often ignored, is the integrity of the navigation solution. Integrity relates to the level of trust that can be placed in the reported position and position error. If the reported error is less than the actual error then it can be said that there is a loss of integrity as the positioning system is reporting misleading information. Maintaining integrity is key to the correct functioning of a positioning system and has been at the forefront of the Sonardyne DP-INS design philosophy since day one.

The 1DRMS error estimates associated with the position information reported by DP-INS are intentionally conservative as a result

4.3 IN-BUILT FAULT DETECTION

DP-INS has its own, inbuilt, integrity monitoring. One of the unique advantages of the Lodestar INS is the independent attitude, heading reference system (AHRS) and INS algorithms running in parallel within the unit. Both of these algorithms compute pitch, roll, heading, with the AHRS algorithm being extremely robust and immune to bias that can be introduced by external aiding sources. The difference between the two pitch, roll, heading calculations are continuously monitored in real time and any divergence is a potential indication of loss of integrity. A further check is the on-going monitoring of the gyro and accelerometer biases. Bias estimates that exceed fixed thresholds based on the gyro manufacturer's technical specifications could potentially indicate a loss of integrity or a hardware fault. The internal integrity monitoring can be viewed on the detailed diagnostic pages in some system variants and alarms are automatically raised in the system where integrity is lost.

A sensor capable of independently monitoring and communicating a loss of integrity to the control system will facilitate fault detection. Although there are limits to the extent that this is possible (there will be failure modes that cannot be internally detected) it increases the likelihood of appropriate action being taken, either on-board or remotely.

5. REMOTE CONTROL AND MONITORING

Given that much of the data required to support high integrity navigation and increased levels of autonomy are already available on-board from existing sensors a further consideration is the value of that data remotely for trouble shooting, support and decision making operations. Various stakeholders have different requirements:

5.1 STAKEHOLDERS & APPLICATIONS

Vessel owners and operators invest in the equipment in the first place in order to meet charter requirements. Maintenance of this equipment is often on a scheduled service or ad-hoc in case of failure and not necessarily aligned with operational and planned downtime.

Asset owners and Charter holders contract in services against specifications which are mandated from a variety of sources but performance against these specifications is not measured, except for infrequent audits and through contractual compliance rules. Reports produced by owners from data analytics will allow them to demonstrate a much higher level of compliance and individual case by case job performance. Early indication and pro-active correction of non-conformities can also be demonstrated.

Crew and Operators need emergency support on the systems they are responsible for when failures occur or when a deeper level of configuration is needed. They hold log books of vessel hours but cannot demonstrate their actual system experience and usage of equipment. There is no feedback loop from the data to evaluate crew performance, training needs and efficiency gaps

5.2 SUPPORTING DATA

Extensive quality metrics related to DP position like measurement noise and associated error metrics are available which, if recorded and processed through a data analytics approach, allow sensors to become truly smart and self-learning. These sensors store what accuracy they can achieve and how good their performance and they access that data not just on an instantaneous basis but also on an average basis and for a particular set of environmental conditions or geographic location. Historic quality maps for all operations would allow the DP system to be smarter about sensor weighting and selection.

Sonardyne system data such as sensor calibration status, USBL transceiver performance, transponder battery status and overall acoustic and / or inertial positioning system performance can be pushed to remote support centres worldwide.

The performance of solid state sensors found in ships gyro's, motion reference units and inertial navigation units can be monitored over the long term to detect changes in bias and drift rates indicting the need for re-calibration. A move away from time based maintenance would offer significant logistics savings and provide fleet wide sensor performance statistics.

Changes in USBL transducer health can be measured on-board and any degradation highlighted in advance of any material loss of performance. Transducer inspection and cleaning can be carried out during convenient periods of downtime rather than as unscheduled maintenance.

System performance and the integrity of the acoustic / inertial position can be monitored over time and degraded / suboptimal performance due to inappropriate system configuration of transponders, environmental factors, etc can be highlighted on board and onshore in near real time. The vessel can then be advised of changes needed to optimise the system performance following arrival on location or following environmental and / or operational changes.

Transponder battery status can be monitored remotely and replacements ordered by a support centre to coincide with port calls and supply vessels.

The long term trend of GNSS signal-to-noise (SNR) ratio on a vessel could also be monitored to let the operator know that the antenna cable needs to be changed out.

5.3 REMOTE CONTROL ARCHITECTURES

It would be inefficient for individual sensor and sub-system manufacturers to implement their own end-to-end remote monitoring solution including on-board data collection, communication system and shore base data centre functions. Fortunately, the major DP manufacturers and some vessel owners existing infrastructure has open interfaces that can be used by 3rd parties. However, the details of interfaces, networks and data centres are different for each provider even where the overall aims and approach are similar.

At the moment a rigid architecture may not work, and some flexibility is needed in the absence of relevant industry standards. In order for the full value to be realised, standards need to be developed that will allow sensors, subsystems, vessels and data centres to share data effectively and securely.

6. SUBSEA OBSTACLE AVOIDANCE

One of the main differences between the high integrity positioning requirements of the offshore oil and gas industry and what is required for smarter, more autonomous ships is that the offshore oil and gas operations do not consider the voyage to and from the site of work to be part of the operation and the automated station keeping only begins once a vessel has arrived on location.

Existing sensors such as radar, ECDIS and echo sounders can be augmented by LIDAR and cameras to provide sense and avoid information as levels of automation increase. However, there is also a need to deal with an uncharted, poorly charted or changing subsea environment. Increasingly commercial and military vessels, private yachts, cruise ships and manned underwater vehicles are fitting forward-look sonar as an important aid to vessel navigation and collision avoidance.

Sonardyne has developed a solution to this challenge, the Navigation and Obstacle Avoidance Sonar (NOAS) provides sonar navigation to 1,500 metres and 3D seabed mapping ahead of the vessel up to 600 metres is possible. The system alerts the operator to shallow water and potential hazards as well as maintaining historical data of the vessel passage for manoeuvring ahead of a vessel to detect potential underwater hazards. This sonar will be an effective tool for remote operation by providing pilots onshore with situational awareness of the surrounding seafloor. It will also serve as an enabling input for an autonomous pilot.

7. CONCLUSIONS

Station keeping systems already in use today offer some level of autonomous operation as they are designed to operate for months at a time with limited human intervention.

The hidden value in the sensors used in a state of the art station keeping system is the additional data available for extended integrity monitoring. When an extended set of measurement and quality data is used the DP systems can more appropriately weight the position information received and more successfully detect faults and raise alarms if accuracy degrades or integrity is compromised. Combined with the in-built fault detection that good quality position reference sensors now support it is possible to extend the level of navigation system integrity beyond current levels, further reducing the need for human intervention.

In parallel remote data analytics based on systems and sensors that are already in service today is becoming accepted and provides significant support, learning and understanding advantages that will improve performance, reduce costs and help make faster and better operational decisions.

Unmanned vessels pose difficult technical, commercial and legal challenges to the shipping community. The combination of increased automation and data analytics based around existing navigation sensors and systems in use in the offshore oil and gas sector today is therefore of significant commercial value, both to the oil and gas sector and to the wider shipping community as we move towards unmanned vessels.

8. ACKNOWLEDGEMENTS

Jan Grothusen, Guidance Marine, UK

9. REFERENCES

1. D.L. Buechler & J.T Hanna, Inertially-Aided Dynamic Positioning. 8th Offshore Technlogy Conference, Dallas, TX, US, 1975.

2. M.R. Carter, DP INS – A Paradigm Shift, Sonardyne International Ltd, 2011

10. AUTHORS BIOGRAPHY

Mark Carter is the Business Manager for Vessel Equipment at Sonardyne International Limited. He joined Sonardyne in 2011 and has 20 years of navigation systems experience across the maritime and aviation sectors. Previously, Mark worked at QinetiQ as engineering manager for the navigation group. This involved working for military and commercial customers to develop GPS anti-jam systems and high integrity aided inertial navigation systems. Mark has a background in Systems Engineering and a BSc in Surveying from the University of Newcastle-Upon-Tyne.

A SIMULTANEOUS OPTIMIZATION METHOD OF THE DIRECTION AND SPEED FOR SHIP ROUTE PLANNING

K S Kim, M I Roh and **S M Lee**, Seoul National University, Republic of Korea
H R Jung, J J Park, D Y Lee and **B K Kim,** Samsung Heavy Industries, Republic of Korea

SUMMARY

The aggravation of financial condition of shipping companies in recent years and the emphasis on reducing greenhouse gas emissions have resulted in renewed interest in further optimizing ship performance through the development of green-ship or smart-ship technologies. A recent IMO study has indicated that ship route planning can achieve reductions in fuel consumption and the associated greenhouse gas emissions through technical and operational measures such as speed management and route planning. Related to these issues, a simultaneous optimization method of the direction and speed for ship route planning is proposed in this study. For this, an optimization problem for finding an optimal ship route is mathematically formulated considering fuel consumption and weather condition.

1. INTRODUCTION

1.1 LITERATURE SURVEY ON PREVIOUS STUDIES

Most of studies for ship route planning separated path optimization and speed optimization. Joo et al. [1] found the optimal path with the Astar algorithm in state of the fixed speed and they modified the path considering weather condition with the Dijkstra algorithm. Afterwards, they adjusted speed on the path. Similarly, Bang et al. [2] found the optimal path with the Astar algorithm in state of the fixed speed, they modified the found path considering weather condition with the evolutionary strategy, and finally, they adjusted speed on the path. Choi et al. [3] targeted ship route planning for the arctic sea. Since the weather condition including ice changes hour after hour, they developed an ice model which has uncertainties of ice behavior. They proposed a method to find an optimal path under the time-varying weather condition with the Astar algorithm. However, they did not perform speed optimization on the path. Park et al. [4] found the optimal path with the Astar algorithm in state of the fixed speed and then they adjusted speed on the path with the geometric programming. The studies of Joo et al. [1], Bang et al. [2], and Park et al. [3] took action in speed optimization with a separating stage. The studies used different optimization methods each other, but they conducted their study with finding the shortest path in the first stage, and then adjusting the speed on the path in the second stage. Vettor et al. [5] formulated ship route planning as a multi-objective optimization problem which has positions and speeds as design variables. They mainly used the evolutionary algorithm for making the route which is comprised of several nodes, and specifically, Dijkstra algorithm was used to calculate objective functions between neighbor nodes. Therefore, their method called the ES-Dijkstra method is similar to the Dijkstra algorithm which is possible to change the speed at each segment between nodes. They also designed a model of weather condition to be used in ship route planning problems.

In addition, several methods which correspond to the cell-free method have been studied. Roh [6] proposed an improved isochrones method which can consider obstacles such as the land in ship route planning. Wei and Zhou [7] proposed a method for ship route planning with the use of the 3D dynamic programming which is similar to the isochrone method. Similarly, Chen [8] developed a program based on the 3D Dynamic Programming. ENIRAM [9] is a commercial program which uses the empirical data based on actual route of ship.

2. MATHEMATICAL FORMULATION OF A SHIP ROUTING PROBLEM

2.1 SPEED-ROUTE SIMULTANEOUS OPTIMIZATION PROBLEM

A ship route planning problem is to determine how to reach arrival port from departure port called the route. The route can be referred to as the combination of path and speed. In ship operation, the path is determined by a series of ship directions and the speed is also determined by a series of engine RPMs. And it is assumed that one route consists of a series of nodes. Nodes mean intermediate points between departure and arrival ports. Thus, design variables of the optimization problem are direction (or heading angle, θ) and engine RPM (rpm) of a ship at each node, as shown in Fig. 4.3.1. In addition, it is assumed that the time between two adjacent nodes is given as a constant value (e.g., 6 hour). As shown in this figure, if the directions are determined, the positions of nodes can be determined. The direction is related to the path and the engine RPM is related to the speed of the ship. If the number of nodes is n from P_1 to P_n, n directions from θ_0 to θ_{n-1} and n+1 RPMs from rpm_0 to rpm_n should be determined as the design variables. The direction at the last node P_n does not need since the arrival port ($P_{arrival}$ or P_{n+1}) is given and the direction from P_n to $P_{arrival}$ is automatically determined. Thus, total number of design variables is 2n+1 in this problem. By connecting all nodes including the departure and arrival ports, the route (path and speed) can be obtained.

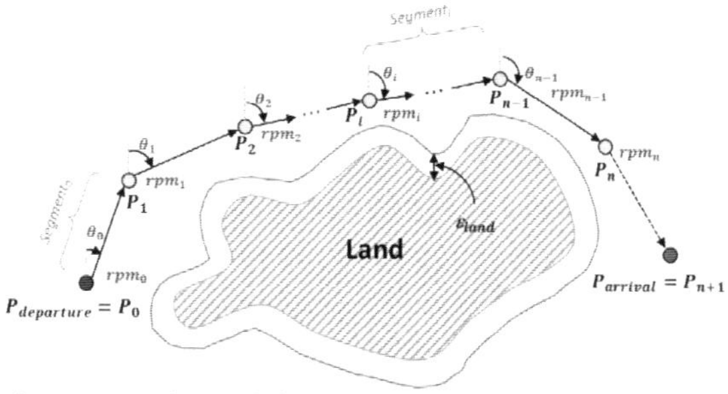

Figure 1: Design variables in the ship route planning problem

The TFOC is used as an objective function which is a criterion for selecting the best one from a number of alternatives. This TFOC depends on the route, that is, values of design variables, including weather condition. Two constraints are used in this study. First, the ship should arrive at the arrival port until the latest time of arrival (ETA_{max}) which is given by a designer. Second, the route for the ship should not pass through obstacles such as the land. Thus, this optimization problem is a kind of a constrained optimization problem which has 2n+1 design variables, one objective function, and two inequality constraints.

2.2 ESTIMATION OF FOC

For calculating fuel oil consumption of the route, a regression model that is made from a number of ship operation data and is a type of black box. This model gives the fuel consumption per unit time (UFOC) and the reduced ship speed if the weather condition.

2.3 OPTIMIZATION METHOD

Optimization algorithms are generally divided into global and local optimization algorithms, and several types of algorithms are available for global optimization, including a genetic algorithm (GA). Global optimization algorithms may be suitable for large-scale problems that have many local optima, but they require more iterations in order to obtain an acceptable optimum in comparison with local optimization algorithms. Several classes of local optimization algorithms also exist, including sequential linear programming (SLP) and sequential quadratic programming (SQP). Each of these local optimization algorithms can be used to effectively find the optimum. However, in some cases, these local optimization algorithms find an optimum that is relatively close to the starting point. Therefore, in this study, GA is the global optimization algorithm that was selected to solve the optimization problem for ship route planning. In this study, NSGA-II [10], which is a well-known and proven GA, algorithm was used to solve the optimization problem formulated in each step.

3. APPLICATION

3.1 STUDY ON THE GUI

A prototype program is developed based on the proposed method in this study. For this, C # language is used on the environment of Microsoft Visual Studio 2015. Fig. 4.3.5 shows a screenshot of the prototype. As shown in the figure, the program consists of ribbon menu, planning view, and map view.

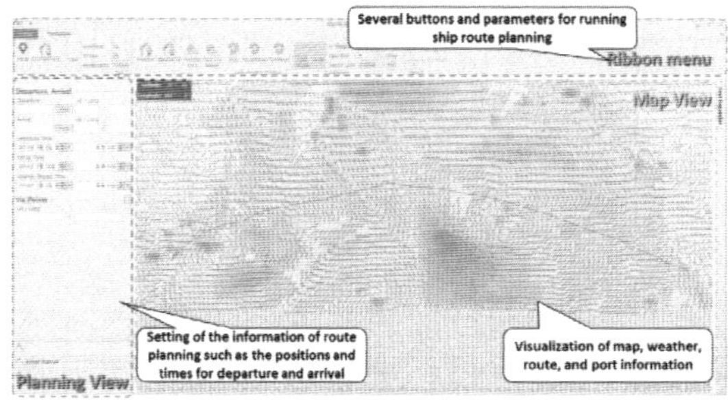

Figure 2: Screenshot of the program

3.2 CASE STUDY

The proposed method and the prototype program were also applied to the route Busan, Republic of Korea to Long Beach, U.S.A.

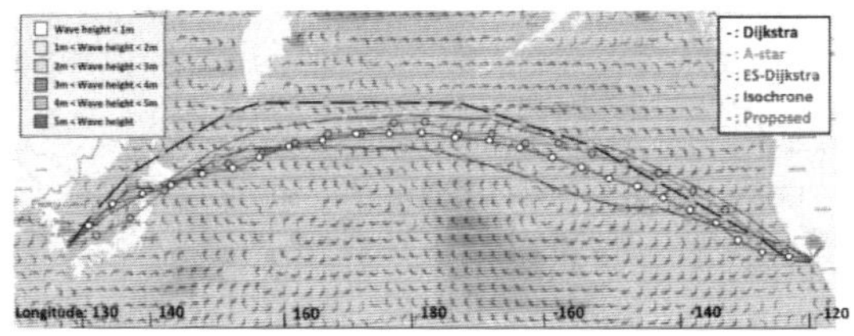

Figure 3: Results of the ship-route planning

Figure 3 shows visualization of the results of the ship-route planning for Case 1 according to the five methods, as follows: the Dijkstra algorithm, the A-star algorithm, the ES-Dijkstra method, the isochrone method, and the proposed method. In the case of the isochrone method and the proposed method, the time duration of each segment is 12 hours.

Figure 4 shows a comparison between the results. The proposed method reduced rpm in relatively bad weather. To compensate it, it increased rpm in relatively good weather. By doing that of strategy, the proposed method reduced total FOC in the hole route consequently.

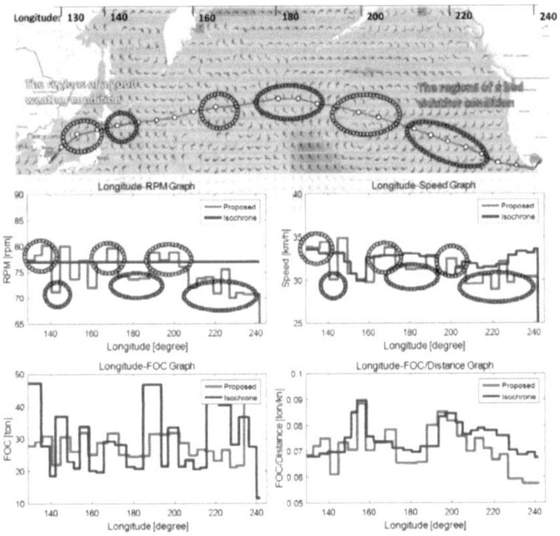

Figure 4: Comparison the results of the isochrones method and the proposed method

4. CONCLUSION AND FUTURE WORKS

In this study, an efficient method for ship-route planning was proposed through the mathematical formulation of a ship-route-planning problem as an optimization problem, whereby a simultaneous optimization of the path and the speed of the ship was enabled. The performance of the proposed method was evaluated alongside those of the other methods according to a comparative test. The results show that the proposed method can yield the route with the lowest TFOC, although the total distance is not the shortest.

In the future, various objective functions, such as sea-keeping and operability, will be included besides the TFOC in the optimization problem. Further, the results of the proposed method will be validated through a comparison with the actual operation data of a shipping company, and the post-treatments will be developed to improve the quality and the smoothness of the routes.

5. ACKNOWLEDGEMENTS

This work is an expansion of the author's previous study (Kim et al., 2016) and was partially supported by

a) BK21 Plus, Education & Research Center for Offshore Plant Engineers (COPE) of Seoul National University, Republic of Korea

b) LRF-Funded Research Center for FSI (Fluid Structure Interaction) at Seoul National University, Republic of Korea,

c) Samsung Heavy Industries, Co., Ltd., Republic of Korea,

d) Research Institute of Marine Systems Engineering of Seoul National University, Republic of Korea.

6. REFERENCES

1. JOO, S.Y., CHO, T.J., CHA, J.M., YANG, J.H. and KWON, Y.K., 'An economic ship routing system based on a minimal dynamic-cost path search algorithm' *KIPS Transactions on Computer and Communication Systems*, 2012.

2. BANG, S.H. and KWON, Y.K., 'Economic ship routing system by a path search algorithm based on an evolutionary strategy', *The Journal of Korea Information and Communications Society*, 2014.

3. CHOI, M., CHUNG, H., YAMAGUCHI, H. and NAGAKAWA, K. 'Arctic sea route path planning based on an uncertain ice prediction model', *Cold Regions Science and Technology*, 2015.

4. PARK, J.M. and KIM, N.W., 'Two-phase approach to optimal weather routing using geometric programming' *Journal of Marine Science and Technology*, 2015.

5. VETTOR, R. and SOARES C.G. 'Development of a ship weather routing system', *Ocean Engineering*, 2016.

6. ROH, M.I. 'Determination of an economical shipping route considering the effects of sea state for lower fuel consumption', *The Society of Naval Architects of Korea*, 2013.

7. WEI, S. and ZHOU, P. 'Development of a 3D dynamic programming method for weather routing' *International Journal on Marine Navigation and Safety of Sea Transportation*, 2012.

8. CHEN, H. 'Voyage optimization supersedes weather routing' *Jeppesen Commercial Marine Journal*, 2013.

9. ENIRAM, 'Proven fuel savings', *Eniram Fleet, http://www.eniram.fi/product/fleet-performance-manager*, 2016.

10. DEB, K., PRATAP, A., AGARWAL, S. and MEYARIVAN, T., 'A fast and elitist multiobjective genetic algorithm: NSGA-II', *IEEE Transactions on Evolutionary Computation*, 2002.

AUTHORS' CONTACT DETAILS

Alessandro Boveri
DITEN University of Genoa
alessandro.boveri@edu.unige.it

Anna Swider
Rolls-Royce Marine AS
anna.swider@rolls-royce.com

Carl Hunter
Coltraco
cmehunter@coltraco.co.uk

Christiano Garau
Parker Kittiwake
nick.arthur@blue-comms.com

Christos Gkerekos
University of Strathclyde
christos.gkerekos@strath.ac.uk

Demetres Armanes
ABS
DArmanes@eagle.org

Esa Takkinen
Napa Ltd.
Esa.Takkinen@napa.fi

Hannah Stones
University of Southampton
H.Stones@soton.ac.uk

Hidenori Makino
Osaka University
makino@naoe.eng.osaka-u.ac.jp

Jesús A. Muñoz
SENER
jesus.munoz@sener.es

Joe Beel
Cisco Systems, Inc.
jbeel@cisco.com

Justin Paul
QintetiQ
JOPAUL@qinetiq.com

Ki-Su Kim
Seoul National University
kisu2511@snu.ac.kr

Lokukaluge Prasad Perera
SINTEF Ocean AS
prasad.perera@marintek.sintef.no

Luca Agnello
National Research Council of Italy – ICAR Institute
luca.agnello@icar.cnr.it

Mark Carter
Sonardyne, Guidance, Veripos
mark.carter@sonardyne.com

Ørnulf Jan Rødseth
SINTEF Ocean AS
OrnulfJan.Rodseth@marintek.sintef.no

Stelios Koroneos
Soldecom
skoroneos@soldecom.com

Yiannis Raptodimos
University of Strathclyde
yiannis_rapto@hotmail.com

The Royal Institution of Naval Architects
8-9 Northumberland Street
London, WC2N 5DA
United Kingdom

ISBN 978-1-5108-8299-7

CURRAN ASSOCIATES INC.
proceedings
.com

Highway Design 2018

Transportation Research Record: Journal of the Transportation Research Board Volume 2672, Issue 39

SAGE Publishing